EISENHOWER PUBLIC LIBRARY

D1216196

12/10

$30.00

A Neighborhood That Never Changes

Eisenhower Public Library
4613 N. Oketo Avenue
Harwood Heights, Il. 60706
708-867-7828

DEMCO

FIELDWORK ENCOUNTERS AND DISCOVERIES

A Series Edited by Robert Emerson and Jack Katz

A Neighborhood That Never Changes

Gentrification, Social Preservation, and
the Search for Authenticity

JAPONICA BROWN-SARACINO

The University of Chicago Press
Chicago and London

Japonica Brown-Saracino is assistant professor of sociology and a faculty fellow in the Center for Urban Research and Learning at Loyola University Chicago.

The University of Chicago Press, Chicago 60637
The University of Chicago Press, Ltd., London
© 2009 by The University of Chicago
All rights reserved. Published 2009
Printed in the United States of America

18 17 16 15 14 13 12 11 10 09   1 2 3 4 5

ISBN-13: 978-0-226-07662-1 (cloth)
ISBN-13: 978-0-226-07663-8 (paper)
ISBN-10: 0-226-07662-8 (cloth)
ISBN-10: 0-226-07663-6 (paper)

Library of Congress Cataloging-in-Publication Data

Brown-Saracino, Japonica.
    A neighborhood that never changes : gentrification, social preservation, and the
search for authenticity / Japonica Brown-Saracino.
        p.   cm.—(Fieldwork encounters and discoveries)
    Includes bibliographical references and index.
    ISBN-13: 978-0-226-07662-1 (hardcover : alk. paper)
    ISBN-10: 0-226-07662-8 (hardcover : alk. paper)
    ISBN-13: 978-0-226-07663-8 (pbk. : alk. paper)
    ISBN-10: 0-226-07663-6 (pbk. : alk. paper)   1. Gentrification—United
States—Case studies.   I. Series.
    HT175.N3975   2009
    307.3´362—dc22

                                                          2009009851

♾ The paper used in this publication meets the minimum requirements of the American National Standard for Information Sciences—Permanence of Paper for Printed
Library Materials, ANSI Z39.48-1992.

For my parents,
Pamela Brown and Michael Saracino

Contents

Preface

I was born in the 1970s near Boston, where my father worked as a pharmacist and my mother as a social worker. In 1979, my parents, nature enthusiasts in search of a "simpler" life, relocated us to the rural, western portion of Massachusetts. A few years later, they purchased several acres of woodland and pasture in Leyden, a hill town on the Massachusetts/Vermont line with a population of just over five hundred. At the time, dairy farmers, factory workers who commuted to work in a neighboring town, and a growing assortment of hippies—many of whom had attended one of the neighboring five colleges—populated Leyden. Like many of their peers, my parents spent our first years in town planting gardens and finishing a Cape Cod–style house complete with a wide expanse of south-facing windows and a greenhouse. When we were not gardening and staining wood floors, we went for wildflower hikes, attended potlucks, divvied natural food from the co-op truck at the town hall, and swam in the Green River. Outside school, mine was a world of newcomers: of bearded men, women in long skirts or overalls, and children with botanical (Alyssum, Japonica) and other such names (Flora, Harmony).[1]

In several senses, this project began there, in the hills of Leyden. In the second grade, I became acutely aware of my position as a newcomer. My family had just returned from a trip to my grandparents' home in western New York. On Monday, a classmate and friend, Maggie, whose family owned a cattle farm, reported that her family had spent Saturday fishing in a dammed section of the stream that crossed my parents' land. "Your dog wouldn't stop barking at us," Maggie complained for the benefit of our classmates. "Maybe it didn't want you on our land," I retorted (my parents opposed fishing, and I concurred). In response, Maggie said, "My father built the dam when he was a little boy." At her words, I knew I had lost our small fight and felt growing remorse for claiming that the land was ours. With a sinking feeling, I also realized that Leyden belonged more to Maggie than to me, that, no matter the number of afternoons I spent in the pool where she had fished, it would never echo with memories of my father's childhood or of his father's before him. Although I might try, I could not envision my father at ten years old, following the banks of Glen Brook in search of rocks suitable for the dam. My father belonged somewhere else, in some other stream. As a native of Leyden (albeit a young one), Maggie had a claim to the pool, to our small schoolhouse, and to many of our classmates with which I felt I could not compete. Thus began my curiosity about the relationships between new and old residents of changing communities.

Ten years later, a student at Smith College, I enrolled in Rick Fantasia's urban sociology course. We read about gentrification, and descriptions of the urban process immediately struck me as parallel to my observations of Leyden. Might gentrification exist in a rural context? Was my parents' move to Leyden less innocuous than I thought? What was it like for Maggie and others to trespass past hippie gardens and barking dogs on land that once belonged to her family? I approached Rick and my adviser, Nancy Whittier, with the thought of writing about rural gentrification for my honors thesis, and they encouraged the idea, quickly convincing me to undertake a study of Leyden.

Much to my surprise, and somewhat to my chagrin, given my growing personal criticisms of gentrification, my study revealed discrepancies between the disposition of some rural gentrifiers and the literature's description of their urban counterparts. While the literature suggested that gentrifiers embrace improvements and welcome others of their social class, some Leyden newcomers sought to prevent change and preferred the presence of Yankee farmers to other affluent newcomers. That is, while my interviews revealed newcomers who expressed animosity toward old-timers and others who seemed relatively indifferent about the presence of retired dairy farmers—focused, as they were, on the town's natural rather than social attributes—others expressed an admiration for old-timers and a self-consciousness about their contribution to the transformation of Leyden that the gentrification literature did not fully account for.

I left college wondering whether my findings were particular to Leyden or to rural gentrification. I wondered whether they would have been the same had I not studied my hometown or had I been a more experienced ethnographer. On arriving at Northwestern for graduate school, I found I was not alone in my curiosity. Faculty encouraged me to further explore the similarities and differences between urban and rural gentrification, and I undertook the study of two Chicago neighborhoods and two New England towns, to which I devote this book.

In graduate school, I decided that I could no longer study Leyden. While I had apprehensions about objectivity and intellectual distance, the presence of three additional sites in my research design largely countered these. Rather, the slow transformation in my mind of Leyden from hometown to research site primarily concerned me. It was better, I decided, to wonder what Maggie made of my family than to ask her myself. Still, as with any intellectual project, my biography followed me into each site: to the small town of Dresden, Maine, where longtime residents admired my rural pedigree ("A town of five hundred? That *is* small!"), and to Chicago, where my time in Northampton, a gentrified factory town with a large population of lesbians, encouraged my curiosity about Andersonville, a traditionally Swedish neighborhood with many lesbian and, later, gay newcomers.

One might surmise that this text, like a first novel, seeks to answer questions I might have asked my parents: Why does one choose to relocate to the central city or a small town? Once there, what sense do newcomers make of their place in the village or neighborhood? How do longtime residents receive them? What is particular about places like Leyden and Andersonville? That is, are the city and country as diametrically opposed as we often presume them to be, or, as Raymond Williams (1973) and David Hummon (1990) might suggest, do we emphasize their distinctions to articulate our place in a greater array of social and geographic categories? While the following pages can only begin to answer such questions, I have chosen as my focus the peculiar finding I uncovered a decade ago in Leyden: that some newcomers, whom I have come to call *social preservationists,* are highly self-conscious about their role in gentrification and, as a result, work to minimize its risks for certain longtime residents.

Acknowledgments

Countless individuals have offered insights that have shaped this book. Among these, several deserve special recognition. For reading papers and chapters and offering helpful conversations about the research, I thank Jean Beaman, Aaron Beim, Ellen Berrey, Henry Binford, Pam Brown, Brooke Brown-Saracino, Jocelyn Brown-Saracino, Corey Fields, Gary Alan Fine, Annie Garvey, David Grazian, Wendy Griswold, Geoff Harkness, Al Hunter, Maria Kefalas, Eric Klinenberg, Peter Levin, Katherine Lieber, Richard Lloyd, Lyn Macgregor, Jeff Manza, Lida Maxwell, Terence McDonnell, Melinda Milligan, Juan Onesimo Sandoval, Mary Pattillo, Krista Paulsen, Mikaela Rabinowitz, Mathew Reed, Angela Steward, Judith Wittner, and the editors of and anonymous reviewers for *City and Community* and *Theory and Society*.

I have also benefited from feedback from organizers and participants at conferences at which I presented chapters, including meetings of the Midwest Sociological Society, the American Sociological Association, and the Social Science History Association as well as at the Chicago Ethnography Conference and the Society for the Interdisciplinary Study of Social Imagery. Members of the Northwestern University Culture and Urban Workshops offered helpful comments. Presentations at Loyola's Center for Urban Research and Learning, the Flying Culture Workshop at the University of Wisconsin, Madison, Cornell's Department of Development Sociology, and the Kahn Institute at Smith College also shaped the book.

Financial support from the Northwestern University Graduate School, Northwestern Department of Sociology, and the Dispute Resolution Research Center of the Kellogg School of Management aided the research.

I thank Wendy Henderson, Brooke Brown-Saracino, Jocelyn Brown-Saracino, Shamsher Virk, Nathan Shepard, and especially Daniel Grossman for their careful transcription of interview tapes. Allyson Bogie and Lisa Magged took careful field notes and wonderful photographs of the 2001 Provincetown Portuguese Festival when I was unable to attend. I also thank Jean Beaman, Dara Lewis, Erin Tracy, and especially Cesraea Rumpf for their research assistance and Dennis McClendon for developing maps of the research sites.

Much of this manuscript was written in the friendly confines of Ithaca, New York. I am indebted to Cornell University's Department of Sociology, where I was visiting assistant professor, for providing me with financial and intellectual support as well as with space—complete with views of the Finger Lakes' rolling hills—in which to write. I also thank my Ithaca friends for providing both intellectual support and necessary distraction.

After a year in Ithaca, I returned to Chicago for a faculty position in the Department of Sociology at Loyola University Chicago. I could not have asked for a better home in which to complete my manuscript. The Department of Sociology provided, not only necessary practical support, but also a convivial atmosphere in which it is a pleasure to work. I have benefited greatly from conversations with my Loyola colleagues and students, especially from their insights about the consequences of the orientations to gentrification I have identified.

As I detail in the preface, the questions behind this study first developed more than a decade ago when I was an undergraduate at Smith College pursuing a senior thesis. Faculty in the Department of Sociology embodied a curiosity about the social world and a passion for its study that was infectious. I remain extremely grateful for the intellectual freedom and encouragement that my thesis adviser, Rick Fantasia, and committee members, Nancy Whittier and Mark Steinberg, offered. They taught me to be true to my data even when it seemed to take me off course. If Rick had dismissed my early findings or offered less guidance as I entered my first field site, it is unlikely that I would have pursued the questions at the heart of this book.

I was lucky enough to receive similar latitude and encouragement in graduate school at Northwestern. I did not imagine drawing on my thesis research after college in large part because I worried that, in the context of America's oldest social laboratory, faculty would be unenthusiastic about the study of small towns. Wendy Griswold quickly proved my anxiety unfounded. First-year graduate students in her classical theory seminar were asked to pick a topic and write a series of essays on it from the perspective of each thinker we read. I chose to write about the topic of my senior thesis: gentrifiers in a small New England town who were deeply unsettled by the changes they wrought. I was heartened by Wendy's interest, and we began meeting to discuss the design of the study that led to this book.

As my dissertation adviser, and even after I completed my degree, Wendy has been a tremendous source of intellectual support. Her insight and unflinching expectations have influenced every page of this book. I am much indebted to Wendy and her family for inviting me to stay in their Maine home while I collected data in Dresden. This allowed me to conduct research without going too terribly into debt as well as to perfect my wood-stacking, fire-starting, and leaf-raking skills. Wendy graciously encouraged visits to her home by my family, which helped alleviate some of the isolation of ethnography. Her "open house" policy is just one indication of her remarkable generosity. I cannot thank her enough for her countless e-mail exchanges, conversations, and readings of papers and chapters.

Like Wendy, other faculty in Northwestern's Department of Sociology offered a rare blend of encouragement, high expectations, and intellectual autonomy. I am especially grateful for the support and encouragement the other members of my dissertation committee—Mary Pattillo, Albert Hunter, and Henry Binford—provided. Each offered helpful feedback on chapter drafts as well as thought-provoking conversation. Together, they guided me through the literatures—on gentrification, urban change, and community—that are this book's foundation, and I am tremendously grateful to have had such inspiring and committed guides. I also extend thanks to the directors of the Northwestern workshops of which I was a part, particularly to Wendy Griswold for her Culture and Society Workshop and to Gary Alan Fine for his Ethnography Workshop as well as for his mentorship in graduate school and beyond.

I owe more than I can express to the unflagging commitment of my editors, Douglas Mitchell, Jack Katz, and Robert Emerson. Jack's initial comments on my early chapters and feedback on a subsequent draft indelibly shaped this book. Bob Emerson read multiple iterations of the full manuscript; not a page is untouched by his insights and suggestions. It has been an honor to work with two ethnographers I so admire. I am tremendously grateful for the trust and encouragement that, with Doug, they extended to me from the beginning. Doug's interest in the project buoyed me through the revision process, and I am indebted to him for, among other things, soliciting tremendously helpful feedback from several anonymous reviewers. While any errors herein are my own, the reviewers' comments, with those of my editors, doubtlessly improved this book. Finally, I thank Timothy McGovern for his able editorial assistance and Joseph Brown for his thoughtful copyediting.

Of course, intellectual endeavors are supported, not only by colleagues, but also by friends. I am indebted to one of my oldest friends, Liz Titus, for an impromptu invitation twelve years ago to share her Provincetown cottage for a college summer. I may not have otherwise had the opportunity to become entranced by the multiple social worlds that compose Provincetown and likely would not later have had the pleasure of studying one of the most unique towns in the United States. I also thank the other steadfast friends who patiently witnessed this project's development. Among these I owe special thanks to Corrina and Angela Steward and Alisa Shor.

Graduate school friends, especially Amit Nigam, Kendra Schiffman, Terence McDonnell, Carla Shedd, Corey Fields, Jean Beaman, and Mikaela Rabinowitz, have provided tremendous intellectual support and delightful levity since my first days at Northwestern. Many have read portions of the manuscript. Of these, Terry McDonnell stands out for having read several chapters multiple times and for always offering spot-on suggestions.

I thank my immediate and extended family for their remarkable love and support. From my earliest memories, my parents, Pamela Brown and Michael Saracino, were attentive to my questions and speculations about the world and, just as important, shared their own queries and suppositions with me. For this, as well as for the support and sacrifices they offered throughout my life, I am unspeakably grateful. I also thank my grandparents, Arlene Filosi Brown, Margaret Vergamini Malley, Frank and Rose Saracino, and the late Robert Brown and Jim Malley, for their collective hope that their children and grandchildren might follow their minds and hearts. I am blessed with two wonderful sisters, Jocelyn and Brooke Brown-Saracino, who have lent their support and intelligence to this project from day one. Likewise, I am lucky to count among my family members Annie Garvey, David Nelson, Shamsher Virk, and many aunts, uncles, and cousins as well as members of the Maxwell and Cunha families—especially Carol and Phil Maxwell. Together, they made this book possible.

Of course, I reserve special appreciation for Lida Maxwell, who couldn't possibly have known what she was signing onto when we met seven years ago. She has offered all manner of practical and intellectual support: she has read and reread chapters and even the full manuscript, endured (and, perhaps, sometimes enjoyed) my extended absences when I was in the field, traveled to visit me in my sites, listened to countless speculations about my findings, clipped relevant newspaper articles, and generally lent her impressive mind to this endeavor. Many of the insights contained herein are Lida's, and I thank her for generously offering them over breakfast, walks, and drinks. Just as important, I owe her a debt of gratitude for encouraging me to put my pen down from time to time. Lida is a constant source of love and optimism, and any endeavor is lighter and better with her at my side.

Last but far from least, I thank my informants for their participation. I remain awed and touched by their willingness to open their lives and thoughts to me. Obviously, I cannot thank them by name, but I want each of them to know how glad I was to meet them and how grateful I am for their choice to share their perspectives with me and for the invitation to enter their homes, offices, and meetings. It is my great hope that they will find that their trust was well placed.

Chapter 5 utilizes—with permission—material that previously appeared in Japonica Brown-Saracino, "Virtuous Marginality: Social Preservationists and the Selection of the Old-Timer," *Theory and Society* 36, no. 5 (2007): 437–68.

Introduction

Mary

I first met Mary—a Portuguese American woman—during a college summer spent in Provincetown on Cape Cod. We frequently shared a shift at a dress shop, an odd couple to the say the least. Mary was working-class and middle-aged, with dark hair and olive skin. Nineteen, I arrived at work either with my skin freckled and my red hair lightened from a day at the women's section of Herring Cove beach or visibly tired from late nights with friends in one ramshackle cottage or another after we cashed out the registers at the shops where we worked to earn tuition money. I relished bike rides to work, while Mary always drove the mile from her apartment. I half dreaded summer's end, when I would return to books and papers and a work-study job at Smith College, while for Mary Labor Day and the break from tourists it promised could not come soon enough. Mary frequently chided me that I should forget my girlfriend (who, as I complained to all, was thousands of miles away in California) and marry a man. I responded by insisting that she hear about one crush or another, and, in turn, she recalled a bygone romance of her own. An unlikely pair, by July we were friends. In an album amid photographs of my college friends is an image of Mary and me in front of the library, my arm tossed casually across her shoulder, her head leaning ever so slightly toward her young friend.

While at nineteen my mind probably should have been on the beach or phone calls to the West Coast, I was intrigued by the larger pattern I recognized in the incongruity between Mary and me, between the Portuguese old-timer and the college girl who shared a counter at a Provincetown dress shop. We were much like the town was in the summer of 1997: a gay cabaret adjacent to a Portuguese bakery; a lesbian poet living next door to a fisherman; a photograph of a statue of Saint Peter on the shoulders of Portuguese

men processing to the harbor, a rainbow pride flag on a house taking posses-
sion of one corner of the frame.

While we did not often speak of it, that summer Provincetown was moving rapidly from one stage of gentrification to the next. The town had already
gentrified, but in 1997 it was not immediately apparent (at least not to this
college sophomore) how complete its conversion would be. In the decade
between that summer and this writing, it has become increasingly upscale,
and, with rising property values, rents, and taxes, one is much less likely to
encounter people like Mary or even college students like me as I was then
(few old-timers can afford the cost of living, and few college students can pay
rent for a summer season while also saving tuition money).

In the decade since I met her, Mary has remained a friend. As her life
changed dramatically with Provincetown, and as I transformed from a cu-
rious college student to a sociologist and ethnographer formally studying
gentrification, her story came to serve as a powerful reminder of gentrification's risks. Even after I began to study Provincetown in 2001—along with
two Chicago neighborhoods (Andersonville and Argyle) and rural Dresden,
Maine—and heard many other working-class residents' accounts, Mary's
story remained forcefully emblematic of gentrification's costs. For this reason, and because Mary's life changes parallel my entrée into the field, with her
permission I briefly share her encounter with gentrification.

During my first Provincetown summer, I came to know much about
Mary and, in turn, about Provincetown. Mary's father was a Portuguese old-
timer who owned a clothing store. After his death, her mother operated a restaurant in her home. Until her mother's death (some years before I met her),
Mary shared a small apartment with her and supported herself with long
hours as a clerk at two stores. In the winter, when most businesses close, she
relied on occasional, untaxed work and unemployment checks. Mary sometimes enrolled in community college extension courses, and she once sent me
a short story she had written about her grandfather's seafaring adventures,
the writing crisp yet sentimental.

As Provincetown rents increased, Mary worried about her ability to remain in town. She told me: "I don't see, unless my business takes off, how
I can make on a regular job enough to live alone, and I don't want to live
with anybody. . . . I'm just too set in my ways now." And, in fact, Mary's fears
were confirmed when several months after the 2002 death of her landlord—a
family friend who kept the rent below market rate—her heirs raised the rent
and readied the house for sale. Mary had no idea where she would find an affordable rental in Provincetown (the only place she had ever lived), and there
were no affordable housing units available for her.[1]

In 2003, six years after our dress shop summer, and after nearly a year without contact, Mary e-mailed, explaining that, some months before, after receiving an eviction notice, she began to scream uncontrollably and was admitted to a psychiatric hospital: "My year has been Eviction, Bankruptcy and slight breakdown. . . . I was to the point of calling shelters and then someone at the psych hospital I was in called the place I am staying in now. . . . It's such a long complicated mess, but I am . . . in a rest/retirement home with about thirty-five people. Most of the staff is from Brazil and so speak Portuguese. It is so nice to have that familiar sound around."[2] Today, Mary remains in the facility she describes above, several hours from Provincetown. A few residents—an aunt, an uncle, and a former coworker—have visited her, but she has not yet returned to Provincetown.

Two years before, Mary had told me of her fear that she would have to leave town to find housing: "I can't imagine not being able to walk down to the post office or seeing a dozen people you know. . . . In the winter you go to [the grocery store], and sometimes . . . you want to go in and get a loaf of bread, and you end up spending half an hour because you run into ten people." While Mary's Provincetown life was never ideal—threaded as it was with financial struggle and loneliness—gentrification strained the ties and place-based identity she had always had. Just as powerfully, preceding her eviction, it colored her life with anxiety about potential displacement.

In a sense, Mary and I knew each other in the last years before gentrification would alter us and our relationship to Provincetown. Mary came to realize her worst fear and perhaps one of the greatest losses any of us can experience: displacement from the only place and people she knew. My loss was much less profound, yet it shaped the choices that color this book. For me, "P-town" is not as it was before I met Mary. Because of Mary and many others I have seen displaced since, I cannot walk Provincetown's breakwater without noticing the rotting frame of a small boat on the harbor floor, and I cannot fall asleep beneath the exposed beams of a cottage without imagining those who once used it as a shed or the nets and traps it once stored.

Given my concern for Mary and others like her in my sites, it may surprise the reader that this book is primarily about gentrifiers. While half the 160 formal interviews I collected in Provincetown, Dresden, Andersonville, and Argyle were with longtime residents, and while I believe that no account of gentrification should neglect their perspective, this book is more about the relationship of people like me to Mary than it is about Mary's relationship to the affluent people who have taken her place in Provincetown.

On hearing a story like Mary's, one braces oneself to hear about newcomers' singular attention to progress and grave indifference to longtime residents, about "pioneers" who seek to transform their place of residence with little regard for those who lived there before them. Indeed, the gentrification literature has devoted much of its discussion of gentrifiers to those who find pleasure in taming the "wilderness" and hope to earn a profit in the process. While my research has uncovered those who fit the pioneer prototype, I was surprised to discover that most of the gentrifiers I interviewed were more complicated characters. Almost all recognized that they were participants in a process that leads to longtime residents' physical, political, and cultural displacement. Many were concerned with preserving one form of local authenticity or another, and few supported wholesale transformation. I entered the field expecting to tell a story about gentrification like those many have written—about growth-machine politics (Logan and Molotch 1987) that conspire with gentrifiers' culture to ensure the displacement of those like Mary—only I wished to demonstrate the similarities between urban and small-town gentrification.

Instead, my interviews and observations continually uncovered those who depart from the literature's account of the ruthless pioneer. They revealed that a diversity of practices and attitudes characterizes gentrification. For this reason, I have a different story to tell about how political economy and culture conspire to shape gentrifiers' residential choices and daily practices and how these, in turn, alter the lives of those like Mary. To tell it, I must begin where my research did, with a review of the literature on gentrification and the actors who drive it.

Gentrification and the Urban Pioneer

For more than three decades, sociologists, planners, and policymakers have paid attention to *gentrification:* "an economic and social process whereby private capital (real estate firms, developers) and individual homeowners and renters reinvest in fiscally neglected neighborhoods [or towns] through housing rehabilitation, loft conversions, and the construction of new housing" (Perez 2004, 139). Importantly, gentrification is also supported by public investment of funds preceding or following the moving in of the *gentry:* typically young, highly educated individuals. Gentrifiers' motivations for moving to the central city or small town are varied, but some do so as an investment strategy or to find affordable housing, employment, or cultural amenities. The gentry's residential choices, together with public and commercial investment, result in the economic, political, and cultural transformation of neighborhoods and towns. Most notably, gentrification breeds rising housing costs and infrastructure transformations geared toward gentrifiers. These contribute to physical displacement and to what Michael Chernoff terms *social displacement:* "the replacement of one group by another, in some relatively bounded geographical area, in terms of prestige and power" (1980, 204; see also Spain 1980, 28).

While nearly all studies of gentrification take the central city as their subject, some document the gentrification of rural areas.[3] Attracted by landscape or other natural amenities, historic properties, or location, rural gentrifiers, like their urban counterparts, contribute to the transformation of their place of residence and to displacement. While scholars identified rural gentrification in the 1970s (associated with the back-to-the-land movement), many agree that baby boomers' retirement, increasing rates of second-home ownership, and technologies that permit telecommuting encouraged the escalation of the gentrification of some small towns.

In this sense, demographic and economic trends—such as the baby-boom generation's effect on housing markets in the 1970s and 1980s, an expanding service sector, increasing economic emphasis on tourism, and low interest rates—encourage the gentrification of urban neighborhoods and small towns (see Long 1980; Spain 1993; Smith 2002). As Logan and Molotch (1987) illustrate, "boosters" with an interest in economic revitalization, such as business, the media, politicians, universities, and cultural institutions, also encourage gentrification (see also Gotham 2005).

However, many scholars agree that demographic, economic, and political factors alone cannot explain gentrification's breadth or individual actors' choices, arguing that cultural tastes also influence gentrifiers' residential preferences. Scholars like Sharon Zukin argue that culture, along with concern for housing affordability and financial gain, contributes to many individuals' decision to invest in gentrifying space.[4] Specifically, early gentrification research identified an ideological orientation among gentrifiers that supported their engagement in the process: the "frontier and salvation" mentality (Spain 1993), which glamorized personal sacrifice and sweat equity as methods for "settling" the untamed city. Economic boosters and the popular press credited gentrifiers with restoring deteriorated and often historically significant housing stock and "infus[ing] moribund communities with new health and an appreciation for cultural activities" (Spain 1993, 158) and with spurring an "urban 'renaissance'" (Zukin 1987, 130). Scholarship also suggested that gentrifiers celebrate neighborhoods for what they might become, rather than for what they are or were. For such individuals, "the prospect of changing a neighborhood was both a challenge and a social responsibility" (Levy and Cybriwsky 1980, 144).

Scholars and others paint gentrifiers as similar to nineteenth-century pioneers on the plains, who had little regard for those there before them. For instance, Jim Stratton draws a direct parallel between frontiersmen and 1970s loft dwellers: "From the moment the earliest settlers learned they could profit by evicting the present tenants and replacing the old housing stock with log cabins the New World was locked on a course of Out with the Old, Up with the New" (1977, 7).

Recent scholarship argues that this ideology persists even as the frontier recedes, sustained by the press, the marketing of "frontier kitsch," and even an "entire cinematic genre that makes of urban life a cowboy fable replete with dangerous environment, hostile natives and self-discovery at the margins of civilization" (Smith 1996/2000, 15, 13). According to the geographer Neil Smith, such tropes tap into imperialist fantasies about recolonization "from the neighborhood out," a consequential fantasy as "urban pioneers seek to scrub the city clean of its working-class geography and history." From this perspective, pioneers engage in gentrification for financial gain and because they regard it as "an expression of personal activism . . . their personal triumph of culture over economics" (Smith 1996/2000, 16, 27, 43; see also Levy and Cybriwsky 1980).

Indeed, accounts suggest that the prospect of transformation still entrances gentrifiers. For instance, Monique Taylor writes that Harlem's gentry "identify themselves, implicitly and explicitly, as gritty and determined settlers, or as homesteaders and urban pioneers." Embedded in this self-image is a belief in the virtue of economic and cultural change: "The stories that middle-class black newcomers to Harlem tell make use of its spaces and other residents to present and explain themselves as actors returning to the ghetto and rescuing it" (Taylor 2002, 98, xi).

Importantly, scholars suggest that gentrifiers particularly celebrate the reclamation of space from *certain* longtime residents: "Just as the frontier thesis in US history legitimized an economic push through 'uncivilized' lands, so the urban frontier thesis legitimizes the corporate reclamation of the inner city from racial ghettos and marginal business uses" (Zukin 1987, 141). This reclamation is often quite intentional: the literature suggests that real estate agents and gentrifiers seek to strip urban space from its "historical association with the poor immigrants" (Smith 1996/2000, 8), and, in Elijah Anderson's words, "the emerging neighborhood is valued largely to the extent that it is shown to be separate from low-income black communities" (1990, 26). This "sanitation process" (Mitchell 1997) involves physically removing old-timers from public space or from the neighborhood altogether (Rose 2004) and otherwise catering to gentrifiers' "desires for safety and relative homogeneity" (Atkinson 2003a, 1841).

In this sense, the literature suggests that gentrifiers' relationship to long-time residents almost uniformly predicates itself on distance and conflict. Reviewing the gentrification literature, Sharon Zukin writes: "In street encounters, [gentrifiers and longtimers] approach each other warily until familiarity with neighborhood routine ensures politeness. . . . New middle class residents often expect crime to be as prevalent as 'background noise.' For their part, existing residents may resent the superimposition of an alien culture—with different consumption patterns and an accelerated pace of change" (1987, 133; see also Levy and Cybriwsky 1980; and Chernoff 1980).

Where the literature acknowledges ideological variation among players in the gentrification field—that is, where it notes that some may depart from the frontier and salvation ideology—it nonetheless assumes a neat correspondence between an actor's economic or structural position and cultural tastes, particularly his or her ideological orientation to gentrification. For instance, Janet Abu-Lughod writes:

> At one extreme are the diverse players who view the land and buildings of the Lower East Side simply as "property"—as commodities with exchange values only. Flippers are perhaps the purest example. . . . Purchasers and renters of rehabilitated apartments share some of the views of developers, they differ only in terms of their relative stakes in a gentrified neighborhood, with owners more concerned than renters. At the opposite extreme are those players who view the land and the buildings of the Lower East Side exclusively in terms of their use value, since they are outside of or very marginal to the system of property. The homeless are perhaps the purest example. (1994, 337–38)

According to this logic, there is an uncomplicated relationship between one's economic position and one's ideology, and gentrifiers should have little concern for Mary or for the Chicago merchant who, on learning of her impending displacement, cried that gentrification was "cutting her root."

Social Preservationists and Social Homesteaders

Given the literature's emphasis on gentrifiers' frontier and salvation ideology and my own inclination to regard gentrifiers as structurally and culturally at variance with longtime residents' interests and well-being, I was surprised when nearly half the eighty-plus gentrifiers I interviewed and many others I observed in Provincetown, Dresden, and Chicago's Andersonville and Argyle departed starkly from the ideology and practices associated with the pioneer. Specifically, thirty-eight of my gentrifier-informants articulated an ideology quite distinct from the frontier and salvation mentality that scholarship has highlighted. These individuals spoke and behaved in a manner that starkly diverged from the pioneer model, and many of their beliefs and practices failed to correspond neatly with their economic interests, sometimes contradicting them.

For instance, I interviewed a Provincetown homeowner—a middle-class, middle-aged, white gay man—who worked to establish a "cultural sanctuary" to preserve the town's "fertile cultural ecology"—particularly its Portuguese fishermen and artists. For him, what makes Provincetown the place where he chooses to live is the presence of those who were there before him. In Chicago's Argyle neighborhood, which is home to many Vietnamese and other Asian old-timers and their shops, I observed a rabbi who seeks to prevent the disruption of Argyle's Asian streetscape at the expense of his *own* cultural tradition despite the neighborhood's history as a Jewish enclave. He said: "This is an old community built upon various cultures and ethnicities. [We] want to have a kosher delicatessen here—and, God, we need a kosher delicatessen—but it doesn't have to be on Argyle. Let's be perfectly honest: it doesn't belong here."

Meanwhile, in rural Dresden, Maine, affluent newcomers purchased acreage from a retired farmer for whom property taxes had become unaffordable. They did not do so as an investment. In fact, they agreed not to develop most of the land. Instead, their motivations were a desire to preserve land from development and to maintain the presence of a farmer who, for them, lends the land meaning. And, twelve hundred miles away, a white, gay resident of Chicago's Andersonville suggested that his move to the traditionally Swedish neighborhood was equivalent to colonization and derided the increasing presence of gay men like himself.

These anecdotes, among hundreds collected over the course of four years of fieldwork, have led me to conclude that, on the whole, the gentrification literature presents an oversimplified image of the gentrifier as pioneer and overstates the relationship between economic position and ideology. Generally speaking, we have failed to fully document the multiple attitudes and practices that characterize gentrification. Most dramatically, the literature has overlooked what I call the *social preservation* ideology and practices of some gentrifiers.

Social preservation, a heretofore unidentified and little-examined set of beliefs and practices to which I primarily devote this book, is an orientation toward gentrification and longtime residents markedly distinct from that of the pioneer (Brown-Saracino 2004). From a social preservationist's perspective, the central city and the small town are attractive because they possess "old-timers"—the longtime residents whom they most admire and who have, they believe, a rightful claim to their place of residence—and their "authentic" communities. Those who adhere to the social preservation ideology—who, like other gentrifiers, tend to be highly educated and residentially mobile—move to live near old-timers and their communities or, on moving to gentrifying space, come to appreciate qualities that they associate with *un*gentrified space, particularly those embodied by old-timers. As a result, those who articulate the social preservation ideology engage in a set of political, symbolic, and private practices to maintain the authenticity of their place of residence, primarily by working to prevent old-timers' displacement.

Social preservation is in some ways analogous to environmentalism. Like environmentalists, who seek to preserve nature, social preservationists—those who adhere to the preservation ideology and engage in related practices—work to preserve the local social ecology.[5] They combine ideology with practice by working to prevent old-timers' displacement, despite acknowledging the disruption their own in-migration causes. Their practices range from the creation of artwork celebrating old-timers or deriding gentrification to the construction of affordable housing and include myriad private acts, such as frequenting old-timers' businesses. Their concern is rooted in empathy for those whom displacement threatens and appreciation for old-timers' communities, particularly those of working-class ethnics.

From a social preservationist's perspective, gentrifiers fall into two bodies: callous pioneers and those who recognize their complicity in gentrification and seek to undo the harm they cause. Preservationists frequently report that they are among a select few who value local social authenticity. At first, having identified social preservation, I concurred that the field was divided between preservationists (who do not fully recognize themselves as such) and pioneers.[6] However, I came to recognize that the notions *social preservationist* and *pioneer* do not capture the full range of variation among gentrifiers. A careful analysis of my data suggests that, as Damaris Rose (1984) first argued, *gentrifier* is a more "chaotic" concept than most acknowledge and than I first found it to be.[7] Many lobby to narrow definitions of *gentrification,* to retreat from definitional "chaos," but my findings suggest that gentrifiers' ideological and practical orientations are diverse (see also Zukin 1987). Gentrifiers vary in terms of the impetus for their relocation, their ideological orientation to their place of residence, longtimers, and gentrification, and their practices.

Indeed, few of the eighty-plus gentrifiers I interviewed fit the literature's image of the pioneer. While social preservationists are most distinct from the pioneer model, most others also depart from it in subtle ways. Many fall between the preservationist and the pioneer: they neither seek to prevent old-timers' displacement nor wholeheartedly support total transformation. Like preservationists, they deem certain place characteristics, such as houses, landscape, or social character, authentic and work to preserve them, although the authenticity they appreciate is rarely directly tied to longtime residents. I have come to think of these gentrifiers—who are neither speculators nor social preservationists—as *social homesteaders.*

In some scholarship, the term *homesteader* is used synonymously with *urban pioneer* (see Spain 1993; and Taylor 2002). Here, I differentiate between the two and assign new meaning to *homesteader. Social homesteaders,* as I use the term, are, like all other gentrifiers, pioneers on the urban and rural frontier who engage in the transformation of poor and working-class neighborhoods to serve middle-class purposes. However, their interest is less in clear-cutting the wilderness—in unmitigated transformation—than in selectively and somewhat cautiously altering their place of residence to build a home. That is, they are not pioneers who boldly and unapologetically seek to "retake space" (Smith 1996/2000, 45).

However, social homesteaders are also less hesitant than social preservationists about making space for themselves alongside longtimers. While most social homesteaders do not readily articulate self-satisfaction associated with gentrification-related transformation, they do advocate for changes to improve their quality of life. For instance, some lobby for improvements to local schools, while social preservationists worry about the costs to old-timers of tax increases that such alterations require. Similarly, many social homesteaders prioritize the preservation of the built or natural environment, which they believe gentrification threatens, over longtime residents.

While homesteaders' beliefs and practices distinguish them from pioneers and preservationists, they are, nonetheless, internally heterogeneous. The motivations for their relocation and practices vary within a set of parameters. Once in a place, they celebrate its characteristics, ranging from landscape to social diversity writ large. What social homesteaders share is a combined commitment to preserving authenticity and ensuring progress. They appreciate authenticity—which they often associate with the built or natural environment—but gentrification's cost for longtime residents and their communities are not their central concern. They welcome others who share their cultural and economic traits and articulate a sense of belonging or ownership vis-à-vis their place of residence.

Thus, while social preservationists are not alone in their resistance to gentrification, they nonetheless stand apart from other gentrifiers. They alone relocate to live beside old-timers and put the preservation ideology to practical use. While some homesteaders articulate a taste for diversity or "the cultural practices of the categorical 'other'" (Mele 2000, 4), there is an important difference between this symbolic consumption of diversity and social preservation. Preservationists enact appreciation for difference, as embodied by old-timers, through practices intended to preserve that difference. Such practices are particularly salient given preservationists' self-reflexivity: their cognizance of their impact on their surroundings, sophisticated understanding of political economy, and concern that symbolic preservation may cause old-timers' displacement. Given its grand departure from the pioneer framework—to which the gentrification literature has largely devoted itself—I take social preservation as the primary focus of this book. However, to deny some attention to homesteaders would risk reifying the oversimplifications that characterize the literature and that I seek to correct.

Indeed, I do not want to overstate the differences between preservationists, homesteaders, and pioneers. A gentrifier's ideology or practices may change over a lifetime, as is the case with several preservationists who once encouraged gentrification through the restoration and sale of historic properties. In this sense, we might consider the categories "ideal types" (Weber 1968/1978). However, if we are to think of ideal types in the sense with which the term is often used—to refer to typologies that are unlikely to exist in ideal form in the social world—it would be an oversimplification to suggest that the categories are strictly ideal types (Weber 1968/1978). During four years of fieldwork, the gentrifiers I interviewed and observed consistently adhered to a single set of beliefs that they partnered with corresponding practices. None espoused concern for old-timers one day and appreciation for rising home values another.

I argue that the ideological and practical consistency that I observed during the period of study results from several factors. First, as the following pages will demonstrate, gentrifiers have access to a plethora of cultural images of the gentrifier and gentrification. Gentrifiers—from the pioneer, to the homesteader, to the preservationist—may understand and position themselves in relationship to such images. Indeed, throughout this book, I present evidence of preservationists' acute awareness of images of the callous or hapless pioneer and effort to frame themselves in opposition to such images. In short, such referents may enable the ideological and practical consistency that I identified. Second, the preservation and homesteading ideologies are inherently contradictory—wedding the desire to move to gentrifying space and to prevent some of the transformation that such movement causes—and it may be easier to consistently adhere to such an ideology than one that demands the avoidance of such contradictions. Third, I have identified gentrifiers' ideological and behavioral differences through observation. The categories are not abstract; rather, they are products of a grounded theory approach (Strauss and Glaser 1967).

In sum, I use the categories of gentrifiers I identified to describe the ways in which gentrifiers oriented themselves ideologically and practically to gentrification during the period of study. The typologies capture, not types of people, but rather how individuals position themselves, ideologically and behaviorally, in relation to a political, economic, and cultural process in which they are engaged. In other words, my typologies are grounded in a specific context—gentrification—and should be understood as bearing descriptive weight only in that context. Therefore, as I further specify in the next chapter, I refer to those who, during the period of study, reliably articulated components of the preservation ideology and engaged in corresponding practices as *social preservationists*. Likewise, I identify those who adhered to the homesteading ideology and engaged in complimentary practices as *social homesteaders* and those who articulated the frontier and salvation mentality and engaged in related practices as *pioneers*.

Roughly half the gentrifiers I interviewed articulated components of the social preservation ideology and worked to resist gentrification and its consequences for old-timers. Observations uncovered many others who are engaged in the practices of preservation and who publicly tout social preservation concerns. Given that social preservationists do not have a name for their ideology or organizations explicitly dedicated to their cause, social preservation is surprisingly widespread. It is plausible that, among contemporary gentrifiers, there are more social preservationists than pioneers because social preservation exists in reaction to gentrification. That is, because social preservation predicates itself on opposition to gentrification's risks and the "urban invader," those who adhere to the ideology possess a remarkably coherent set of beliefs and practices (London 1980). Because they know who they do not wish to be (heedless pioneers) and what they do not want the space in which they live to become (gentrified), they are highly conscious of how they "should" think and act in relation to old-timers and, more generally, gentrification. This self-awareness vis-à-vis gentrification is not the sole domain of social preservationists. Some homesteaders demonstrate related self-consciousness, but preservationists are particularly self-critical, and this explains much of the breadth and steadfastness of their ideology and practices.

Table 1 outlines the key differences among gentrifiers.

TABLE 1. Key ideological distinctions among gentrifiers

Type of newcomer	Origins	Vision	Attitude toward newcomers	Attitude toward old-timers
Social preservationist	Desire to live in authentic social space and affordable housing for middle class	Social ecology to be preserved and enjoyed; recognition of old-timers' culture	Dilute the authenticity of space; displace old-timers; threaten real community	Colorful; authentic; desirable
Social homesteader	Desire to live in authentic space and affordable housing for middle class	Improved space that includes embodiments of high culture and certain original features	Potential allies in improvement and/or efforts to preserve the built, natural, or social environment	Appreciation for diversity writ large; objects of uplift or symbolic preservation
Pioneer	Affordable housing for middle class; promise of economic gain; excitement of revitalization	Frontier to be tamed and later marketed	Welcome fellow pioneers; increased safety; rising property values	Threatened by; critical of

Note: Adapted from Brown-Saracino (2004).

Where Were Social Preservationists and Social Homesteaders Hiding? Explaining the Gap in the Literature

Cumulatively, preservationists and homesteaders complicate the typical story about gentrifiers' ideological and practical orientation. They challenge the notion of a monolithic "urban invader" (London 1980) by suggesting that, among gentrifiers, there are multiple orientations to place and gentrification. Gentrifiers do not possess a single value structure.[8] As a few other scholars argue, this suggests that the pioneer framework either is outdated or has always oversimplified the field.[9]

The widespread presence of social preservationists and social homesteaders in my sites is somewhat surprising given that few scholars have devoted serious attention to those who depart from the urban pioneer prototype.[10] While some discuss antigentrification movements (Smith 1996/2000; Mele 2000), these have been associated with "punks, housing activists, park inhabitants, [and] artists" (Smith 1996/2000, 3) and do not identify the teachers, business owners, and Republicans I found to engage in social preservation alongside the more likely culprits listed above.

In the following paragraphs, I offer a few explanations for this gap in the literature, a topic I revisit in the conclusion. These explanations include the possibility that social preservation is expanding in relation to mounting popular self-consciousness about gentrification's risks. Because this only partially accounts for the absence of scholarly attention to social preservation, I argue that several prevailing approaches to the study of gentrification have also contributed to this omission: an emphasis on how gentrifiers' culture serves capital, the notion that demographic traits neatly produce orientations to gentrification, and the assumption that gentrifiers' concern for old-timers is always insincere and never translated into practice.

In exploring the first explanation, I caution the reader that I do not wish to argue that social preservation is an entirely new process. It may, however, be an expanding one, and popular criticisms of gentrification likely contribute to the self-reflexivity that drives social preservationists and discourages homesteaders from participating in the reclamation that pioneers celebrate. Today, public debate about gentrification is ubiquitous (Smith 1996/2000). For example, in 2004, a Chicago bookstore hosted a meeting on community preservation at which, before a white, middle-class crowd, a speaker identified gentrification as a threat to "vibrant, eclectic" immigrant neighborhoods. In 2005, Chicago's Fight Club—a venue dedicated to public debate—hosted a "fight" entitled "Gentrification: Destruction or Creation of a Neighborhood?" The material promoting the fight on the club's Web page suggested that "from its very beginning the term [gentrification] was a pejorative" and included links to scholarship on gentrification as well as "a very in-depth blog post called 'Gentrification: Good or Evil.'"[11] And, in 2006, while flipping through the channels at a Chicago gym, I came across a cable access station's presentation of a study of gentrification by Loyola University Chicago's Center for Urban Research and Learning (see Nyden, Edlynn, and Davis 2006).

As this evidence suggests, the discourse about the threat that, newcomers—or "yuppies"—pose to old-timers is by no means limited to academic tomes. Indeed, it is widespread. In a New York Times article about Brooklyn's gentrification, the director of an organization that works to prevent displacement said of stores catering to yuppies: "'Now you can't swing a cat without hitting one'" (Kirby 2003, B46). Such sentiment is apparent on popular television: in an episode of the NBC program Seinfeld, the character Kramer tells his friends: "You yuppies are destroying the fabric of our neighborhood!" His particular concern is that an upscale coffee shop or bakery will replace a "mom-and-pop" cobbler shop. Such anxiety extends to small cities and even to rural areas. The Indigo Girls' 2006 album, Despite Our Differences, includes a song bemoaning rural gentrification and capturing the self-consciousness of "suburban pioneers." Amy Ray writes:

> It's been you and me on this frontier
> Trying not to be suburban pioneers.
> Fighting off the pavers
> And the associations
> And the covenants against the trailers. . . .
>
> (RAY 2006)

Also in 2006, the Web site of the radio program *A Prairie Home Companion* published a letter decrying the movement of "yuppies" into the letter writer's hometown:

> Everything you talk about reminds me of where I grew up in San Rafael, CA. which was once a "small town" settled by Italians. . . . We knew the milkman, the garbage man and the mailman by their first names. . . . Up until about 1973 I could ride my horse anywhere downtown and tie him up to a parking meter and get an ice cream at Woolworth's for a quarter. . . . Then something happened about five years ago. . . . Snobby young yuppies with their whiney, undisciplined kids have moved in and all everyone seems to care about is how much money you have and what kind of car you drive and how much "stuff" you have. It's all about how good you look in spandex while getting your decaf-soy-non-fat-latte at Peet's Coffee and talking about how sore you are because you just worked out for 2 hours at 24-Hour Fitness. So now we are looking into new places to live—something we never thought about before—but this doesn't seem like "home" anymore.

The show's host, Garrison Keillor, urged the letter writer to reconsider moving, suggesting that she offers value to a town overrun with yuppies: "San Rafael will be poorer if you leave, my dear, and those whiney children will be even less likely to hear about the old Italians . . . which would be a pity."[12]

Thus, in the popular press as well as in the academic literature, gentrification has acquired negative overtones, and it is likely that many gentrifiers are aware of this and that social preservation is, in part, a response to such criticisms of the process. This supports Tim Butler and Garry Robson's argument that "gentrification as a concept may be reaching a 'sell by date'—largely because of its connotations. Nobody is in favour of gentrification."[13] They suggest that gentrifiers are increasingly aware of their "effects on the 'other'" (Butler and Robson 2003, 15, 4).[14]

While increasing popular criticism of gentrification may contribute to the expansion of social preservation and this, in turn, may partially explain why earlier scholarship did not identify the process, it is an insufficient explanation for the gap in the literature. This is not a historical study, and I cannot pinpoint social preservation's origin, but some of my informants report that, as early as the 1970s, the ideology motivated their residential choices

and shaped their practices. Even assuming that social preservation is a bur-
geoning process, this does not explain why other contemporary scholars have
paid it little attention. In short, we must still explain the absence of sustained
examinations of departures from the pioneer model.[15]

A second explanation for this gap is that, despite calls in the 1980s and
1990s to consider the possibility that residential choices are not purely economically rational, most gentrification scholars continue to do so, therefore
limiting attention to ideological variation (Zukin 1987, 142; Caulfield 1994).
This is not to say that such calls went altogether unheeded. They attracted
attention to gentrifiers' culture and garnered recognition that economic interests are not the only motivations for residential choices. However, subsequent work on gentrifiers' tastes and beliefs emphasizes how culture *serves*
gentrifiers' economic interests as well as the interests of local government,
private capital, and other boosters. In short, most have taken gentrifiers' cultural tastes and practices to be the unambiguous handmaiden of local economic change and personal financial and status gain.[15]

For instance, while the sociologist Sharon Zukin's pioneering study of
loft conversions insightfully delineates the role of cultural tastes and actors
in the accumulation of capital in the central city, it also argues that this is
closely linked to elites' economic interests: "The city reaps a benefit—a social
and financial payoff—from loft living's demographic and cultural effects"
(1982/1989, 3).[16] Richard Lloyd tells a similar story about how the avant-garde
culture of neobohemians in Chicago's Wicker Park facilitates gentrification
even as adherents mourn the displacement of urban "grit." Lloyd writes:
"[The] practices of neo-bohemia contribute to other strategies of capital
valorization in the postindustrial city" (2005, 44). While his evidence of how
galleries, bars, and shops encouraged gentrification is careful and convincing, he regards neobohemians' complaints about gentrification as largely aesthetic and as anything but an affront to the neat relation between culture and
capital. This analysis, like Zukin's, arises from the notion that culture serves
capital.[17]

While I do not deny that culture often functions in this way, social preservation suggests that the relation between culture and capital is not always
as neat as the intellectual tradition suggests. Social preservation pushes us to
consider the possibility that culture—even that of relative elites—can complicate processes of capital accumulation, such as those associated with gentrification, even if it does not effectively prevent or stall it. A lack of recognition of this possibility may have hindered recognition of social preservation.

A third explanation is that scholars failed to identify social preservation
because they identify gentrifiers by their demographic traits and assume that

ideology neatly corresponds to structural position. If this is, in fact, the case, then, because social preservationists share many demographic traits with other gentrifiers, and because all gentrifiers move to areas populated by those less educated and affluent, preservationists would not have been immediately distinguishable. Thus, the expectation of neat correspondence between demographic traits and orientation to a process like gentrification helped mask social preservation.

Fourth, researchers may have overlooked social preservation because they conclude that, since places gentrify regardless of some gentrifiers' discomfort with the process, such departure from the frontier and salvation ideology is insincere or otherwise unworthy of attention.[18]

This is striking given that some have acknowledged some gentrifiers' awareness of the displacement of longtime residents or appreciation for urban diversity.[19] For instance, Neil Smith (1996/2000, 32) suggests that, by the 1980s, the antigentrification movement induced the Board of Real Estate of New York to print an ad defending the process. Similarly, Elijah Anderson describes the movement of white liberals into a neighborhood to establish a racially and economically egalitarian community. "Indeed," he writes, "many found inspiration, if not affirmation, in their relationships with blacks of the Village and the nearby ghetto" (1990, 8, 17 [quote]). As was noted above, Butler and Robson identify a taste for diversity and difference among certain London neighborhoods' gentrifiers, who "are, on the whole, a highly reflexive bunch of people who are very aware of the consequences for, and effects of their interactions on, the areas in which they live" (2003, 4). Identifying a more general trend, Richard Florida writes: "The creative class is drawn to more organic and indigenous street-level culture." It is also drawn to a "cultural community" that is often "reviving-downscale" (2002, 182–83). Like Florida, Richard Lloyd notes artist gentrifiers' appreciation for "street level diversity" (2002, 520). While such scholars capture gentrifiers' desire to distance themselves from gentrification and appreciation for diversity, they tend to regard these traits as largely symbolic and/or insincere, and, perhaps as a result, few offer sustained studies of departures from the pioneer prototype.

In short, most scholars have taken such self-consciousness and taste for diversity to be at best ironic, and at worst deceptive, or the prerogative of a particular type of gentrifier, for example, African Americans gentrifying predominately African American neighborhoods or first-wave gentrifiers concerned with preventing their own displacement (see Allen 1984; Rose 1984; Caulfield 1994; Taylor 2002; Lees 2003; and Pattillo 2007). Many presume that, because neighborhoods gentrify regardless of gentrifiers' self-consciousness, such sentiments must not translate into practices that resist gentrification.[20]

For instance, Damaris Rose tells of condominium owners in gentrifying Montreal neighborhoods who "articulate an egalitarian vision of the inner city
while grappling openly with the paradoxes of personally being beneficiaries
of the economic and social dynamics of gentrification." As she found: "Many
in this group are emphatic about their appreciation of the neighbourhood's
current social class diversity and state that they wouldn't want to live there
if a gentrified 'sameness' set in due to the arrival of many more people like
them! . . . [T]his group of interviewees articulate discourses of cosmopolitanism and urbanity that, while superficial and aestheticized, . . . seem neverthe-
less central to their identity constructions." However, she swiftly dismisses
this perspective, writing that it is "no doubt practiced only 'in the mind'"
(Rose 2004, 299–300). Besides pointing to the swift dismissal of departures
from the pioneer prototype, this underlines the limits of the interview and
survey research that is the foundation of much of our understanding of gentrifiers and their motivations. For instance, since Rose's conclusions are based
solely on interviews, we cannot be sure that such beliefs are "practiced only
'in the mind.'" The assumption that such beliefs are divorced from action
and are, therefore, insincere may arise, in part, from the fact that most studies
examining attitudes toward diversity or social mix are not based on observation, leaving researchers with little occasion to test whether belief translates
into practice, and, therefore, allowing cynicism to prevail.

The media mirror this depreciation of gentrifiers' self-consciousness, noting the ironic presence of an antigentrification ideology among gentrifiers.
An April Fool's Day issue of a San Francisco paper "read 'Old Yuppies Decry
New Yuppies' and 'Pot Calls Kettle Black'" (Solnit and Schwartzenberg 2000,
122), and the satirical newspaper *The Onion* published a facetious article that
reads: "A three-year resident of Chicago's Wicker Park neighborhood, lashed
out Monday against encroaching gentrification. 'See that big Barnes & Noble
on the corner? You better believe it wasn't there back in '98,' said Smales, 34,
a finance manager with Accenture. 'This whole place is turning into Yup-
pieville. You can't throw a rock without hitting a couple in matching Ralph
Lauren baseball caps walking a black lab.' Smales then took his golden lab for
a walk" ("Resident of Three Years" 2001). While fictional, such satire notes a
social trend: gentrifiers' wariness of gentrification, which has become a powerful symbol, as well as the tendency to dismiss such resistance as merely
hypocritical (Smith 1996/2000, 34).

What are the consequences of having overlooked social preservation? In
1987, Sharon Zukin noted a "worrisome stasis in the field" (132), and this
book seeks to address her concern some two decades later. It also aims to fill
what Jon Caulfield considers the "silence of 'gentrifiers' themselves in the

scholarly literature" (1994, xi). In so doing, I identify a range of voices, including preservationists' particular perspective, which stands markedly apart from that of other gentrifiers. Despite the work of those like Caulfield, as a result of the literature's focus on outcome and the assumption that gentrifiers are universally economically rational, the intellectual stasis to which Zukin referred has become so powerful that we have missed much of gentrifiers' ideological and practical variation. Indeed, we have overlooked significant attributes of the *process* of gentrification. At this stage, there is arguably little new to be learned about gentrification's outcomes—particularly about its high costs for longtime residents—but our certainty about such outcomes has distracted us from a serious analysis of the motivations behind gentrifiers' participation in the process and of their daily practices. In short, it has prevented us from attending to many of the factors that drive gentrification and the resultant displacement.

Thus, this book offers empirical findings and theoretical perspectives that depart from the gentrification literature in at least three key ways. First, it suggests that few gentrifiers possess the frontier and salvation ideology (Spain 1993) that scholars and the media—and the gentrifiers themselves—typically associate with them. Second, I demonstrate that concern for a sense of place, community, and authenticity and a related self-consciousness about gentrification are not the purview of a small minority of gentrifiers, nor are they limited to given locales or stages of gentrification, as some scholars have argued (Rose 1984; Butler and Robson 2003). Instead, gentrifiers' attitudes about gentrification and related practices are quite diverse. Third, I introduce social preservation. Those who espouse the social preservation ideology and engage in related practices stand apart from others who appreciate diversity and express self-consciousness about gentrification, for they are particularly cognizant of their participation in gentrification and seek to orient their action in opposition to that of the ruthless invader. They further differentiate themselves by unifying their appreciation for diversity and concern about gentrification with practices intended to preserve old-timers' presence and communities.

This book does not celebrate social preservation as an answer to the harm that gentrification causes longtime residents like my friend Mary. While I hope that planners and policymakers will benefit from a consideration of social preservation, I do not frame social preservation as a policy suggestion or otherwise devote serious analytic attention to the outcomes of preservationists' practices. There is much scholarship on gentrification's outcomes, but we know less about contemporary gentrifiers' motivations, beliefs, and daily practices. Thus, I seek to document how gentrifiers think, feel, and behave and do not regard any of their approaches as a simple solution to the inequalities endemic to gentrification. Indeed, I discourage the reader from regarding social preservation as a clear-cut solution, as I have often paused to reflect that the experiences of those like Mary are all the more ominous when we recognize that many of those whose tastes and choices contributed to her displacement recognized this risk and even sought to avoid it.

That said, while I do not wish to celebrate social preservation or position it as a policy solution, I seek neither to unmask the contradictions inherent in preservationists' ideology and practices nor to otherwise debunk them. It is easy enough to point to social preservationists' hypocrisy (of which they are well aware and self-critical), but I have, instead, adopted what I regard as a more challenging task: to understand how gentrifiers conceive of their action and to identify some of the differences (both intended and otherwise) that they make in the places where they live. Sociologists rightfully devote much attention to the processes that produce and reproduce inequality, but, to fully understand the social world, we must also study efforts to reduce inequality, even if they are less successful than their practitioners hope they will be.

Furthermore, many social scientists seek to give voice to the perspectives of those to whom society ordinarily gives less weight. While this is a valuable and admirable task, there is much that we can learn from the study of power holders, of those whose beliefs and choices are of particular import for the less powerful, such as those at risk of displacement. In short, this book suggests that it is not enough to identify actors' structural locations. We must also understand their culture and how ideology both affirms and upsets their political-economic positions.

What Follows

In the following chapters, I propose that the field of gentrification consists of newcomers who possess a variety of attitudes about gentrification and who engage in an array of practices. Their approaches to gentrification range from that of the pioneer, who consciously "retakes" space (Smith 1996/2000, 45) and who has long been the focus of scholarly attention, to the social preservationist, for whom the value of a neighborhood or town is contingent on old-timers' sustained presence. Homesteaders' ideology and practices fall between these ends, and, as I explore in chapter 4, local context shapes the tenor of all gentrifiers' discourse and practices.

The goals of this book are twofold. First, it identifies general variation among gentrifiers. For this reason, I devote a chapter to social homesteaders. I draw attention to how homesteaders depart from the literature's depiction of the pioneer and, in so doing, remind us of the diversity of attitudes and practices that fuel and accompany gentrification. Second, the book primarily devotes itself to the nearly unexamined ideology and practices of social preservation. In so doing, it emphasizes how ideological—rather than merely material—differences shape the lives and practices of a variety of actors in gentrifying neighborhoods and small towns.

In the next chapter, I outline my methods and introduce the sites. The chapter on social homesteaders follows. Amid a book primarily devoted to social preservation, it demonstrates gentrifiers' heterogeneity and reminds us that not all efforts to resist gentrification and preserve authenticity are synonymous. Finally, it places social preservation in stark relief by demonstrating how it is different from the preservation movements to which scholars have traditionally attended, such as those of the built and natural environments. Chapter 3 describes social preservation, emphasizing adherents' motivations for relocation and beliefs about gentrification, old-timers, and community. Necessarily, in this section, I approach pioneers and longtime residents as preservationists regard them: as "objects" against which they understand themselves and their neighborhood or town. In other places, I remind the reader of their subjectivity by devoting space to variation among gentrifiers and longtimers.

The following chapters document the strategies that preservationists employ in each site (chapter 4) and preservationists' selection of old-timers from a pool of longtime residents—for they do not work to preserve all longtimers (chapter 5). Chapter 6 explores the origins of the ideology of social preservation and how some gentrifiers come to adopt it. Chapter 7 takes those whom social preservationists identify as old-timers as its subject, documenting how their reactions to gentrification depart from preservationists'. Finally, the conclusion asks us to consider what social preservation means for those like Mary who experience physical, cultural, or political displacement at the hands of gentrification. In it, I also discuss scholarly traditions—those that encompass and extend beyond the study of gentrification—that account for the gap in the literature and, thus, kept social preservation unidentified and little explored.

The Research Sites and Methods

This book examines the gentrification of four sites: two Chicago neighborhoods (Andersonville and Argyle) and two small New England towns (Dresden, Maine, and Provincetown, Massachusetts).[1] Specifically, it explores the factors that motivate gentrifiers' relocation to the central city or a small town, their practices, and the response of residents—new and old alike—to gentrification. While my curiosity about a particular site (Provincetown) and a semi-autobiographical interest in rural New England influenced site selection, I used census data to select sites that over the past decade had experienced demographic change indicative of gentrification, such as rising property values, median income, and percentage of residents with a B.A. and decreasing poverty rate and proportion of children and the elderly.

I sought communities that were similar in such ways but geographically and demographically distinct. I chose sites that varied in terms of longtime residents' racial, ethnic, and occupational characteristics, stage of gentrification (Clay 1979), and population as well as in terms of newcomers' characteristics. Two are in urban areas, and two are small towns. Two were traditionally white, ethnic enclaves, while one is home to a large population of Asian immigrants. Both Andersonville and Provincetown have experienced a notable influx of gays and lesbians, and Provincetown is a tourist destination. While the sites, which are equally divided between the small town and the urban neighborhood, were not randomly selected, they provide valuable sets of similarity and difference (see table 2).[2] In the sections that follow, I describe the sites and my methods.

TABLE 2. Key characteristics of research sites

	Stage of gentrification	Old-Timers (as defined by community)	Other long-time residents	Newcomers	Location	Urban/rural character
Dresden	Early/mid	Yankee farmers and others who work the land	Russian Americans, factory workers, state employees	Substantial population of white retirees, back-to-the-landers, and dual-income commuting couples	Maine	Rural town
Provincetown	Advanced	Portuguese fishermen and their families and, secondarily, struggling artists	WASPs, descendants of Pilgrims, gays with history of residence, non-Portuguese working class	Large population of affluent, white gays and lesbians	Cape Cod	Small, isolated tourist town
Andersonville	Advanced	Swedish Americans and, secondarily, independent business owners	Middle Easterners, lesbians, Japanese residents, middle- and upper- class longtime residents	Large population of white lesbians, affluent, white heterosexual families, and, more recently, affluent, white gay men and young heterosexual couples	Chicago's North Side	Urban neighborhood
Argyle	Early/mid	Asian American (particularly Vietnamese) business owners and their families	African Americans, Appalachians, SRO residents, Ukrainian Jews, descendants of theater-district past, a handful of Swedish Americans	Predominately white population of middle-class heterosexuals, gay men, and artists	Chicago's North Side	Urban neighborhood

Sites

DRESDEN

Dresden, Maine, is a picturesque town of 1,625 with rolling hills and shady dirt roads located twenty-three miles from coastal Boothbay Harbor, thirteen miles from the Bath Iron Works Factory (a major employer), and eighteen miles from the state capital, Augusta. White Americans first permanently settled the town in 1752, when the land's proprietor induced German and French Huguenot immigrants from Boston to establish homes where Dresden stands today. The promise that any man who settled one hundred acres was entitled to the property encouraged settlement. For a short time, the town, then known as Pownalborough, was the county seat, and residents built a courthouse on the banks of the Kennebec, one of Dresden's two rivers.[3] The courthouse, a massive colonial house built in 1761, still stands today. Containing Maine's only remaining colonial courtroom, the building, which until the 1950s was home to descendants of its first proprietors, attracts visitors who amble through an old tavern and a courtroom where John Adams once argued a case.

Most of Dresden's early land parcels included fertile, tillable riverside acreage. Settlers supplemented farming with fishing; the sale of hay, fish, poultry, and wood subsidized what were subsistence farms. During the eighteenth

FIGURE 1.01. Scenic Waterway, Dresden

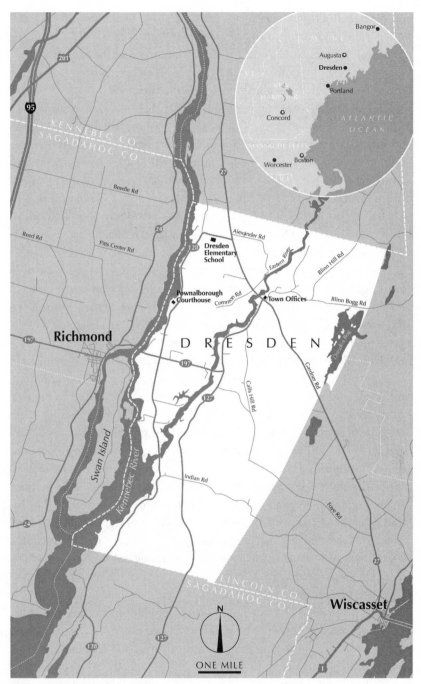

MAP 1. Dresden. Map by Dennis McClendon.

and nineteenth centuries, residents used the rivers to transport farm products. For this reason, the town contained many wharves as well as ferries and private toll bridges, and, in the nineteenth century and the early twentieth, it was a regional shipbuilding center. In 1850, Dresden reached a population high of 1,419, which it did not attain again until the 1990s.

Historically, Dresden's largest industry was that of ice harvesting, which flourished from 1870 to 1920. With the aid of seasonal laborers, residents cut and shipped ice to the American South. Boardinghouses and hotels sheltered seasonal laborers, and, after the ice industry was lost to the electric icebox, summer tourists rented the rooms. Today, all that remains of the town's ice industry is a book written by an old-timer (Everson 1970) and an elaborate wooden replica of an icehouse in a back room of the Pownalborough Courthouse Museum. The failure of the ice industry led Dresden to reach a population low of 620 residents during the 1920s.

Until World War II, there were as many as fifty or sixty dairy farms in town (Bolté 1983, 117). Residents also raised blueberries, strawberries, apples, and potatoes. Despite the fact that Dresden "has 25 percent of all prime farmland soils in Knox and Lincoln Counties combined" (Town of Dresden 1991b, 15), as the twentieth century progressed Dresden families found it increasingly difficult to support themselves through farming, the common means of support throughout New England. Many sons, who provided important farm labor, did not come home after World War II. Some died in the war, but more went to live elsewhere or chose a trade other than farming. Large Midwestern and California farms were tough competition for the small, New England farm (Kramer 1987). These factors, combined with the growing expense of equipment (tractors, tillers, and trucks), led to the closure of most Dresden farms. As residents wrote in Dresden's 1991 Comprehensive Plan: "Many sources of employment outside of Town have become available. Instead of being farmers or fishermen, the majority of the populations work outside of Town and Dresden has primarily become a bedroom community." Referring to the libertarian ideology popular among old-timers, they added: "The independent attitude of long time residents is still evident however" (Town of Dresden 1991a, 4).[4]

In the mid-twentieth century, Dresden received its first influx of middle-class newcomers. A Russian immigrant purchased dozens of Dresden farms, most with a view of the Eastern or Kennebec River, and sold them to other Russian immigrants living in New York and New Jersey, who restored the homes. Some moved permanently to try a hand at farming, and others used the farms as summer retreats (Jaster 1999). However, the Russian settlement was short-lived, and its cultural centers were across the river in neighboring

Richmond. Today, few Russians remain in town, although a few maintain tracts of land in Dresden.

A second wave of newcomers—primarily young, highly educated white couples—arrived in the 1970s as part of the back-to-the-land movement. Many built houses through a local program that provides instruction in timber frame construction and started small home businesses or worked in neighboring towns as doctors, lawyers, or teachers. In 1991, the town noted that "the increase in the Dresden population since 1970 has been great" and that, while the population had yet to surpass that of 1850, "the number of houses in Town has *more than doubled* since" (Town of Dresden 1991b, 5, 6).

However, the most substantial population increase occurred between 1980 and 2000, and the increase continues at a fast pace (see table 3).[5] The town's population grew from 998 in 1980 to 1,625 in 2000, and the number of housing units increased from 552 in 1990 to 739 in 2000, with additional homes built each year since. Several developments have appeared in town, including one composed of log cabins, another of trailer parks, and, more recently, one that includes an airstrip and another with riverside lots listed at $250,000. Recent newcomers include retirees who prefer Dresden's riverfront properties to expensive coastal land and dual-income couples who enjoy the

TABLE 3. Dresden demographic shifts (1990–2000)

	1990	2000
Population	1,332	1,625
Percentage of population under the age of 19	28.6	26.8
Percentage of population 60 or older	14.7	14.5
Percentage of population with bachelor's degree or higher	14.19	21.5
Percentage of work force in professional or managerial position	25.63	30.1
Percentage of work force in farming, fishing, or forestry	5.5	3.1
Percentage of households earning less than $15,000	19.05	15.4
Percentage of households earning $100,000 or more	1.6	4.2
Number of structures built in decade	138	181
Median owner-occupied home value ($)	87,900	97,900
Median contract rent ($)	355	443
Number of people traveling 60 minutes or more to work	44	63

Source: U.S. Census data.

town's location between two major employers: the Bath Iron Works factory and the state government in Augusta. Among recent newcomers and the back-to-the-landers who preceded them there is nearly universal appreciation for Dresden's landscape and a desire to preserve it from development. Many enjoy hiking and skiing across the town's rolling hills or walking back roads, and some boat along the Kennebec River to coastal Boothbay Harbor.

As Dresden's population has increased, the number of residents with a bachelor's degree rose by 7 percentage points. The number of households earning at least $100,000 was more than two and a half times higher in 2000 than in 1990, while that of households earning less than $14,999 has decreased from over 19% to 15.4%. Longtime residents and preservationists note that newcomers increasingly hold positions of power on town boards and committees, a key indication of change in a town governed by town meeting. An elected board of three selectmen oversees town employees and regulatory boards—such as planning, school, and finance—as well as the conservation and solid waste committees.

Today, a gas station, a post office, a volunteer fire department, an antique store housed in a former church, the town office, and a small restaurant (take-out only) constitute Dresden's center. A sign across from the town office points to a used bookstore, while the town library, housed in a private academy that once educated local schoolchildren, sits outside the town

FIGURE 1.02. Town Hall, Dresden

FIGURE 1.03. Barn, Dresden

center, within a mile of the old town hall, where most town meetings are held. While the local historical society has preserved a handful of one-room schoolhouses, the town operates a modern elementary school and buses junior high and high school students to a neighboring town.

By 2000, only a little over 3% of Dresden's workforce listed farming, fishing, or forestry as an occupation on census forms. Yet a number of successful greenhouse, organic, strawberry, and blueberry operations remain in town, and roadside signs advertise eggs, organic produce, potted plants, and pick-your-own strawberries. However, hobby animals, such as llamas and horses, dot far more of today's pastures than do milk or beef cattle. While many longtime residents are manual laborers or employees of Bath Iron Works or the state, social preservationists primarily seek to preserve farmers, their families, and others who work the land.

PROVINCETOWN

Provincetown, Massachusetts, the easternmost town on Cape Cod, was incorporated in 1727. The region was long home to Native Americans, who fished in the natural harbor. In 1620, on their arrival in the New World, the Pilgrims explored the land. However, their sojourn was short-lived; after several weeks, they sailed to Plymouth in search of a more hospitable landscape (Vorse 1942, 64–65). Harsh conditions, especially poor soil, made Provincetown's permanent settlement difficult. In the eighteenth century, "English, French, and Portuguese fishermen and whalers, privateers and smugglers" took up temporary, often seasonal, residence on the tip of the Cape, and neighboring towns—worried about their "heavy drinking, brawling,

FIGURE 1.04. Welcome to Provincetown

FIGURE 1.05. View of Harbor, Provincetown

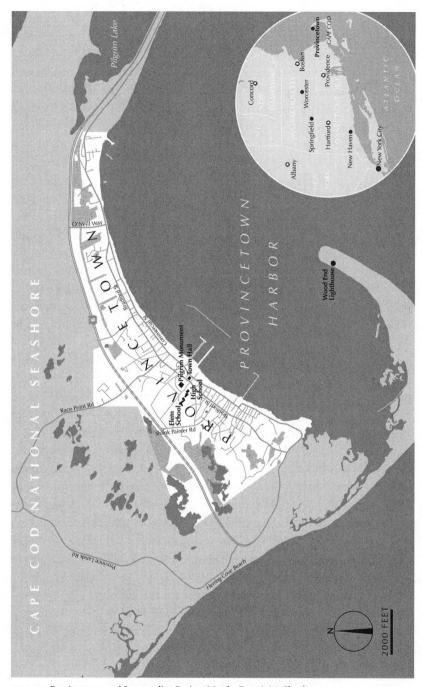

MAP 2. Provincetown and Surrounding Region. Map by Dennis McClendon.

gambling, and other bacchanalian pursuits"—petitioned the state to incor-
porate the town. Today, residents boast of the town's early reputation as "a
place where anything goes" (Shorr 2002, 8).

By the mid- to late nineteenth century, the town had become a world-
renowned maritime center as a home harbor to whaling vessels and cod and
herring anglers. Captains went to the Azores in search of skilled but inexpensive labor, and the town's west end became a Catholic, Portuguese enclave.
Descendants of these sailors remain in town, although, owing to rising housing costs and the fishing industry's collapse (a product of overfishing and increased regulation), many have relocated to the Cape towns of Truro and East-
ham or left the Cape altogether (Shorr 2002, 94; Faiman-Silva 2004, 55–57).

In the late nineteenth century, the whaling industry suffered with increasing reliance on petroleum, and the fresh fish industry surpassed the salt industry. It was during this period that Provincetown first became a tourist
destination. The author Mary Heaton Vorse recalls returning to the town in
1936 after a long absence to find it transformed. She writes: "Since Provincetown stopped making salt, it has had only one crop. That crop was fish. . . .
Now it had another crop—tourists—and so had all New England. . . . The
whole town had become a rooming house. . . . A stream of tourists choked
the streets. . . . Here people congregate like sand fleas on a dead fish" (1942,
293–94). Vorse was herself a part of the transformation of Provincetown into
an arts colony, which was sparked in 1899 by Charles Hawthorn's establishment of the Cape Cod School of Art. Subsequently, at least four other art
schools opened in town, and the Provincetown Players—a theater troupe
composed of the likes of Eugene O'Neill, Susan Glaspell, Louise Bryant, and
Vorse—began productions on a town wharf (Shorr 2002). Provincetown was
subsequently termed "Greenwich Village North," having become one of the
foremost American colonies for art, theater, and writing (Braham and Peterson 1998, xiii). In 1968, Robert Motherwell, Myron Stout, Jack Tworkov, Alan
Dugan, and Stanley Kunitz (among others) established the Fine Arts Work
Center, which continues to provide classes in writing and the visual arts as
well as prestigious residential fellowships.

The theatrical and artistic aspects of the town, and its use as a navy port
during World War II, encouraged the establishment of the town as a haven
for gays and lesbians.[6] In 1950, concerned with the town's reputation for its
"boys," leaders sought to discourage "exhibitionists" by passing regulations
that discouraged employing or renting to gays (Shorr 2002, 96). These efforts proved unsuccessful, in part because working-class Portuguese families
depended on rental income from gay tourists and artists (Gleason 1999; Krahulik 2005).

FIGURE 1.06. Carnival Parade, Provincetown

Except for a brief period during the 1980s when AIDS-related deaths dev-astated the town's gay male residents—claiming in a single decade one-tenth of the town's population—Provincetown has flourished as a gay enclave (Braham and Peterson 1998, xv). Today, bars and stores catering to gay cli-entele outnumber seafood restaurants and Portuguese bakeries. There are an abundance of guesthouses and clubs geared toward gay men as well as some catering to lesbians. Provincetown, with its risqué "dick-dock" and dunes, circuit parties, and themed weeks for bears, lesbians, and same-sex parents, attracts an array of queer tourists.

In general, the town's popularity as a tourist destination continues un-abated. On a busy summer day, the population can surpass fifty thousand, in sharp contrast to a winter population of approximately thirty-two hundred (see table 4). Today, the town is a medley of things: an increasingly upscale tourist destination, a Portuguese community, the home of the Pilgrim Monu-ment (run by Mayflower descendants), a gay enclave, and a place of reclusion for artists and writers.

Over the last fifteen years, Provincetown has experienced significant de-mographic and economic transformation. The number of households earn-ing less than $14,999 has fallen from nearly 36.5% in 1989 to just over 23% in 1999, and in 2000 the median home value was $323,600—more than double 1989's median value of $148,100. In 2005, the *Provincetown Banner*, the town's weekly newspaper, frequently listed one- and two-bedroom condominiums

TABLE 4. Provincetown demographic shifts (1990–2000): U.S. Census data

	1990	2000
Population	3,561	3,431
Percentage of population under 18	12.95	8
Percentage of population 65 or older	18.03	17.8
Percentage of population with bachelor's degree or higher	26.9	37.9
Percentage of work force in professional or managerial position	27.8	30.4
Percentage of population (reporting ancestry) of Portuguese descent	30.84	22.8
Percentage of households earning less than $15,000	36.5	23.3
Percentage of households earning $100,000 or more	1.9	8
Median owner-occupied home value ($)	148,100	323,600
Median contract rent ($)	443	590

for $400,000–$600,000, "cute studios" for over $150,000, and houses for $1–$2 million. As housing prices have continued to escalate, citizens, advocacy groups, and the town manager have worked to establish affordable housing, and the town hired an affordable housing consultant. In 2007, the Web site of the town's affordable-housing task force included a red box marked with bold white letters reading: "Provincetown Homeland Threat Advisory / HOUSING CRISIS / Significant Risk of Loss of Year-Round Community."

Similar concern appears in work by Provincetown artists and poets. For instance, in her poem "On Losing a House," the longtime resident Mary Oliver writes:

Amazing
how the rich
don't even
hesitate—up go the
sloping rooflines, out goes the
garden, down goes the crooked,
green tree, out goes the
old sink, and the little windows, and
there you have it—a house
like any other—and here goes
the ghost, and then another, they glide over
the water, away, waving and waving
their fog-colored hands.

(OLIVER 2002, 32)

Several years earlier, Mark Doty wrote in "Breakwater" of a landscaped sullied by wealth:

> After the crowded houses disperse,
> after the bleached condos
> of the summer millionaires
> accentuating the beach
>
> like stress marks penciled
> on a line of verse, after the squared lawns
> and hedges watered into emerald,
> and the final, fascist architecture
>
> of the motel, suddenly,
> uninterrupted horizon: a long exaltation
> of marsh greening out to
> —there, that far white line.
>
>   (DOTY 1995, 40)

Both authors articulate a concern for place characteristics threatened by gentrification, a concern shared by many Provincetown residents.

In 2000, the town's population was almost entirely white. However, since then, business owners (by some counts as many as 25%) have come to rely on a federal "guest workers" program, hiring seasonal workers from Ireland, the Czech Republic, and, most recently, Jamaica to staff hotels and work in kitchens (see Faiman-Silva 2004, 64–65; and Krahulik 2005, 10). Business owners suggest that they must hire such workers because the college students (like me) who once provided summer labor can no longer afford the local rents. Some proprietors provide dormitory-style housing for foreign workers.

Today, those of English and Irish descent (32.2% combined) outnumber the Portuguese, who composed just over 20% of the town's population in 2000, having constituted more than 30% a decade before.[7] Much to the chagrin of many Portuguese residents, their dominance of the town's government, which, like Dresden's, is run by town meeting, has given way to that of gays and lesbians, who populate many boards and, for a time, made up the full membership of the board of selectmen. Nonetheless, Provincetown preservationists seek to preserve working-class Portuguese (preferably fishermen and their kin) and, secondarily, struggling artists.

ANDERSONVILLE

After the Chicago Fire of 1871, today's Andersonville, a neighborhood on Chicago's North Side, began its transformation from a sparsely populated farm-

FIGURE 1.07. Andersonville Residential Street

ing community, surrounded mostly by celery fields, to a Swedish enclave. The devastation of the fire and the cost of rebuilding with materials required by Chicago's new fire code encouraged Swedes, who before the fire centered in today's Lakeview neighborhood, to migrate outside the city limits to today's Andersonville. In their new location, they could build with wood, and the area promised jobs, as a hospital and a cemetery were under construction (Lane 2003). These residents built the three-flats and large Victorians that still compose much of the housing stock.

Longtime residents recount that, until World War II, Andersonville flourished; most residents lived and worked in the neighborhood. Several Swedish congregations, restaurants, delis, bakeries, and bars served neighborhood Swedes. However, as across the United States, World War II wrought significant change. Soldiers returning from war were anxious to take advantage of the GI Bill and buy a home, and new highway construction enabled a daily commute from a suburban home to a city job (Lane 2003). As a result, the neighborhood lost many young people to the suburbs.

Because of depopulation and disinvestment, by the 1960s working-class Latinos and African Americans were able to take the place of the Swedes who had left for the suburbs. Crime, particularly associated with gangs, and racial tension characterized the commercial center. In an effort to preserve the area's Swedish and middle-class character, in 1964 Swedish business owners, with the help of politicians, christened the neighborhood "Andersonville" in memory of a Norwegian minister who once aided ill Swedish immigrants. While Andersonville is officially a part of Chicago's Edgewater neighborhood and, by the Chamber of Commerce's measures, has begun to extend into the official Uptown neighborhood, Chicagoans know the neighborhood by its symbolic name, which weds the area to its Swedish past.

Beginning in the 1970s, and continuing into the next two decades, lesbians began to move to the neighborhood, an in-migration buoyed by the presence of a renowned feminist bookstore that opened in the neighborhood in the 1980s. By most accounts, lesbians constituted the first wave of Andersonville's gentrifiers. This is notable given the literature's attention to the role of gay men in gentrification's first wave and rare acknowledgment of lesbian gentrifiers.[8] Around the same time, Iranians and others of Middle Eastern descent opened a handful of restaurants and groceries along Clark Street, Andersonville's primary commercial center. By 2003, a (subtly) Swedish-themed streetscape and a variety of small businesses, ranging from Swedish delis to Middle Eastern restaurants, a proliferation of sushi and Italian restaurants, gay-owned antique stores, high-end frame shops, an upscale gym popular with gay men, bars catering to young, straight professionals or lesbians and gays, and clothing boutiques characterized the commercial center.

In recent years, many affluent heterosexual couples and gay men have moved to Andersonville. At the neighborhood's annual summer festival in 2005—"Midsommar Festival"—which was once a Swedish street fair, the Chamber of Commerce sold T-shirts depicting a Swedish flag and a gay pride flag merging above the caption "Swedish and then some . . . " Most residents agree that the neighborhood is increasingly "and then some . . . " and decreasingly Swedish. In 2006, the gay club and hamburger chain Hamburger Mary's opened on Clark Street (inducing a Swedish longtimer to sarcastically ask, "Can hamburgers be gay?"), and the number of neighborhood bars serving a predominately gay population has tripled over the last five years. Indeed, as rising home prices pushed gay men out of the city's "Boystown,"

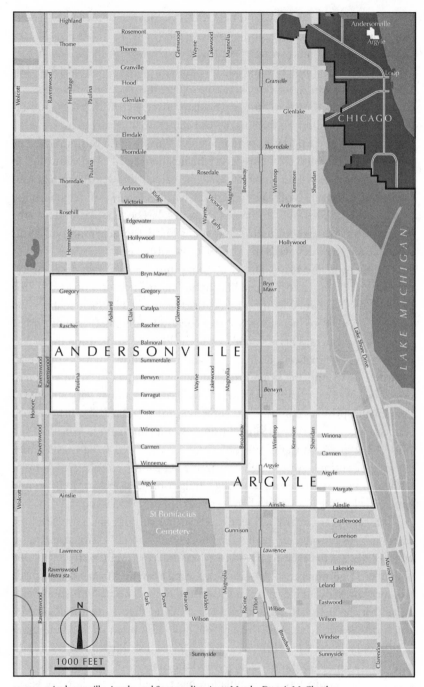

MAP 3. Andersonville, Argyle, and Surrounding Area. Map by Dennis McClendon.

FIGURE 1.08. Swedish Dancers, Andersonville Midsommar Festival

many moved north to Andersonville, thereby contributing to the outpricing of some remaining lesbians (many lesbians had already been displaced).

Like Provincetown, Andersonville is significantly gentrified: 13.6% of households earned $100,000 or more in 1999, compared to 2.1% in 1989. The median home value increased from $150,533 in 1989 to $300,167 in 1999, and recent advertisements frequently list homes for between $500,000 and $1.5 million. As table 5 indicates, Andersonville has also experienced substantial increases in the percentage of residents with a bachelor's degree and in professional or managerial positions, while the proportion of children and of the elderly has decreased.

The gentrification of Andersonville has radiated from its recently designated historic district, Lakewood Balmoral (many of whose residents do not consider Lakewood Balmoral to be a part of Andersonville), with its grand homes and tree-lined streets, next extending into central Andersonville (between the north-south arteries Broadway and Clark and the east-west arteries Foster and Bryn Mawr).[9] However, property and rents in the areas west, north, and south of central Andersonville—home to more modest three-flats and single-family dwellings—are increasingly expensive, as newly landscaped yards, condominium construction, restorations, and high-end vehicles make readily apparent. When I began my research, I collected interviews and observations within Andersonville's boundaries as defined by the Chamber of Commerce: between Glenwood and Clark and between Foster

FIGURE 1.09. Andersonville Chamber of Commerce T-Shirt

TABLE 5. Andersonville demographic shifts (1990–2000)

	1990	2000
Population	14,363	13,390
Percentage of population under the age of 18	21.36	14.7
Percentage of population 65 or older	11.35	10.5
Percentage of population with bachelor's degree or higher	30.67	37.3
Percentage of work force in professional or managerial position	29.58	50.5
Percentage of population Caucasian	60.18%	64.45%
Percentage of population of Swedish descent	3.9	3.05
Percentage of households earning less than $15,000	21.5	11.5
Percentage of households earning $100,000 or more	2.13	13.6
Median owner-occupied home value ($)	150,533	300,167
Median contract rent ($)	436.37	683.33

Source: U.S. Census data.

and Bryn Mawr. Later, in accordance with the Chamber and residents' re-visioning of the neighborhood, I extended my area of study to include the space between Foster and Carmen (adjacent to Argyle) between Bryn Mawr and Hollywood, and between Clark and Ravenswood, and many Andersonville events I observed likely included some residents from beyond this area. However, much of my data is from Andersonville proper.

In 2000, those of Swedish ancestry composed only 3.05% of the neighborhood population. In fact, there are fewer Andersonville residents of Swedish heritage than of German, Irish, English, Polish, or Italian ancestry. Nonetheless, Andersonville social preservationists work to preserve Swedes, particularly those who own small businesses. Recently, they have also sought to preserve independent business owners.

ARGYLE

Settled in the same period as Andersonville, Chicago's Argyle neighborhood has in the past century seen several significant demographic shifts. In the early twentieth century, it housed a vibrant Jewish community and its synagogue. Meanwhile, the larger Uptown neighborhood, of which Argyle is a part, served as summer quarters for wealthy Chicagoans and, later, as the ritzy theater district and home to Essanay, one of the first motion picture studios. From 1907 to 1917, Essanay Studios served as "Hollywood on the prairie" (McNulty 2000, 120). Several noteworthy jazz clubs, including the famous Green Mill, which still operates a few blocks south of Argyle, and cavernous theaters neighbored Argyle. During Prohibition, Uptown was known for its speakeasies. Rumors persist that Al Capone favored certain nightspots and that he built tunnels connecting mob establishments.

During the 1950s and 1960s, much of the neighborhood's affluent population relocated, and Argyle's affordability and the relatively small number of African Americans attracted Appalachian migrants (Bennett 1997, 39). Argyle Street sported numerous bars, remembered by residents for their rowdy clientele. Many thought Uptown to be a white, urban slum, and, in the late 1960s and early 1970s, it attracted the attention of leftists, who sought to organize residents (Maly 2005, 71). Some organizers from that period, particularly one alderwoman, remain active in ward politics, and many social service agencies and residential facilities that developed at the same time continue to serve Uptown's working-class, poor, mentally ill, and immigrant populations.

The 1970s saw significant population change, as Latinos and African Americans moved to the neighborhood. Refugees from Vietnam, many of whom were capitalists who fled communism, followed, some sponsored by national religious organizations. Chinese immigrants also moved to the neighborhood, and some sought to establish a "Chinatown North" on Argyle Street. Vietnamese immigrants simultaneously established a commercial presence on Argyle, attracting the nickname "Little Saigon." While Vietnamese and Chinese residents and businesses are most visible, Cambodian, Laotian, Thai,

FIGURE 1.10. Argyle Street and Pagoda

and Korean immigrants also moved to the neighborhood, establishing a pan-Asian presence. Together, the newcomers rid the neighborhood of its strip clubs and rowdy bars.

Today, Argyle Street is home to several pho restaurants, butchery windows sporting duck and squid, Thai restaurants, social service agencies serving Asian immigrants, and an el stop bedecked with a green pagoda. The area also houses several single-resident-occupancy hotels (SROs), which are home to many mentally ill or otherwise disabled residents. Thus, a walk down Argyle Street and the surrounding blocks is a study in contrasts. One can step into a Vietnamese grocery in which a predominately Asian clientele selects fresh tofu, fish, and produce. Outside, African American teenagers converse on a corner, and, if a train has recently arrived from downtown, a stream of young, white professionals moves east toward residential streets near the lake. It is likely that this stream will be met with requests for money from panhandlers, some of them SRO residents who spend their days on Argyle's streets.

Over the last decade, an increasing number of childless, highly educated, mobile individuals have moved to Argyle, many of whom have joined block clubs that seek to reduce crime and "beautify" the area. Some have worked (thus far unsuccessfully) to displace a long-standing leftist alderwoman whom they believe blocks economic development, while others lobby to protect Argyle from gentrification emanating from Andersonville.

Initially, newcomers primarily purchased condominiums in the eastern portion of the neighborhood, adjacent to a park with access to Lake Michigan, and single-family homes in the neighborhood's western section, in close proximity to Andersonville. They report that the neighborhood's affordability and its grand apartments built for actors and other performers who once worked in the entertainment district—with their wood floors and closets built for gowns—attracted them to Argyle. More recently, newcomers have invested throughout the neighborhood. As a result, between 1990 and 2000, Argyle experienced a decline in children and households earning less

FIGURE 1.11. Argyle Chinese New Year Celebration

than $14,999, a sharp increase in the median rent, and increases in the pro-
portion of residents with a bachelor's degree and professional or manage-
rial positions (see table 6). Over the past few years, gay men have restored
homes and purchased condominiums, in part attracted by a popular gay bar
as well as by two gay-owned restaurants. In 2007, property listings ranged
from $99,000 for a one-bedroom apartment in a co-op building near Lake
Michigan to $465,000 for a three-bedroom condominium along Kenmore
Avenue.[10]

Argyle, even more so than Andersonville, does not have official boundaries. Residents even debate about what to call their neighborhood. My research began, and remains concentrated in, the area bordered by Marine
Drive, Broadway, Foster, and Argyle. However, as block clubs shifted their
boundaries and informants referred me to residents in adjacent areas, I began collecting data between Argyle, Ainslie, Broadway, and Marine as well
as in an area adjacent to Andersonville, between Winnemac and Lawrence,
Broadway and Clark.[11] These shifting boundaries are themselves evidence of

FIGURE 1.12. Chinese Mutual Aid Association, Argyle

TABLE 6. Argyle demographic shifts (1990–2000)

	1990	2000
Population	19,587	18,853
Percentage of population under the age of 18	20.77	16.59
Percentage of population 65 years or older	16.3	14.76
Percentage of population with bachelor's degree or higher	23.5	31.93
Percentage of workforce in professional or managerial position	23.13	38.2
Percentage of residents listed as Asian	23.5	22.8
Percentage of residents listed as Vietnamese American	N.A.	7.2
Percentage of residents listed as Caucasian	42.3	45.1
Percentage of residents listed as African American	24.3	22.08
Percentage of residents of Hispanic origin	16.63	13
Percentage of households making less than $15,000	49.3	32.94
Percentage of households making $100,000 or more	1.47	7.2
Median owner-occupied home value ($)	225,333	182,933[a]

Source: U.S. Census data.

[a] This figure reflects a drop in median owner-occupied home value in one of the three census tracts that compose Argyle. In the other two tracts, the median value increased by nearly $100,000 between 1990 and 2000 (from $106,750 to $202,350), and more recent data suggest that home values have increased throughout the neighborhood.

Argyle's gentrification, as residents of central Argyle began to identify with the slower-to-gentrify area between Ainslie and Argyle, and as residents of the more stable area adjacent to Andersonville, as well as of Margate and Castlewood Terraces, began to identify with the "improved" Argyle.

In 2000, Asians constituted 22.8% of Argyle's population, while African Americans made up 22.08% and whites 45.1%. Social preservationists seek to preserve Asian business owners, with a particular emphasis on Vietnamese residents, who constitute 7.2% of Argyle's population.

Methods

This book does not seek to document the municipal, national, and even global political and economic trends, institutions, and structures that facilitate gentrification. Nor does it seek to systematically identify the costs of gentrification to original residents.[12] Many others have posed such questions, and their work informs this book. Rather, I reclaim an earlier approach to the study of gentrification by focusing on the in-migrants who participate in the process.[13] This book captures the sites in the midst of gentrification, albeit at different stages thereof. It asks why individual gentrifiers choose to move to gentrifying spaces and how informants—from pioneers to homesteaders, preservationists, and longtime residents—react to, speak of, and participate in gentrification. Secondarily, it examines the relationships between such diverse residents, including their struggles for power.

To answer such questions, this book draws from forty-eight months of observation in the four sites. I took field notes at church services, political assemblies, protests, safety, town, and block club meetings, and plenary sessions and more informally observed residents' daily life on the street, in parks, in stores, and in other public places. I became a regular at many meetings, so much so that during a 2004 meeting a member of a Provincetown board asked me to publicly recount details of an earlier meeting and several years into my research members of a Chicago alderwoman's staff began greeting me by jokingly asking, "Why haven't you finished yet?" On a few occasions, Dresden officials glanced around their meeting room to ensure that a reporter who covers local politics was not present before discussing an agenda item, no longer regarding my presence (or that of my blue notepad) as out of the ordinary.

As I detail in appendix 1, my ethnographic approach was three tiered. First, I sought to observe the organizations, events, and informal public interactions of every population and interest group in each site. However, given the constraints of studying four sites and my interest in newcomers' reactions to community change, I later prioritized the observation of settings and events in which residents discussed gentrification or, more generally, change processes or mechanisms such as economic redevelopment or landscape preservation. These included block club meetings at which residents established plans to increase safety, protests against upscale housing, and school committee meetings at which longtime residents rallied to preserve local schools. I did not prioritize the observation of events or organizations at which I was likely to encounter people with a particular point of view (whether for or against gentrification); I sought simply to conduct an ethnographic census of spaces in which residents discussed gentrification (directly or indirectly). Finally, I observed events that drew a cross section of residents, such as festivals, street life, and town meetings. This allowed me to observe the interactions of a diverse array of residents.

Given the study's comparative nature, my field notes reflect periods of focused and consistent observation in a site, interrupted by periods of observation of another. For instance, my notes contain extensive observation of Provincetown in the summer of 2001 and winter of 2004. In the intervening period, when I concentrated on Chicago and Dresden, I relied on newspaper reports, the observation of community events (e.g., the 2002 Portuguese Festival and the town's 275th birthday celebration), and interview questions about events that occurred during my absence.

I collected ethnographic data in Chicago over the course of four years (2001–5); in Provincetown for five months, with several repeat visits for community events; and in Dresden for three months in 2003, with four additional extended visits for observation of town events in 2004.[14] In addition, I observed seventeen community festivals, such as Provincetown's Carnival Parade and Argyle's Chinese Lunar New Year celebration.

I supplemented observation with interviews with 160 individuals: 45 residents of Provincetown, 40 of Dresden, 39 of Argyle, and 36 of Andersonville.[15] In each site, roughly half the interviews were with newcomers and half with longtime residents. Newcomers included social preservationists, homesteaders, and pioneers. Most interviews with longtime residents were with those whom most residents, and particularly social preservationists, deem old-timers: those longtime residents popularly imagined to have the most legitimate claim to the local past, such as Dresden farmers and Andersonville Swedes.

Thirty-eight (about half) of the newcomers interviewed were later coded as social preservationists and most of the rest as homesteaders. I categorized an informant as a social preservationist when he or she stated a desire to prevent or stop gentrification, expressed concern that gentrification threatens social authenticity, articulated appreciation for old-timers and concern that gentrification will lead to their displacement, and associated his or her neighborhood or town with at least some longtime residents (e.g., referred to Provincetown as a Portuguese fishing village or complained that Andersonville is no longer a full-fledged Swedish enclave). Those coded as social preservationists also engage in political, symbolic, or private practices to prevent old-timers' displacement and/or gentrification. They range from those who advocate for old-timers professionally (e.g., through housing advocacy) to those who express appreciation and concern for them through private practices (e.g., by purposefully patronizing their stores).

I coded an individual as a homesteader when he or she articulated awareness of or self-consciousness about gentrification's risks for markers of authenticity and concern for the preservation of place attributes such as the built or natural environment or aspects of local social character. However, despite their cognizance of gentrification's risks, homesteaders, unlike social preservationists, believe that "responsible" gentrification can benefit a neighborhood or town and that they can personally contribute to positive change, whether through historic preservation or increasing safety. Homesteaders report that they welcome other newcomers, often as potential preservation allies. In contrast to social preservationists, homesteaders do not center their preservation aims on longtime residents. Thus, while homesteaders share with preservationists some concern about the risks gentrification poses to authenticity, they are neither as resolutely critical of gentrification nor as concerned with longtime residents as preservationists.

While also cognizant of gentrification's risks, pioneers articulate unmitigated excitement about the transformation of their place of residence and a commitment to gentrification-related progress. Pioneers speak of longtimers with indifference or animosity and welcome other newcomers as partners in the taming of the frontier. The language that they use when speaking about their property and their place of residence is akin to that of speculators; they regard their participation in gentrification as an investment that will benefit, not just themselves, but the greater good.

Interviews—which were semistructured, recorded on audiotape, and transcribed—varied in duration from half an hour to three hours, with most lasting at least one hour. Most were held in informants' homes, although, in some cases, they were held in public locales such as a café, restaurant, park, or pub. The interview protocol was nearly identical for new and longtime residents. I asked the informant to describe his or her basic demographic characteristics, such as occupation, education, age, ethnicity, and sexual identity, and to give a brief family and residential history. I then asked about his or her decision to move to or stay in the site, civic involvement, and local conflict as well as for an account of community change and his or her hopes and concerns for the neighborhood or town's future. Importantly, I rarely asked direct questions about gentrification, allowing my informants to broach the topic if they chose to. One-third of the way into my research, I began to close interviews with newcomers with questions about their politics and cultural tastes, such as, "What are you currently reading?" and "What place would you most like to visit?" (For a list of interview questions, see appendix 2.)

On entering my sites, I sought interviews with the community leaders—for example, block club presidents, members of political committees and boards, and merchants—who populated the events I observed. Thus, my sample contains a high proportion of civically engaged residents of every ilk. However, I recruited informants from organizations popular with a diverse array of residents, such as streetscape committees, block clubs, and the board of selectmen.

Later, realizing that my sample contained a disproportionately high number of residents engaged in formal organizations, I consciously sought those with limited involvement in local politics or social clubs. I began asking informants for references to those less or uninvolved in civic or community life. I also approached residents for interviews outside formal events, such as at social gatherings and at museums or shops. I sought a sample reflective of local demographics in terms of informants' race, class, length of residence, gender, sexual identity, and age. While my sample of newcomers is by no means random and is more representative of those formally engaged with local organizations and institutions than those who are not, it otherwise approximates the ideological and demographic character of the newcomers in my sites.

My sample of longtime residents is intentionally less representative. While half of my interviews were with longtime residents, I primarily sought interviews with those whom social preservationists seek to preserve and whom I identify as old-timers, for I was particularly interested in documenting the perspective of those on whom newcomers focus attention. For instance, I wished to know what Swedes made of the Andersonville Chamber of Commerce's effort to mark the neighborhood as Swedish with a streetscape. Thus, my sample of original residents is weighted toward old-timers. However, I interviewed some who fall outside the old-timer category, such as Andersonville Lebanese merchants and longtime Provincetowners who are neither Portuguese nor artists, as well as some whom gentrification had already displaced. Throughout the text, I use *old-timer* as social preservationists do and *longtime resident* to refer to the full pool of those who resided in the space pregentrification.[16]

As I hope the following pages make evident, two features of the research design proved particularly beneficial: the combination of ethnographic observation with interviews and the comparative study of four sites. Interviews provided an opportunity to document informants' explanations of their observable behavior as well as to capture their private beliefs about community change. In turn, the observation of community life allowed me to determine the relation between private beliefs and public practices. Without the combined methods, we would not have a full understanding of, for example, Argyle social preservationists' beliefs about affordable housing and, thus, could not fully explain the motivations behind their participation in protests against condominium conversions. In the same sense, interviews alone would have left us with questions about whether social preservationists translate their beliefs into practice.

Second, the study of four distinct sites reveals, not only the breadth of social preservation, but also the factors that shape preservationists' practices and rhetoric. For instance, had I studied only Dresden, we might believe that social preservation is characterized by careful avoidance of the public sphere to preserve old-timers' political power and traditional practices. Similarly, had Argyle been the sole site, we might imagine that all preservationists engage in protests for affordable housing and vociferous public debate with other gentrifiers. More important, had I studied only rural or urban sites, we might have come to think of social preservation as an exclusively urban or rural phenomenon. And, finally, had social preservation been observed in only a single site, I might have taken my findings to be anomalous, rather than evidence of a widespread but largely overlooked set of beliefs and practices. Indeed, I might have missed the fact that nearly half the newcomers to four very distinct gentrifying communities are acutely uneasy about the changes associated with their residential choices and take it on themselves to counteract the impact of their presence on old-timers. From this discomfort arise the practices and lifestyle choices to which the following chapters turn.

2

Beyond Pioneering:
Social Homesteaders as Uneasy Gentrifiers

Much of the gentrification literature presents a single image of the gentrifier as an urban pioneer. As gentrification scholarship richly documents, pioneers celebrate gentrification's benefits and unabashedly welcome the transformation of the wilderness. They seek financial gain as well as the less tangible rewards associated with the social, cultural, and physical transformation of the place in which they live. Pioneers welcome others who share their demographic and cultural traits and the restaurants, shops, and institutions that suit their tastes and satisfy their needs. Except for those who regard historic restoration as the key to economic revitalization, they express little interest in preservation.

While I argue that most gentrifiers depart from the frontier and salvation framework (Spain 1993), I encountered some whose beliefs and practices fit the literature's description of the pioneer. I begin this chapter by introducing two pioneers as background and contrast to the social homesteaders who are this chapter's focus. The first is Fred, a white gay man in his fifties who owns several prominent Provincetown businesses that serve gays and lesbians. Fred, who is athletic and classically handsome, moved to town in the 1980s with his partner and certainly fits the pioneer prototype. Like other pioneers, he was drawn to Provincetown by the excitement and sense of promise he associated with gentrification. Fred sold his law practice outside Boston so he "could start a whole new life, a new business. That was exciting." He was also drawn to Provincetown because of qualities he associates with newcomers. Specifically, it was a place where he and his partner could "be free, express who we really are. We don't have to worry about who's looking at or who's talking to you or affecting your business in any way."

Through their businesses and engagement in local politics, Fred and his partner helped institutionalize Provincetown's gay and lesbian character. The Website for their guest house boasts photographs of muscled, shirtless men and of the renowned gay and lesbian entertainers who perform there. Indeed, Fred celebrates "the fact that we've calibrated the town into kind of a resort destination, as opposed to just being a quaint little fishing village." Not only have Fred and his partner benefited financially from this transformation; they have also become figures in local and national gay circles, hosting fund-raisers for national gay advocacy groups, and in so doing have marked, not only their place in such circles, but Provincetown's as well.

Fred is aware that most longtime residents do not share his enthusiasm for gentrification. He said: "There are many local people that don't like the way the town has turned into a very commercialized tourist destination spot." He is also aware that he is able to participate in town politics in part because longtime residents have retreated from the public realm: "Many local people don't want to volunteer and be chastised in the press or to be ridiculed or second-guessed by business owners, so it's very difficult. Many of us that moved here are ex-professionals: hospital administrators, doctors, dentists, lawyers, you name it. CEOs of companies."

This awareness of political displacement does not change his positive assessment of Provincetown's gentrification. Indeed, Fred is fairly indifferent to longtime residents' displacement, saying: "Nobody's put a gun to anyone's head to buy their piece of property. Many of the old-timers' families or estates would rather move away, would rather not move back to town." That said, Fred supports the construction of affordable housing for those who are "middle and median income": "*Those* are the people. *That's* your workforce. *That's* your labor force." In keeping with these sentiments, he attended hearings on affordable housing, and his partner was one of the realtors who served on the town's affordable-housing task force. However, Fred regards affordable housing as a tool for ensuring, not the continuity of local character, but, rather, his own and the town's continued economic success.

Fred's appreciation for progress and transformation may be rooted, in part, in his understanding of himself as a person who is anything but nostalgic. When I asked him (as I asked all informants) whether, given the power, he would freeze the town in a particular historic moment, Fred answered: "You know, I'm not one to say the '70s were better than the '80s, were better than the '90s. I tend to want to adapt to the times. My favorite time is between 3:30 and 5:30, sitting on my deck looking at Long Point. I freeze that in time before I get back to the madness. I also love walking home at 1:30, quarter to 2:00 in the morning and seeing the throngs of people, then walk to the West End, and it's quiet and I can go to sleep."

What Fred did not say is that, when he is not walking quiet, nighttime streets or sitting on his deck, he is orienting the town toward the future. Like other pioneers, he directs himself toward progress and transformation, toward the town's conversion from quaint fishing village to resort destination. Implicit in this is his desire to live in a place in which he and his partner can be open about their sexual identity and an attachment to place that neither references the past nor rests on a sense of authenticity. P-town is precious to Fred for reasons other pioneers prioritize: the promise of change, of profit, and of a community composed of others like oneself.

Similarly, Richard, a white attorney in his late thirties who has lived in Chicago's Argyle neighborhood for several years, expresses nearly singular appreciation for gentrification-related transformation. Despite his awareness that such transformation leads to displacement, Richard talks about his gentrification efforts with moral certitude, arguing that Chicago benefits from the taming of the frontier and that such benefits outweigh the cost of displacement.

Having observed Richard at safety and beautification meetings, I scheduled an interview with him. I subsequently learned that, like other pioneers, he describes gentrification as a natural and, therefore, inevitable economic process and says that he moved to Argyle because he believed that it provided an opportunity for homeownership and a chance to enjoy the thrill of living on the urban frontier. Having witnessed friends benefit from rising property values in other neighborhoods, he decided that purchasing in Argyle was a wise financial investment: "It was incredibly bad on the street where I bought. [But] the deal I got on a condo was unbelievable. I just couldn't pass it up. . . . It was a renovated building. It was really spacious. I got a five-bedroom place. My mortgage right now, I couldn't even rent a studio apartment for what I pay for the mortgage, . . . and I had known that progress kept coming north up the lake." He soon found himself caught up in the excitement of early gentrification: "When I moved in here, it was like the wild, wild west. It was, literally. You couldn't sleep at night."

Richard expresses little concern for the people or businesses that populated Argyle before he arrived. And, despite his acknowledgment that gentrification leads to displacement, he believes that neighborhood improvements are for the greater good: "Some of the people feel threatened. Like it's the white people coming in here, kicking them out, moving them out, and to some degree it's true. It's gentrification. . . . We tried passing the word to all these people that we're just trying to make it a cleaner, safer place for everybody." While Richard says that he would like to make the neighborhood nicer for everybody, he openly admits that he wishes to rid Argyle of "certain elements" and is proud of his work to ensure the displacement of those he identifies as prostitutes, drug dealers, and even renters: "This street had the reputation of being the place to come on the North Side to buy drugs, score prostitutes, and drink and party. . . . We're changing that. We're screaming for some development. We need some more property owners on this block, and hopefully that will come about through our actions."

While Richard appreciates the "cultural diversity of the city," he hopes — in stark contrast to social preservationists, who lobby to preserve Vietnamese, Chinese, and Thai businesses — that some will leave: "Personally, I'd like to see something around here besides Chinese restaurants. . . . There aren't enough restaurants around here convenient to just go out and have breakfast somewhere in the neighborhood. . . . It's got to be a multiuse neighborhood for everybody."

Richard is aware of criticisms of his approach to gentrification and ex- presses some concern for the "deserving" poor: "That's a fact that's going to happen [i.e., people will be displaced]. But the fact is, I think it's a good thing in a way. I mean, I feel sorry for the poor people who are decent people who won't be able to live here, but, you know, I hope there will be places to go for them." For Richard, gentrification-related displacement is justified by the in- evitability of economic change, his certainty that revitalization is for the com- mon good, and his personal interest in financial gain. His words almost echo early Chicago sociologists' beliefs about natural competition and expansion: "This is the nature of how the city has been. It was once a nice neighborhood, it fell into disrepair, and now people are claiming it back. I can't feel sorry for everybody because what do we live with: a horrible neighborhood or a clean, nice neighborhood? It's a tricky gray area: social responsibility versus gentrification. Like I said, I don't feel bad." In the end, while Richard is aware of gentrification's risks, he regards his participation in the process as ethical and even laudable.

This perspective and related practices — litter removal, daily 911 calls to report African American men gathered on corners — are affirmed by other gentrifiers, including one whom Richard regards as a teacher: "There's this other guy. . . . He's been here sixteen years. . . . And he was like my mentor . . . my coach." This suggests that some explicitly learn to be pioneers. While Richard reports that he prioritizes gentrification over social responsibility, he and others like him evoke an ideology that regards the taming of the city as a social responsibility and the bulk of original residents as adversaries. In their minds, to stand in the way of revitalization is to prevent progress and economic gain, and they express little interest in preserving authenticity.

At a 2001 Argyle streetscape planning meeting composed of gentrifiers, a political aide, and a Vietnamese business owner, Richard argued that local African American youths threaten gentrification efforts, that streetscape plans are superfluous without crime reduction. "Gangbangers" whom Richard suspected cut street lamp wires to facilitate drug deals particularly incensed him. When a social preservationist argued that what worried him about Richard's anecdote was, not the presence of drug dealers, but the risk that wire cutting posed to their lives, Richard retorted: "Then we'd be rid of them."

While a few of the assembled nodded in agreement, most shifted in their seats, peered studiously at the table, or otherwise communicated discomfort with Richard's perspective. A few openly concurred with the social preservationist, expressing concern for the safety of neighborhood youths. On the basis of my interviews with several members of the group, I can surmise that, beyond finding this particular comment offensive, several gentrifiers consider Richard's orientation to gentrification discordant with their own.

Such moments reveal that, while the gentrification literature has generally devoted itself to those like Richard and Fred, only a minority of gentrifiers shares their beliefs or engage in practices singularly dedicated to transformation. As the anecdote also reveals, some gentrifiers' ideology and practices depart sharply from those of the pioneer. Those whose departure is most stark are social preservationists, who seek to prevent gentrification and the displacement of old-timers and who shape their ideology and practices in opposition to the Richards and the Freds that they encounter.

However, preservationists are not the only gentrifiers who depart from the urban pioneer prototype. Many at Richard's table at the Argyle meeting fall—ideologically and practically—between the preservationist and the pioneer. Among them are *social homesteaders.* Unlike Richard and Fred, who openly take pleasure in the settling of the wilderness, most homesteaders seek to unobtrusively carve out a space for themselves. To an extent, they share social preservationists' awareness of gentrification's risks and their interest in authenticity. However, they wish neither to charge onto the frontier nor to stand in the way of progress.

This chapter demonstrates that social homesteaders share a set of ideological traits and related practices that distinguish them from social preservationists and pioneers. They have a more ambivalent relationship to gentrification than do other gentrifiers, regarding it neither as inherently problematic nor as a simple solution to the problems of urban or rural disinvestment. They regard themselves as potential agents of positive change through building or landscape preservation, but not as those who will unambiguously save the central city or small town either through gentrification or by resisting it. In this sense, they are less self-conscious about their place in their neighborhood or town than are either of their counterparts. They neither celebrate their presence in the way of pioneers nor criticize it as preservationists do.

As a sort of shorthand, we might think of social preservationists as those who regard their neighborhood or town as a place that is primarily old-timers' "home." They perceive themselves and other gentrifiers as uninvited guests. Some even suggest that they are invaders. In contrast, while mindful of longtime residents, homesteaders regard their neighborhood or town as *their* home. They neither tiptoe through the space in the manner of preservationists nor regard it in the speculative terms of the pioneer. Homesteaders engage in the preservation of place characteristics that they deem authentic but are fairly unconcerned by longtime residents' displacement. Their sense of home depends on place characteristics that they regard as authentic as well as on the notion that their presence does not automatically alter that authenticity. For pioneers, the alterations they make to a place are what make it home; for them, home is an amalgamation of market, comfort, and the allure of transformation.

This chapter is organized around the place characteristics homesteaders most appreciate, those they deem authentic and seek to preserve. Homesteaders' appreciation for certain local attributes contributes to their vision of the future of the place and shapes their practices. Some engage in landscape preservation, while others cherish a "melting-pot" atmosphere, embodied by the presence of residents of diverse backgrounds. Still others seek to return the place to a historic or natural heyday.

Like the ones to follow, this chapter reminds us of the intellectual and pragmatic risks of assuming uniformity among gentrifiers. It does so by calling attention to variation from the pioneer prototype and by attending to homesteaders' internal diversity, for the motivations for homesteaders' relocation, experience of their place of residence, and practices are expansive. They regard distinct place features as authentic, and their practices range from efforts at crime reduction to historic preservation. Some build relationships with old-timers as part of an uplift strategy, and many wish to preserve landscape. What binds homesteaders is a worldview that rejects unmitigated gentrification in defense of particular place characteristics but, nonetheless, finds something redeeming in gentrification.

The Past: Saving Sticks and Stones

Social homesteaders often regard architectural preservation as the cornerstone of revitalization and as essential to maintaining or creating local character. In this, they locate the authentic state of their neighborhood or town in the past and in the built environment. As a result, some regard the historic restoration of their home as part of a broader reclamation of local historical authenticity. Some participate in historical societies or block clubs that seek to limit development or protect housing stock. For most, a taste for historic houses is tied to a broader esteem for the local past, often for a social or cultural heyday such as Argyle's theater district past or Provincetown's art-colony history.

While some discover an appreciation for historical authenticity after moving to a gentrifying neighborhood or town, others relocate in search of such authenticity. This is true of Helen, who moved to Dresden as a young woman in the early 1980s. Seated in the kitchen of her nineteenth-century Maine farmhouse, she recalled why she and her husband moved to town: "We were working in a historical museum. Coming up to Maine was kind of like a real-life museum in a sense. . . . We really had a feeling about this area. I don't know if it was just Dresden, but kind of this general area [seemed like] something out of the past."

Helen expressed a particular appreciation for local architectural integrity, which "hasn't taken the direction that [it has] in Massachusetts." Of her 1830 house, Helen said: "You can walk into the cemetery, and other people who have lived [here] are buried out there." Indeed, her house is adjacent to a cemetery perched on a wide expanse of hillside; a line of trees at the top of the hill frames the gravestones. From Helen's yard, one can view the trees, the gray tips of the stones, and the white fence surrounding them.

Gathering knowledge about local history affirms her sense of Dresden's authenticity. Indeed, Helen seeks information about the history of her own home: "We have a picture of someone who was adopted and lived here in the 1860s . . . and [someone] who founded the Augusta library. . . . Just moving into an old house was something we really wanted to do." Helen is not alone in this. Social homesteaders frequently recall the families who once lived in their homes, crops farmers grew in Chicago soil, or a woman who waited at a window for her husband's ship to appear in Provincetown's harbor.

For those like Helen, the preservation of architecture is also about the maintenance of social history. Helen is drawn to the past because she believes that prior generations had stronger communities: "I have a real strong interest in community and the idea of being part of community, being part of a group. In the past, we can see how people have done it." She relishes old-timers' stories about the past: "We were very fortune to hear some of the stories about the early 1900s. . . . A lot of those [old-timers] have started to go." Thus, Helen's restoration efforts are rooted in a desire, not for personal economic gain, but, rather, for connection with a particular social history that, for her, lends her home meaning.

However, despite her attention to old-timers' passing, Helen does not regard gentrification as a threat to local history (although she does believe that uncontrolled growth will taint the landscape). Indeed, she believes that gentrification is of little consequence: "I don't see that big of a change in the twenty-some years we've been here." For her, the simple passage of time is what threatens the vitality of the past, not the "tasteless" symbols of gentrification such as condominiums that other homesteaders, like George, a Chicagoan, decry.

I first met George—an athletic forty-year-old—during his Andersonville block club's garden tour on a patio behind the Victorian house he and his partner, a doctor, had owned for four years. George explained that their home once stood in the midst of a celery farm. While he spent years working in the arts, George is now a realtor and, perhaps as a result, is knowledgeable about Andersonville's housing stock and history.

After meeting me at the tour, George agreed to an interview, suggesting that we meet at Andersonville's Starbucks, which is very popular with many of the neighborhood's gay newcomers. While deeply concerned with preserving buildings from gentrification-induced development, he isn't opposed to all gentrification-related change. He told me that his decision to buy his house was affirmed when Starbucks opened in Andersonville just weeks before his closing. "I thought this'd be the next great neighborhood," he said. He explained: "I'd been having a lot of first-time buyers move up here, and [they] kind of energize the market. They bring it all up. So I had been bringing my Lakeview clients who'd lived in Lakeview, but couldn't afford to buy in Lakeview, up here and saying, "This is the next alternative.' You know, you can walk around at night, feel comfortable, and you're close to the lake, and public transportation is good."

George reports that he purchased for similar reasons: "It was what we could afford, it was the closest to the lake, the best neighborhood to get the most house we could." In fact, the affordability has allowed George and his partner to contemplate purchasing a summer home in Saugatuck, Michigan, a beach town popular with gay men that is sometimes lauded as the Provincetown of the Midwest.

While he celebrates the past, George appreciates the fact that Andersonville has become more upscale since he purchased: "From a real estate standpoint, the house has . . . more than doubled in worth." He hopes that gentrification will extend beyond Andersonville: "I would like to see Broadway, and Winthrop, and stuff east of us get a little bit nicer. . . . It would be great if you could really attach that neighborhood and go all the way down Clark Street. 'Cause right now there's kind of that wasteland between Montrose and Argyle."

Despite his enthusiasm for such expansion, George recognizes that it will lead to the displacement of independent businesses, saying: "It's too bad because it's smaller businesses, but I think they're kind of out of sync with where the neighborhood's at. . . . In ten years, this is gonna be as congested as Lakeview. The taxes will have gone up, and a lot of these smaller stores won't be able to survive. This little hardware store won't be here any more. . . . I go there every chance I can just to try to be supportive of them, but, you know, the rents are just gonna be so high that you're gonna have to be an Einstein's or Chipotle." He regards independent businesses as necessary casualties, expressing greater concern for the destruction of the historic-building fabric.

In fact, George holds himself partially responsible for the demolition of a "gorgeous house," believing that the price he paid for his own house encouraged a neighbor to view his home as an investment and to tear it down to build condos. "We were sick about it," he said. He regards the replacement of houses with condos as a worrisome trend: "Even as a realtor [I think] keep the nice big houses; there's plenty of gross stuff to [tear] down instead." While he has not advocated for the codification of his concerns (he lives just outside a national historic district), he openly shares them with neighbors and frequently boasts about his home's historic importance.

Indeed, history is of the utmost importance to George: he restored his home, framed historic photographs of the house, and often converses about local history. While he recognizes that gentrification will alter Andersonville, like Helen he prioritizes physical symbols of the past rather than residents. Yet, despite his interest in and attention to rising home prices, he values history over profit. He decries his neighbor's choice to demolish a home to construct lucrative condos, and, despite his realty practice, it has not occurred to him that he could do the same.

While an interest in the past motivated George's and Helen's relocations, other homesteaders developed an appreciation for the past after moving to their neighborhood. This was often the case for Argyle homesteaders, several of whom moved because of its affordability and later found pleasure in local history. During interviews, several offered tours of their condominiums, typically in gracious three-flats. They pointed to photographs documenting Argyle's entertainment-district past and proudly pulled open heavy oak doors to reveal closets made to accommodate top hats. A few ushered me into spacious dining rooms, and one described the elegant meals that, she imagined, early-twentieth-century residents hosted in the space.

Even a homesteader who does not live in such a building expressed a great appreciation for their presence in the neighborhood, saying: "The buildings you see on Argyle, my God, they're so nice! They have so much architecture. . . . They're so beautiful: their architecture, their layout. They're spacious. . . . The closet in the [building right there] has a fold for mink coats. That is one of the antique ones. There are pictures of [Charlie] Chaplin and Gloria Swanson when they used to live here." He wishes to preserve— through buildings—a connection to this past, a history more connected to a bygone elite than to contemporary Argyle.

Some bring concern for historical authenticity to formal organizations. In Provincetown, several serve on the Historic District Commission and re- gard themselves as custodians of the past. For instance, at a contentious 2004 meeting, a thirty-year-old commission member railed against the aesthetics of a 1960s ranch house, arguing that it should be replaced by a home that reflects the town's distant past: "We have a historic district to protect—a nineteenth-century fishing village. This is McDonald's. Go anywhere, you'll see this. We need more Provincetown." At other meetings, members voiced their belief that details, like windows, shape town character. In fact, the com- mission codified such concerns by restricting those architectural choices or remodeling practices—often initiated by newcomers—that it regards as in- different to local history. Commission members voted, for instance, that the new owners of a nineteenth-century home must preserve its farmhouse aes- thetic while dividing it into condominium units. In other words, they believe that the best way to maintain ties to Provincetown's authentic small-town past is through preserving buildings rather than people.

Such concerns sometimes pit homesteaders against longtime residents. For instance, over sandwiches in her eighteenth-century home, Gloria, an elderly Portuguese woman, recalled her tense interactions with the com- mission. In so doing, she revealed how, "in a subtle way, the ideology of historic preservation facilitates [or at least threatens] the removal of a pre- gentrification population" (Zukin 1987, 135). Gloria recalled that the previous winter she noticed her curtains moved with the wind. She set about finding affordable replacement windows, a difficult task for an eighty-year-old who struggles with rising property taxes. After installing them, she was asked to appear before the commission to explain why she had not replaced them with a historically accurate (and more expensive) set. She was shocked that the town could tell her what type of windows to install. At the meeting, she was so upset her body shook uncontrollably, and she warned the commis- sion of her heart problem. Her story illustrates how homesteaders' attempts to maintain authenticity can place them in conflict with longtime residents, especially those for whom ordinances pose a financial burden.

Thus, efforts to preserve the built environment pit some homesteaders against longtime residents. However, this is not uniformly true. While Helen does not resist gentrification or worry that gentrification will contribute to displacement, she enjoys longtimers' stories of Dresden's past. For her, oral history lends meaning to buildings and encourages deference to those who were there before her. And even George—who welcomes the economic benefits of gentrification—appreciates the historical authenticity he associates with longtime residents' mom-and-pop businesses. This suggests that, while an orientation to historic authenticity can and often does encourage conflict with longtime residents, appreciation for the built environment sometimes (although not always) results from appreciation for those residents and their history. Just as pressingly, homesteaders' interest in historic authenticity sometimes leads to conflict with pioneers, such as Provincetown homeowners who wish to build an addition to their home or divide it into condominium units.

A Natural or Agrarian Landscape

In Dresden, the most rural site, several homesteaders recalled that their move was motivated by an appreciation for landscape.[1] This orientation to nature was also present in Provincetown, where gentrifiers take pleasure in ski treks over snowy dunes and walks along the shore. However, in Dresden, such sentiments were of greater influence, explicitly shaping homesteaders' residential choices and encouraging efforts to preserve the town's natural and agrarian landscapes.

My conversations with Dresden newcomers often turned to the landscape. For instance, on a cold winter afternoon, a pair of retired professors explained that they moved to Dresden out of appreciation for the natural scenery. The wife said,

> [We like] the immediate physical surround of our house and the fact that when winter is over, which is followed by deerfly season, when that's over, we can walk in the woods and hardly ever come across anybody. We love these boulders, as nutty as that sounds. . . . I take [the dog] every morning up past this blueberry field, and it changes all the time, and it's so beautiful. And then [the] farm on the right-hand side—I mean every morning, no matter how bad the weather is, I think every time it's just so beautiful to see that. And then . . . driving up this hill you come through this tunnel of trees when we're coming home. I just love that part . . . God it's beautiful.

Her husband agreed, acknowledging that he was surprised to develop an appreciation for Dresden's winter landscape: "Once the snow had covered everything and the trees had shed their leaves, you could just see everything in a new way. The woods became much more open. To me I guess my favorite thing is just looking out the window."

While the walks through woods and along dirt roads that such sentiments inspire are largely solitary, the attitude this couple expressed is widespread. Several other homesteaders referred to wildlife spotted or beloved walks. For instance, another retired academic couple repeatedly interrupted our interview to call attention to birds outside their large dining-room window, including an eagle that flew over the Eastern River as we talked. They kept a pair of binoculars within reach and seemed as familiar with local wildlife as with their new neighbors.

This appreciation for nature is of consequence for local politics and interactional dynamics. It contributes to conflict between homesteaders and longtime residents over land use and preservation. For instance, the retired academic couple who cultivated a love for Dresden's winter landscape recognizes that this makes them critical of longtimers. Indeed, they openly acknowledge a distaste for some—particularly those who hunt. Over coffee, their dog resting at their feet, they complained that longtime residents make it difficult for them to enjoy their passion for walks along Dresden's scenic roadsides: "It's alarming—Budweiser must be the beer of choice around here—when you stop to think: why are beer cans being thrown out of the cars? Because they're drinking and driving. And you think: they drive too fast around here, drive too fast . . ." They acknowledged that such suspicion is probably mutual. Laughing, the wife said: "People probably resent that we have a [pedigreed dog]. In the winter we put a coat on her. They probably want to kill her for that." While such homesteaders obviously do not romanticize longtime residents' practices, they are more self-reflexive than pioneers about their relationship to longtimers.

Conflict between longtime residents and homesteaders over preservation of an authentic landscape is rooted, not only in different ideological perspectives, but also in related practices. Helen, the Dresden newcomer whose appreciation for local history I discussed above, is at the forefront of efforts to restrict growth and enforce existing land-use ordinances through town boards and even legal action.

When I interviewed her in 2004, she expressed concern that several developments were planned on prime agricultural land: along the fertile banks of the town's rivers and in a vast blueberry field. She noted that, while the Planning Board works to ensure that new construction complies with ordinances, the board of selectmen (composed of two longtime residents and one relative newcomer) sometimes hesitates to enforce ordinances when they conflict with local land-use traditions. Indeed, on several occasions I sat on a wooden bench in the town office while the board urged a native to close his illegal scrap metal dump. While interactions between the selectmen and the dump owner—a large man who typically arrived at meetings unshaven and wearing a plaid shirt, jeans, and work boots—were tense and the selectmen threatened to close the operation, they did not do so during my period of study.

Inactivity of this kind incensed Helen, as did a proposal to build a motocross racetrack in her part of town. Helen worried that the track would disrupt Dresden's landscape and destroy markers of the agrarian past she loves. In this, her views—embodied in a lawsuit she filed with other residents to prevent the construction of the track—conflicted sharply with those of many natives, some of whom welcome the leisure opportunities the motocross would create, and many more of whom believe that those who own the land should be able to do with it as they wish.

Concerns about development have even led Helen and her husband to contemplate moving out of town. She explained: "Part of it is lack of planning. Anything might come up." "There's land right over there," she said, pointing out the window to a green pasture adjacent to her acreage, "that's been in [the neighbor's] family for a long time, but one generation has died, and I'm not sure what will happen to it in the future."

Helen worries that she cannot prevent her neighbors from selling parcels of their land to newcomers or allowing their children to place trailers on it. She can and does restrict her own development practices: while she and her husband are far from affluent, she does not contemplate selling her own land. Dresden's landscape is of greater value to her than profit from a land sale. For this reason, if she could, Helen would ensure that her neighbors could not develop their land. Thus, her appreciation for Dresden's landscape places her in an adversarial relationship with many longtime residents. As an advocate for development restrictions, she has been at the center of several conflicts. Rather than deferring to the libertarian perspective on land use held by many native Mainers, she has been at the forefront of efforts to preserve those qualities that she deems authentic: Dresden's rural past as embodied by landscape.

A desire to restrict development and limit population growth is common among Dresden homesteaders. At a small gathering at the home of a top state official, Andrew—a successful attorney who has lived in Dresden for more than two decades—said:

> It has grown a lot. I guess what concerns me is that we still have a minimum lot size of one acre. That means that . . . the signature landscape that we think of as rural countryside [is threatened]. . . . [There's the potential] to create

sprawl. . . . We build houses in the country and then wonder and regret the
decline in agriculture. If we don't pay attention to fertile soils, we'll shoot
ourselves in the foot. We've got a few farms on great prime farmland. If we
chop that up . . . if we throw that land away, it just doesn't make sense. . . .
Thank God for [a farmer who put his farmland into a trust that forbids future
development].

Those in attendance listened carefully as Andrew—a much-respected town
figure—spoke, frequently nodding in agreement.

Several homesteaders who share Helen and Andrew's concerns are members of the town's Conservation Commission, which serves to protect local
natural resources. Meetings, held in the small town library, often begin with
chatter about wildlife. At one meeting, commission members traded stories
about rare chestnut trees hidden in the region's deep woods. Each month, the
commission secretary publishes an essay in the town newsletter—the *Dresden Communicator*—listing wildlife residents have observed, and the organization sponsors an annual wildlife photography contest. In this, it celebrates
and even cultivates an appreciation for nature. It also actively recruits new
residents, whose appreciative conversations about pheasant and deer contrast
with the hunting stories many Dresden old-timers shared with me. Indeed, at
the same table in the library around which the commission congregates, an
old-timer complained to me about the increasing hunting restrictions that
newcomers have placed on their land.

Like Helen and Andrew, commission members lobby for policy changes
to sustain Dresden's rural character. At one meeting, a member proudly noted
that a recent analysis of current development patterns found that Dresden
stands out among other local towns in terms of its rural character: "Dresden
is a lovely little island in the middle [of more developed towns]." Another
member concurred, also enjoying Dresden's distinction: "That was the best
part. Dresden was the only rural town around. It was the yellow hole in the
donut based on growth factors." "Yes," the chair added, "Dresden has some
very large undeveloped areas. [We] could download the information into
town computers. This is a valuable planning tool." Another member agreed,
saying: "It's a good way to allow towns to develop while respecting habitat. . . .
It's what we all, I think, value about the town." Later at the same meeting a
member said: "We don't want to encourage people to move to Dresden!"

Attention to the natural character of the Dresden landscape frequently
draws homesteaders into politics. They lobby to preserve the town's authenticity, which for them rests on the maintenance of a rural New England landscape, a landscape one part wooded and wild and one part agrarian. Unsurprisingly, of my informants, Dresden homesteaders most frequently correlate

authenticity with landscape. While one might anticipate a similar association in Provincetown, my other small-town site, homesteaders there more frequently commit to sustaining the town's historical authenticity and its current character as a gay and lesbian enclave. This may be because Provincetown's character as a gay and lesbian town and destination is more popularly celebrated and neatly encapsulated than Dresden's social identity and because narratives about the landscape may be more readily available to Dresden homesteaders than to those in Provincetown. Furthermore, Provincetown's large population of gay, lesbian, bisexual, and transgender newcomers is particularly appreciative of the social, economic, cultural, and political opportunities their town affords. Perhaps for such reasons, in the minds of many Provincetown homesteaders the town's social traits outfigure its landscape.

The Future: Cultivating the True Potential of Place

In contrast to those homesteaders who are oriented toward the past in terms of historic or landscape preservation, some appreciate the sense of progress or potential that they believe gentrification can achieve for their neighborhood or town. Some such homesteaders—like many other gentrifiers (particularly pioneers)—purchase in a place because they believe it to be gentrifying and want to be a part of its transformation. Others, once established, orient themselves to their new place of residence's future, specifically, to its potential as a site for community (often composed of newcomers), commerce, and sustained residence. For such homesteaders, authenticity rests in the true potential of their neighborhood or town, and they take pleasure in cultivating this in a manner akin to those who relish preserving embodiments of the past or landscape. Among homesteaders' perspectives and related practices, this falls closest to that of the pioneer. What divides the two orientations is homesteaders' interest in authenticity and their concern (even if limited) for gentrification's consequences for longtime residents.

Some homesteaders, when asked about their residential choices, report that they selected a place because it was "improving." This is especially true in Argyle. For instance, a writer who works from his Argyle condo said: "[Argyle] seemed to be going through a period where . . . [it] was—I kind of want to say gentrifying, but that's not quite it. . . . It seemed there was more of an investment in the community. People seem to care about it more than in the past." An Andersonville realtor similarly reported: "I thought it would be the next great neighborhood to live in. . . . I kinda knew that this was the next neighborhood that was gonna kind of pop."

Such words indicate that homesteaders are attracted to economic poten-

tial associated with gentrification, and some are quite comfortable acknowledging this motivation for their relocation. For instance, a woman explained
why she moved to Andersonville: "I liked the . . . space I could get for the
money . . . and the fact that I could still get downtown very quickly several
different ways. . . . I wanted to be near the lake."

However, such homesteaders are not merely attracted by the prospect of
financial gain. The excitement of living in a place that is about to "pop" also
appeals to them, and they invest significant resources to help the place reach
its zenith. Neighborhood boosterism and even protests sometimes take on a
festive air. For instance, a coalition of Chicago gentrifiers held upscale wine
and cheese fund-raisers and encouraged residents to attend theater openings
and to participate in lively protests (e.g., a New Orleans–style funeral procession to the alderman's office calling for support for new development). In
2006, the same group held a progressive dinner party in three historic homes
for $100 per plate. In this way, a financial return is not the only reward for im-
provement efforts. They also offer nonmaterial recompense, often in the form
of camaraderie, an amity that sometimes unites homesteaders and pioneers.

Some homesteaders suggest that longtime residents appreciate improvement efforts and that their favorable response serves as an important affirmation of progress for those self-conscious about gentrification's risks. For
instance, an Argyle social worker said: "[Longtimers] are all excited because
they've seen what it was. They've seen it go through so many changes. They're
just so thrilled that it's getting better. That's exciting to me. They are amazed
that young people are moving back into the neighborhood. Amazed. In terms
of the amenities, they just can't get enough of the changes." In this sense,
progress-oriented homesteaders are not indifferent to the past or to longtime residents, and this distinguishes them from pioneers. Some believe that,
through sweat equity and fiscal investments, they can return the space to its
authentic and often more prosperous past. It is, perhaps, unsurprising that
a social worker is among those progress-oriented homesteaders concerned
with longtimers' response to gentrification.

Other homesteaders justify their orientation to progress by referencing
markers of a "dangerous" local past. This is true of Michael, an Argyle block
club leader. On a winter evening, Michael welcomed me into his warm and
expansive condo, which sits on a newly rehabbed but still gritty Argyle block.
His wife greeted me from behind a granite counter, where she stood sipping
red wine. After conversing with the couple about their home, I followed Michael into the living room, which, with its fire and suede couches, seemed
worlds apart from the not-yet-polished street outside the floor-to-ceiling
windows.

Seated in an armchair, Michael began to answer my questions. The son of working-class parents, he was the first of his family to attend college and worked his way first through college—at a large state university—and then through law school. Today, he and his wife, who are in their forties, enjoy beach vacations and fine wine and food.

During our interview, Michael relished recounting stories of Argyle's "dangerous" past: "It used to be really bad: a lot of drug dealing, a lot of gangs, a lot of drinking." He told of deliverymen who hesitated to enter his building, only to discover, once inside, that it had transformed from a crack den into high-end condominiums. Michael enjoys speaking of such transformations because he helped usher them in. An active block club member, he meets regularly with politicians about revitalizing Argyle's commercial district, where Asian-owned businesses cluster, and posts messages on his block club's Web site about "nuisances" such as graffiti and low-income developments. Most weeks, he devotes hours to neighborhood concerns and cultivates ties with other gentrifiers. He reports that local diversity, as well as affordability and proximity to Lake Michigan, attracted him to Argyle. However, he has come to think of such diversity as more problematic than he imagined it would be. While he enjoys the local Asian restaurants, he believes that Argyle has more than its share of Section 8 housing and mentally ill residents and worries for his wife's safety when she walks alone after dusk.

Perhaps as a result of such concerns, he celebrates signs of gentrification: "I like the up-and-coming area where we're getting restaurants, we're getting theaters. . . . Everybody wants to be in this neighborhood, and it's nice." In contrast to Richard, the pioneer who hopes "American" businesses will replace Asian shops, Michael hopes Asian-owned concerns will "improve" and, in so doing, attract affluent tourists. In 2005, he participated in an effort to encourage Asian-owned businesses to remove burglar bars:

> You know, if a place is barred up at the end of the night, it basically sends a message that when we're not here we really don't care what happens in the street. . . . And, giving something to the Asian community, they were here when it was hell, and those bars had a purpose. And what we're just trying to show them . . . is that it's no longer necessary. Look down your street: you don't have gang bangers and thieves everywhere. You have a lot of people that are owning their places, that are really caring about the neighborhood. The neighborhood's just not what it used to be. There's no longer the crack house down the street.

With further improvements in mind, he organizes dinners at local restaurants that have removed their burglar bars, taking pleasure both in the promise of restoring Argyle to its heyday state and in the camaraderie such events cultivate among like-minded newcomers. At community meetings, when other gentrifiers express concern that his proposals will displace old-timers, Michael responds that *all* residents will benefit from a safer, cleaner, more upscale neighborhood.

Michael and other homesteaders like him evoke an ideology that regards revitalization as a socially responsible act that will help a place recover from blight and return to its true nature. In their minds, to stand in the way of revitalization is to prevent progress and to squelch the sense of movement and potential that they believe lends their place of residence economic and cultural value. They celebrate progress in a manner that makes some other gentrifiers and many longtime residents uncomfortable, but their vision of the future of their place of residence is not as indifferent to longtime residents as that of the urban pioneer, and it rests on a desire to return a place to an authentic past or heyday that few pioneers share.

Newcomers: Community and Character

Place character or the promise of joining a community draws homesteaders to a neighborhood or town. However, what attract many homesteaders, and what they seek to preserve in the face of gentrification-related change, are the character and sense of community associated with newcomers, not with longtime residents.

For instance, a semiretired homesteader explained what made him and his wife decide to buy property in Provincetown: "I came in the mid' to late '60s, and I was enamored with the town, I loved its character. That's the right word I'm looking for. I always said if I could have a second home this is where I'd like [it to be]. . . . It was very different exposure even though I'd been all over the world at that point [with the military and as a sales representative]. . . . I think Provincetown was like no other city. It had a lot of characters if you will. . . . [It was] far more open, it was far more sexual." The "characters" to whom he refers are gay and lesbian newcomers. For him, their presence and institutions separate Provincetown from other, less authentic places. As a person who strongly values freedom, he finds gays' and lesbians' political power and sexual license refreshing.

The story of a Provincetown lesbian homesteader also reflects this perspective. It reveals that an orientation to newcomers produces or is produced by an indifference to longtime residents much like that which pioneers articulate as well as by the progress orientation common among Argyle homesteaders.

For Sue, a middle-aged lesbian with a lucrative high-tech job who splits her time between Boston and Provincetown, the town's value and authenticity—those things that distinguish it from other places—rest on the presence of gay and lesbian newcomers.

In the summer of 2002, over dinner at a restaurant on a terrace on Commercial Street, Sue recalled her first trip to the town when she was in her teens. Traveling with her first girlfriend, she immediately fell in love with the town's beaches and with aspects of the town associated with newcomers: "I loved it. I mean it was free. It was gay. It was fun. I love the beach, so anything on the beach is going to be great. It was . . . diverse. It was . . . open to the gay lifestyle."

These are traits that Sue continues to appreciate. Indeed, Provincetown's fishing village past and Portuguese character draw little of her attention. Between bites of chicken, she mused: "It must have been [more of a fishing village when I first lived here], though I wasn't too tied into it. Everything you read says that [it was]. There are still boats hanging out there." Her world is one of newcomers. She explained: "I don't really know many people who grew up here."

While Sue has spent the summer in Provincetown for the last thirty years and is conscious of its rapid economic change, few changes bother her, least of all declining affordability. For instance, while she observed that working-class residents have relocated, she did not associate their relocation with gentrification. She casually mentioned: "The woman who cleans my house decided to move from here to Wellfleet [a more affordable outer Cape town]. I'm not exactly sure why. I don't know why she would. Maybe they needed more space?"

This is not to say that Sue is indifferent to protecting what she values about the town from gentrification; indeed, she seeks to preserve certain aspects of its character. For instance, she was conscious of the increasing presence of affluent gay tourists, particularly those whom she identifies as "circuit boys," who fill sidewalks on summer days and spill onto the streets and beaches outside clubs at night, a presence tied to Provincetown's increasingly upscale character. She suggested that this trend stands in stark contrast to the Provincetown of the 1980s, when AIDS dramatically reduced the number of gay male tourists and business owners. Like several lesbians I interviewed, Sue has conflicting feelings about this demographic shift. She celebrates the renewed health and extended lives of her gay friends while also mourning the decline of what she recalls as Provincetown's brief hiatus as a "women's town." In this sense, she appreciates a *particular* culture and aesthetic that she associates with certain gays and lesbians, for her pleasure does not extend to the circuit boys. Indeed, she worries that circuit boys and the upscale resort town that they embody threaten the place characteristics she most appreciates.

Similarly, while she welcomes the capital that funds the upscale restaurants and bars she enjoys, she finds some markers of change less tasteful. Sue recalled: "There was a time when Burger King tried to come in here. . . . If that happened, Provincetown would lose its charm." I was familiar with this issue, despite the fact that it occurred before I began fieldwork in Provincetown. During the college summer I worked retail there, my Portuguese coworkers often referred to the debate over Burger King. Unlike Sue and many other newcomers, they welcomed the affordable meals that the fast-food chain sells. However, for Sue, Provincetown's charm rests in the fact that "it's quaint . . . eclectic . . . not middle America, not white bread, boring," and Burger King, like circuit boys, threatened Provincetown's distinction. Unsurprisingly, my Portuguese coworkers were on the losing side of the conflict (although, several years later, despite similar resistance, a Subway sandwich shop opened on Commercial Street). In short, according to Sue, Provincetown's authenticity or uniqueness is tied to its gay and lesbian residents and their culture and institutions, not to its longtime residents.

Provincetown homesteaders are not alone in their attention to and appreciation for newcomers. Most homesteaders associate community with those like themselves. For instance, an Argyle professional described her community as follows: "Mostly professionals. People like us who moved [here] to see it get better. That is the majority of people that we see. . . . Those are the majority of the people we're friends with."[2]

This attention to newcomers, as well as other homesteaders' progress orientation, suggests that the line separating homesteaders and pioneers is sometimes thin. While far from universal, celebration of gentrification's successes, particularly the presence of newcomers and evidence of progress, is seen among some homesteaders. Living in a place characterized by newcomers can create a sense of community or affirm one's personal identity. For gay and lesbian gentrifiers, a town composed of others like themselves may be of particular import as it creates a sense of safety, identity, belonging, and power. However, homesteaders' celebration of newcomers is distinct from pioneers', for homesteaders express concern for preserving place characteristics that they believe gentrification, or perhaps better put *re*gentrification (Lees 2003), threatens.

Diversity

Urban homesteaders often report that diversity is part of the appeal of their neighborhood. For instance, a homesteader said that he enjoys Andersonville's "unique mix of diversity": "I mean, this is probably one of the most diverse neighborhoods in the city . . . you've got Korean, you've got Italian, you've got Mexican, you've got Japanese, you've got French, you've got all-American." Another indicated that he values a combined sense of diversity and safety, saying: "There's a little bit of everything."

As such statements suggest, homesteaders generally refer to what I have come to think of as diversity writ large, to a cosmopolitan and distinctly urban heterogeneity that includes both longtime residents and gentrifiers. In a sense, homesteaders celebrate eclectic neighborhoods or towns made diverse by their own presence.[3] The fact that, despite their appreciation for neighborhood diversity, many homesteaders do not see gentrification as a threat to place character affirms this. For instance, a thirty-something Argyle social worker explained: "I always categorize the neighborhood as a big melting pot, with wonderful neighbors and lots of diversity that is still on its way. That's what I always say. . . . I hope it still keeps its diversity; that's one reason why we like living here so much. But I hope that it's a little bit cleaner. I hope that there's more retail opportunities that make it a nice place to live." Another resident expressed appreciation for diversity but reported that she hopes to open a coffee shop and wine bar in a storefront that currently houses a grocery store serving a working-class clientele. While those who move in search of diversity may be disappointed by gentrification's homogenizing effect, they only selectively rally against certain gentrification-related changes. In this sense, they seek something of a middle stage of gentrification: a place largely composed of newcomers that also contains remnants of its social past.

Homesteaders' appreciation for diversity is also sometimes part of a vision of transformation and the related uplift of original residents. They imagine a gentrified space peopled by longtime residents, but their vision includes the transformation of those longtime residents. In other words, they wish both to avoid displacing and to "refurbish" the original residents.

Charlotte, an African American woman who grew up in a middle-class Southern family, best embodies this perspective. After college, she moved to Chicago: "It represented the bigger picture of what I was always looking for in a community and city. I love the city . . . [and] diversity, especially of minorities and African Americans who are doing positive things and held positions of authority and were involved in creating their community." However, several months of living with relatives on the South Side taught Charlotte that she is most comfortable in "*really* diverse communities." The South Side with its predominately African American population was too homogeneous for her.

A decade after she arrived in Chicago, Charlotte and her husband—a successful businessman—moved to Argyle: "We decided to look for a place to invest in. One of my priorities was to live close to the lake." She and her husband were attracted to Argyle's most upscale section, Margate Park, which Charlotte jokingly calls the "platinum area of Uptown." With a park on two sides of a tree-lined cul-de-sac, the neighborhood was attractive to Charlotte because it represented "the new and the old at the same time": "I loved that it was a cul-de-sac street." However, she also believed that it was a place in which she could cultivate community: "I always said that, wherever I invested, I wanted to be part of a community. . . . I'm totally into it now. It didn't take long." For her, home ownership was key to community involvement: "I realized that, once we had our home, it could build a foundation of building and growing into the community." Indeed, today she says: "I love walking within our area and seeing people on the street and knowing their names. . . . I probably know most of the children and the dogs in the neighborhood, and I don't have one animal or a child." She fondly describes Argyle as "vibrant, eclectic, and totally diverse."

During my fieldwork, I sometimes encountered Charlotte at community events and meetings. She could frequently be found gardening in a park near her building, sometimes with local schoolchildren or single-residency-occupancy hotel (SRO) residents. Several residents suggested that I interview her, referring to her leadership role in the Margate Park Block Club. On the afternoon when I interviewed her, Charlotte sat behind a countertop in her sunny kitchen, the open expanse of her living room behind us. Plant-filled balconies stood beyond floor-to-ceiling windows. A professional vocalist, she had just returned from an audition. Still in striking interview attire, she offered me a soda and patiently answered my questions.

On moving to Argyle ten years ago, Charlotte immediately became involved in local organizations. While she wished to become a part of the community, she suggests that her initial involvement was accidental. On a walk, she discovered that the alley next to her condominium was full of litter. After collecting the trash, she wondered whether planting a garden would discourage littering. A few phone calls to city officials later, Charlotte was planting flowers in several public spaces near her home, and she soon became the informal custodian of a play lot in the park next to her building. She regards beautification as a necessary device for improving the neighborhood: "In a way it's kind of selfish because I'm trying to create what I want to see. But in another way I wish that other people would just create what they want to see and then . . . it can manifest into a space that's safe, creative, artistic." She recognizes that her efforts have likely contributed to rising property values but contends that she "didn't set out" to have that effect.

Charlotte's civic work does not stop with beautification. She also volunteers at the elementary school near her home: "I would go to [the elementary school] and read to a fourth-grade class on a regular basis. And that really established a nice connection with the school. Because I live across from the play lot, I recognized that it's probably better to get to know these children and let them know that you're not intimidated by them and you're going to call them out on the good and the bad." Charlotte recognizes the distinction between her own and many other newcomers' relationship to the children of poor and working-class longtime residents, which she characterizes as antagonistic and rooted in fear. She said: "There's so many of our neighbors who are so fearful of the youth." Indeed, just a few blocks from Charlotte's condo lives an Andersonville gentrifier who takes comfort in the fact that his dog scares local young people. He said: "When we'd lived here for like three months, I went to take the dog for a walk. The minute I came onto the front porch, these three little [African American] kids ran up and said, 'Oh my God, it's that dog!' and then just ran away, which I didn't mind because it keeps people away."

Perhaps because of her appreciation for diversity, Charlotte consistently works to transform her own fear of certain neighborhood populations. For instance, on joining a local theater troupe in which she performed alongside mentally ill community residents, she "started realizing that we live in such a diverse community and you really can't stand in judgment, you just have to be who are you and as authentic as possible and to create your own space." Indeed, this mirrors the philosophy of the theater troupe, which uses interviews with neighborhood residents for script development. One year, I attended a musical production by the troupe, which was about a garbage man who often dressed in drag. The show depicted the community slowly rallying around the man, and, in the final scene, the entire cast—consisting of halfway house residents, recent immigrants, children from local schools, and upper-middle-class newcomers like Charlotte—joined the garbage man (dressed in gown and heels) in song. The friends who accompanied me to the performance agreed that it had the aura of an Uptown-style *Sesame Street* scene, for it simultaneously embodied a wholesome celebration of community and the enigmatic union of disparate residents around a transvestite character.

Charlotte is well aware that, if anything unites Argyle's residents, it is the fact of gentrification. However, she also recognizes that gentrification is a divisive issue. Recent debates over proposed upscale developments revealed to her that some longtime residents and newcomers alike are unhappy with gentrification: "Some people would say that more yuppies are moving in [and that's a bad thing], but I don't look at it as yuppies or even dinks or binks. I see it more as a progression." She positions herself between newcomers who wish to see mentally ill residents of halfway houses displaced—"people who want to put out another group of people and have Starbucks on every corner"—and those who resist change. Charlotte admits: "I like development. I like nice shops. My attitude is I'd rather have it created in our neighborhood and support our community than take [my money] to Lincoln Park or up to Evanston. . . . I like mom-and-pop situations, but then of course I would love to have a Target real close to us."

Indeed, she believes that old-timers who own businesses should accommodate new residents. "I wish he would change his decor a little bit," she says of the owner of a local Chinese restaurant. "If you make it pretty, you can attract more of the people who might be intimidated coming into your store. . . . I think that as Argyle changes . . . you can change the aesthetics to be more pleasing." She applauds Vietnamese business owners for slowly welcoming newcomers into their establishments, for recognizing that "the money being spent in their establishments can only help their establishments" and that "they don't have to just cater to their own": "[They should be] welcoming when you come into those shops and not suspicious. . . . I enjoy going to some of the stores and enjoy so much of their cultural diversity within their community." In general, she envisions the neighborhood "*moving to another level* of the entertainment area, of diversity, of just such a good mix of people and businesses" (emphasis added). In this, her appreciation for diversity and her appreciation for transformation go hand in hand.

How does someone like Charlotte reconcile these dual impulses? The potential displacement of businesses and residents concerns Charlotte, but, ultimately, she believes that gentrification alone is not to blame: "I don't have pity, but I have empathy. . . . But I know as people we create our lives. And we create it through our choices and our thoughts, and what we think is what we create. . . . I don't know, I don't want to see people pushed out because of taxes and property value. But at the same time . . . it's up to the individual to take responsibility for some of the choices that they make." She is frustrated when people see her as an urban pioneer: "Some . . . [think] I am one of those yuppie buppie dinks who has come to change things and kick out the poor people."

Indeed, Charlotte stands firmly between the pioneer and the preservationist. As a leader of Argyle's beautification movement, a principal member of her block club, an affluent homeowner, and a supporter of economic development projects, she sides with many urban pioneer–style gentrifiers. Yet, unlike the pioneers I encountered, she devotes significant resources to helping those residents less privileged than she is: the children she reads to at the elementary school and the SRO residents who help her with community gardens. Still, her approach to original residents, while rooted in respect and concern for diversity and individual longtimers, is close to the uplift model associated with urban reform movements and social service organizations. Charlotte's vision for Argyle is neither the neatly transformed upscale neighborhood of which pioneers dream nor a carefully preserved, rough-around-the-edges space. She wishes for a melting-pot amalgamation, and this requires both preservation and transformation. In both her appreciation for diversity and her position between the two key typologies, she is not alone.

Identification with Longtimers

Some homesteaders share with longtimers an appreciation for certain place characteristics. This is particularly the case in Dresden, where some newcomers and longtime residents possess a libertarian perspective that encourages resistance to efforts to restrict development and other protective covenants. In this sense, not all gentrifiers have an adversarial (like pioneers) or protective (like preservationists) relationship with longtimers. Some homesteaders side with them in debates with gentrifiers, yet their perspective is not rooted in empathy for or the romanticization of old-timers that social preservationists express. Instead, it emerges from shared political or material interests.

This was true of Randy, a retired military officer, who moved to Dresden in 2004. I interviewed him on a rainy fall afternoon in the shell of the house he was building on a bank of the Eastern River. While he had yet to take up residence in town (he and his wife, a physical therapist, were living in a neighboring town during construction), Randy had already joined the old-timer-dominated volunteer fire department. While many newcomers spoke of their affection for cross-country skiing and walks on back roads, Randy looked forward to finding others with whom to hunt and snowmobile.

He said that it was property, not people or history or culture, that brought him to Dresden: "We weren't looking so much for a town. We were looking for property. Waterfront was one of them. Acreage was another. And I wanted raw land. Raw land rather than an existing building because I wanted to do what I wanted to do and not have to adapt myself to what was there already." Having moved to Dresden and enjoyed the freedom to build the retirement home he and his wife had envisioned, Randy wishes to preserve that freedom for others, and for him this means accepting that Dresden will inevitably change: "In some of the small towns I've lived in [in Massachusetts] for the last thirty years its always, 'Yeah, I'm here; now let's close the door.' So I know what they're talking about [when they talk about controlling growth], and I can see their point. I like the way it is now, but you have to realize that things don't stay static; they change. Right now we're getting bigger."

Another newcomer's occupation—as the owner of a greenhouse plant business—encouraged his alliance with longtimers after conflicting with other newcomers about his proposed business. I spoke to Charlie, who enjoys reading the Bible and socializing with his employees, in the office of his large greenhouse operation. Outside, his dog barked beside his farmhouse, and employees prepared for spring planting.

Although he had long worked in the greenhouse industry, Charlie did not plan to open a greenhouse business when he moved from Rhode Island to Dresden. Instead, he moved to Dresden to work for a religious organization in a neighboring town but was convinced by friends that Maine was an ideal location for a greenhouse operation.

However, his neighbors expressed much resistance when Charlie applied for permits for greenhouse construction. All newcomers, some of whom Charlie estimates have lived in town for twenty years and "some less," his neighbors worried he would disrupt their view and pose health hazards. He recalled the basis of their complaints: "I think it all boils down to the fact that they're afraid of change. And they all wanted Dresden to stay the way it was when they got here. . . . They raised a lot of havoc with me. . . . They thought I should just keep it open. They want to use it as their personal parkland. That's the thing around here that I find with people that move here—they buy land. They call it a farm, but it's not really. It's just open land." While Charlie believed that his neighbors' concerns were frivolous, the permit process was anything but. He reports that fourteen state and federal agencies had to approve the operation and that he appeared at a town hearing on the matter. His neighbors, who distributed fliers prior to the hearing and hired a lawyer, initially generated sympathy among residents but were unable to block construction. When I conducted my interviews, many longtime residents and social preservationists had come to support Charlie.

Charlie's conflicts with his neighbors did not end after the hearing:

They told me at one point that this wasn't agriculture, that it's a factory. And I said, "Well, what is a farm?" And they said you have to have cows. So I went out, and I bought two cows and put them in my yard. . . . I had two steers, and they both bellowed loudly. I was their mother, and they'd follow me. . . . One of them was loose all the time, and he'd come down to the greenhouse or go for a walk in the woods, and you wouldn't find him more than a few hundred yards away. He followed me wherever I went. He'd go down the road if I went to visit somebody, and, if I wasn't back here in time to feed him, he'd come find me. . . . [The neighbors] thought I was crazy. The neighbor across the road, their dog used to come and poop in my yard all the time, and I never said anything, but my cow went over there and pooped in her yard, and of course she yelled about it.

As his words illustrate, Charlie was aware that newcomers' fondness for long-time residents is often limited to a certain type of resident who engages in certain kinds of practices (e.g., raises cattle). As a college-educated newcomer whose farming practices were hardly old-fashioned, Charlie did not fit their image of the idyllic farmer, nor did his greenhouses fit their image of the idyllic farm landscape. His experience is reminiscent of the words of a dairy farmer from my pilot study of Leyden who quipped to his daughter that new-comers only want "cardboard cutouts of cows in the fields."

Perhaps because he had been on the side of a conflict that longtime residents often occupy, when I asked him what the town might be like in a de-cade, Charlie sounded much more like a libertarian Mainer than a newcomer. He said: "Well, certainly, it will be grown up a little bit more with houses, . . . but that's fine. They'll buy my plants. I think there should be some type of controlled growth, but . . . there's fifteen hundred people that live here, . . . and there's fifteen hundred different opinions on what should happen here." Likely as a result of the conflict, Charlie believes that his vision for Dresden should not have greater influence than anyone else's.

This suggests that occupation and even circumstance (e.g., finding oneself in conflict with newcomers) can shape homesteaders' sympathies and iden-tifications and that, despite their shared economic position, gentrifiers do not automatically ally with one another. Indeed, just as many longtimers do not find their gravel pits or scrap metal piles (against which gentrifiers often lodge complaints) aesthetically displeasing, Charlie admires his greenhouse, set against open fields: "Right now there's not much in it. But in the spring-time in the middle of February or March you come here, and it's a big square of green amongst the snowy environment."

Despite having been at the center of what residents recall as one of Dres-den's most acrimonious conflicts, Charlie is fond of the town. He said: "It's kind of like that old show *Mayberry RFD*. I was down at the town office so often for so many weeks or months [during the conflict]. They're open every Thursday night. I call [the selectmen] the town fathers. They hang out there and answer people's questions about building permits or whatever. Well, anyway, I'd been there so much, and then after the hearings were over I didn't go. And a few weeks later I got a phone call one Thursday night, and it was them. They wanted to know if I was OK." Recalling this, Charlie smiled and looked at his boots, seemingly smitten with Dresden.

Thus, Charlie and Randy fall outside the pioneer and preservationist models. While they possess economic and educational capital that separates them from most longtime residents, they do not dote on them, and they do not have an adversarial relationship with them or seek to uplift them or celebrate their authenticity. Instead, they rarely distinguish themselves from longtimers and side with them on issues related to gentrification and development.

This perspective may be rooted in political orientation (e.g., shared libertarianism), a common working-class identity (as tradesmen or merchants), or even circumstance (a conflict with other newcomers). It may be more likely in a place like Dresden, where longtime residents' identity is closely tied to a political ideology (libertarianism) and many newcomers share natives' racial and regional identity. Regardless of its origins, this perspective stands in opposition to the pioneer model, which assumes antagonism rather than identification.

Conclusion

This chapter drew our attention to social homesteaders' orientation and practices. Homesteaders do not seek to stand in the way of gentrification as social preservationists do. Nor are they indifferent to its direction or as certain of its inevitability as pioneers. In a sense, they are like the traffic cops who stand on Provincetown corners during the busy summer season. Rather than seeking to stop traffic or to hurry it on, homesteaders seek to direct the flow of capital, culture, and politics in particular directions. They wish to avoid a complete standstill, for they want to feel at home in their neighborhood or town, and this requires certain "improvements." Yet they seek to preserve characteristics that they worry gentrification threatens and, therefore, direct change in a manner in which they hope will protect the authentic or fragile.

Social homesteaders remind us that the literature, with its focus on the pioneer, has neglected the full range of orientations to gentrification. They also demonstrate that even a group like homesteaders is internally diverse. In fact, while this chapter identified the ideological traits that bind homesteaders, it also demonstrated their heterogeneity. Individual homesteaders emphasize distinct place characteristics and engage in different practices. Some homesteaders' beliefs and practices are more aligned with those of pioneers and others with social preservationists.

Place variation explains some of the differences among homesteaders. For instance, Dresden's rural terrain attracts those who emphasize landscape authenticity, while Argyle's relatively early gentrification encourages appreciation for progress. That said, homesteaders' attitudes are not uniform within any site, so place alone does not explain their diversity of approaches.

An individual's personal traits also shape his or her approach to gentrification. For instance, some gay, lesbian, and bisexual homesteaders celebrate Provincetown's gay-resort identity. Many wish to preserve the rough-around-the-edges gay enclave they discovered in the 1980s or even the 1990s. Again, personal traits do not ensure a particular view, for the queer gentrifiers in my sample are actually *more* likely to engage in social preservation than to be homesteaders, and, in Provincetown, many gay and lesbian preservationists work to save Portuguese families rather than gays and lesbians. Nonetheless, place and personal traits contribute to homesteaders' beliefs and practices even if they do not determine them.

Homesteaders' nearly universal appreciation for local authenticity stands as evidence that an interest in authenticity, preservation, and self-reflexivity about gentrification is not limited to atypical gentrifiers, such as African Americans or first-wave artists. Nor is gentrification only an "avenue of identity construction" (Taylor 2002, 68) for some newcomers; it functions that way for all. This will not surprise students of place and identity, for, as David Hummon (1990) and others demonstrate, community identity is a vital mode of self-definition, and narratives about place are an important resource for identity construction and articulation. This chapter calls gentrification scholars to reclaim this perspective as nearly all gentrifiers celebrate what they find to be distinctive or authentic about their neighborhood or town. Indeed, this book will continue to reveal the striking effects of definitions of *authenticity.* In the next chapter, and throughout the rest of the book, we will see how a different orientation to authenticity—one that emphasizes old-timers' character and community—shapes social preservationists' residential choices and practices.

3

Social Preservation

It took two lengthy interviews for Tom to recount his two decades of work resisting Provincetown's gentrification. In fact, after the first interview—conducted at his cluttered kitchen table amid art books and paperwork for his massage business—he e-mailed me to say that our two-hour interview had only scratched the surface. For instance, we had not discussed the Community Compact, an organization he cofounded in 1993 that provides temporary housing for artists and raises funds for AIDS care. Nor had we ventured into his small backyard to examine the septic tank he transformed into a performance space to comment on the town's limited number of affordable rentals. This is not to say that we squandered our first interview. We had already discussed his art installations that parodied gentrification, such as the "Last Alms Museum," established in a soon-to-be-bulldozed house, and his "Sold to Cape Cod" musical performance, chastising the town for, in his mind, selling its soul for tourist dollars. Through our interviews and subsequent observations, I sought to pinpoint the beliefs and sentiments that drive him. I also wanted to chart their relationship to his personal history and to the beliefs and practices of others who resist gentrification in Provincetown and beyond.

In this chapter, I describe what I learned from Tom and thirty-seven others who articulate the social preservation ideology and engage in related practices. While the difference between Tom's orientation to gentrification and those of other gentrifiers may already seem self-evident, I carefully delineate the specific beliefs that distinguish social preservationists from homesteaders and pioneers. First, I document the accounts social preservationists provide of the impetus for their relocation, which they typically locate in appreciation for ungentrified space, namely, space marked by old-timers' presence. I also explore social preservationists' vision of the future of the space in which they live, which emphasizes old-timers' sustained physical, cultural, and political presence, and their more general concern that gentrification will destroy the authenticity of their place of residence and, therefore, threaten the distinction between their home and other, less authentic places. In concert with such concerns, social preservationists bemoan the cultural, social, political, and aesthetic implications of gentrification even as they acknowledge (and criticize) their participation in the process.

Examining all this material, I document preservationists' attitudes toward their neighbors, specifically, their antipathy toward other gentrifiers and their romantic attachment to old-timers. The chapter reveals the significant ideological differences between social preservationists and other gentrifiers. Not all preservationists are as bold or as committed as Tom, but their orientation to gentrification, old-timers, and place is, nonetheless, remarkably different than that of the pioneers who dominate the gentrification literature and even from that of the social homesteaders we met in the previous chapter.

Origin Stories

One way in which social preservationists articulate what they value about their place of residence and explain how it differs from other, less authentic places is through what I have come to term *origin stories:* elaborate and rehearsed narratives of how they came to live where they live that rest on the notion that some places are more authentic than others. These origin stories contrast the spiritual, political, or aesthetic qualities of the place to which they moved with the soulless, capital-driven, homogeneous place from which they moved. Such "other" places serve as a reminder of the value of their new place of residence as well as a warning of gentrification's risks. Alternately, their stories reveal how preservationists came to value the authenticity of their place of residence after relocating.

For instance, sitting on a picnic table beside his garden, Robin, a bearded Dresden preservationist studying local architectural history, recalled his decision to move to town:

> I grew up in the suburbs . . . just north of Boston . . . very urban or suburban the whole time, and I hated every minute of it. I always wanted to go out to the country. . . . When we drove into this town, we thought: Whoa, this is a rural town; this is the real thing. . . . It still makes me giddy to think about it now. . . . We thought: yeah, this is it; you can't get any more authentically New England than this. This isn't the tourist version; this is the real thing. . . . And there's nothing virtual about this. There's no suburban imitation here. This is the genuine artifact.

For him, Dresden's authenticity rests on landscape and "the people who had been here before us." Importantly, he contrasted Dresden with his experience of Vermont, which he found unsatisfactory: "You could see the neighbors in that particular community were quite close, but they were all summer houses. They were all summer homes." For Robin, this community of summer residents was not as "real" as that to which Dresden old-timers belong. From his perspective, the old-timers' community lends the town authenticity.

Others use origin stories to explain how they came to think of their place of residence as home, rather than how they came to live there. Unlike Robin, whose preservation ideology informed his relocation, these individuals adopt a preservation orientation after moving. This was true of Leslie, who, like many other preservationists, used an origin story to illustrate how gentrification threatens place characteristics that led her to make Provincetown her permanent home.

At her suggestion, I met Leslie, a slender, middle-aged woman with a tangle of curly hair, on a Saturday in Provincetown's center beside the white clapboard town hall. A leftist long active in local politics, she stood among a small group protesting the Iraq War. After she finished distributing antiwar fliers, we walked through slushy February streets to her home.

Once inside her small house, an old Cape Cod with white walls, sparse furnishings, and gleaming wood floors, Leslie introduced me to her husband, who was at work in his studio, and boiled water for tea. A few minutes later, mugs in hand, we retired to her study, which she has lined with Victorian novels and furnished with two sturdy armchairs.

We began talking about her house: of the work that went into its renovation and of her disapproval of its growing value. We also spoke of her children and her great love for the Cape Cod seashore. Shortly, the conversation turned to Provincetown. Leslie spoke vividly of her first winter in town: of small rented rooms, potlucks, and the surprising discovery that, despite the fact that she is an artist, much of what she loves of the town are traits she associates with its Portuguese-fishing-village past and present, rather than with its reputation as an artists' colony. She said: "When I came here, it was like, 'Oh my God, it's home. I really have a home for the first time in my life.'" Leslie recalled what made her decide to stay permanently in Provincetown:

> Water. I never even knew I had such a deep-seated desire to live by water. . . .
> The fishermen. I mean we really had a thriving fishing community. I don't
> know how thriving it was, but it felt thriving. It felt very much like it was a
> huge part of the town. . . . There was this mix of washashores and the Por-
> tuguese fishing community. . . . It was really like a true mix of people. It was

amazing. . . . To me it was a very remarkable place. . . . And, by the time I came
[in the early 1980s], they were probably [already] weeping about the changes.

Leslie has herself wept about the town's changes. Indeed, she cried during
our interview. Her sadness is rooted, in part, in empathy for those she has
seen displaced: "I've seen so many people have to leave. Not only because of
the job I have now [as an affordable-housing advocate], but before that. . . .
Many of those people are gone: the artists. . . . They were a real part of this
town for a century. For more than a century." However, she is also sad because, as the town gentrified, it lost much of what led her to call it home and
distinguished it from other places she has lived, such as Northampton, Massachusetts, which she found to be newcomer dominated and oppressively
politically correct. She recalled what she loved of Provincetown: "There were
kids. It's so sad; if you look at the annual report, I think we had two years in a
row where we had one child born here. I miss kids terrifically. It started changing really fast toward the late '80s, and then 1990, '91, man it just went crazy.
That's what changed it. What basically changed it the most was the condoiza-
tion of Provincetown." She elaborated: "[The town] just feels soulless. . . .
It's like this place feels really soulless. Why would I want to stay here?"

Yet Leslie remains in Provincetown. Indeed, she makes several appear-
ances in my field notes. She is in the front row at a crowded affordable-
housing hearing, where she lists the number of residents waiting for housing,
her voice alternately rising with anger and breaking with sorrow. She is also
at a town meeting, voting for the Community Preservation Act, which set
aside funds for affordable housing. Like many others, she dedicates much
time to ensuring that the place in which she lives reflects her tastes and values,
and, for her, the social character of the town—as constituted by old-timers'
presence, culture, and community—is what makes her think of it as a place
worth living in, albeit a place that is truly home only to fishermen, their children, and the struggling artists who arrived before her.

Preserving People, Place, and Community

As the above passages suggest, social preservationists emphasize the social
rather than the physical amenities of their place of residence. While, like
many other gentrifiers, they are concerned with the preservation of community character, they believe that this is best accomplished by preserving
old-timers' presence and practices. Some, like Robin, relocate to live beside
old-timers, while others, like Leslie, move to live near another social group
(e.g., newcomer artists) only to find that they are drawn to old-timers. Still

others move out of appreciation for the built or natural environment, later turning to old-timers for a sense of local character and community.

The latter was the experience of one Dresden resident, Sarah—a Republican and avid tennis player married to an affluent lawyer. When I arrived at her home to interview her, I parked my car and paused to enjoy the view that inspired her move: rolling green hills crosscut by a river with New Hampshire mountains on the horizon. Petite, with sandy hair in a neat bob, Sarah welcomed me into her restored farmhouse with a serious manner.

Seated on a couch beside windows overlooking hay fields and a weathered barn, Sarah recalled that her young family moved to Dresden in the 1970s in large part because the town promised boating and hunting opportunities, proximity to the coast and the state capital, a well-regarded elementary school, and an "agrarian landscape" that she loved. On her arrival, Sarah threw herself into community-building activities, organizing cultural events, volunteering with the schools, and engaging in local and state politics.

However, after several years of residence, she noticed that some old-timers resented what they regarded as newcomers' intrusion into local politics and community. Sarah laughed and said: "I always say we just went up the coast until we ran out of people, and that brought us to Maine." But, in fact, she has come to love the presence of local characters, particularly farmers and their families. She explained: "I have a real fondness in my heart for the people who live off the land." She particularly appreciates a certain life attitude that she associates with farming: "People who live their own lives and don't try to live other people's lives for them. You know I'm very fond of that. It's always been very endearing. Even in politics, I just loved it."

Sarah worries gentrification will lead to old-timers' displacement. Musing about riverside lots listed at $250,000 (a previously unheard-of price in Dresden), she said: "People from away bought this and probably have the finances to support it. There are a lot of families who had this in their family for three hundred, two hundred years, and they are making a living off of that land, cutting wood, plowing driveways, whatever, and they don't necessarily have the finances to be able to support the increase in taxes, and so they're going to be driven off, and that means you get away from your agrarian, your rural feeling." As mentioned in the introduction, because of such concerns, Sarah and her husband purchased acreage from their neighbor, a retired farmer for whom property taxes had become too expensive. They did not do so as an investment. In fact, they agreed not to develop the fields that constitute much of the acreage. In so doing, they sought to preserve an old-timer—a part of Dresden they have come to love—and the landscape that brought them to town in the first place.

This belief that old-timers are primarily responsible for lending space value—for making, for instance, a landscape truly rural or agrarian—can place social preservationists in conflict with social homesteaders, who wish to preserve the natural or built environment. For instance, not only does Sarah forgo the economic gains she could receive from developing her acreage, but, to preserve old-timers' presence, she also refuses to act on her desire to use zoning restrictions to prevent farmers from developing the landscape she loves, a tactic many of her homesteader friends and neighbors rely on to restrict growth and longtime residents' practices that they believe threaten the integrity of the local ecology and landscape, such as operating gravel pits and scrap metal dumps. Thus, despite the fact that she loves Dresden's landscape, at this time Sarah prioritizes the preservation of people and what she regards as their traditional way of life.

Many social preservationists give similar precedence to the preservation of old-timers over historic buildings. According to a man who resides near the University of Illinois Chicago campus (several miles from my Chicago sites), historic preservation encourages rising property values and displacement and, therefore, is antithetical to *real* preservation:

> These 7 old buildings . . . they're only saving the shell of the buildings. . . . The real big tragedy here is that while I think saving buildings [is] important . . . people are more important than buildings, and to have a huge building saved where the people who lived here are all gone is to me not real preservation. What to me real preservation is about—and a lot of even preservationists don't understand this—is preservation of a place, which is the building fabric, but it's also the people too, and the culture, and the food, and the signage, and the music, and the interaction. That's a place, that's culture, and this was a great place, and it should've been saved. (Eight Forty Eight 2002)

Similarly, at a Chicago discussion of gays' and lesbians' role in gentrification, a well-dressed, middle-aged gay man said: "Historic preservationists try to preserve houses when we know that *people are the culture*. We have to figure out ways to codify it." Along the same lines, an affluent Provincetown innkeeper said: "If we're trying to preserve something of the community, the way to do it isn't through preserving wood shutters on buildings. It's through things like affordable housing initiatives and *who really lives here.*"

This dedication to preserving old-timers' presence is particularly notable given that, as Sarah's story illustrates, many preservationists share homesteaders' appreciation for the built or natural environment. For instance, the Provincetown innkeeper I just quoted focuses on preserving residents, but she and her partner carefully restored their bed-and-breakfast and decorated

it with antiques. Thus, social preservationists do not entirely divorce themselves from the tastes and concerns of homesteaders.

Yet, despite this dual appreciation for physical and social amenities, when
push comes to shove, social preservationists demonstrate that they value the
social character of their place of residence, particularly that which they associate with old-timers, over the built or natural environment. For instance,
several Argyle social preservationists lobbied in support of a proposed high-
rise assisted-living facility for the elderly despite pioneers' and homesteaders'
complaints that it would destroy the street's aesthetic and historical integrity.
A social preservationist focused on the social benefits the facility would provide for old-timers: "Many of our seniors in the area live just above poverty
level. . . . This is very much the case in the Asian community. . . . They have
the additional plus of Argyle Street where they can visit . . . shops that carry
the foods and things they are accustomed to and be served by individuals
who speak their language." Underneath his concern for old-timers is an interest in social and cultural preservation, in maintaining old-timers' presence
and traditions. In contrast, an Argyle homesteader said of the same facility:
"Let's face the fact that low-income, elderly housing is rarely architecturally
stimulating (to say the least) or reflective of its surroundings, and, moreover,
it seems to quickly depreciate."

Social preservationists' prioritization of old-timers' sustained presence
over the preservation of the built or natural environment rests on their deep
appreciation for old-timers and for the sense of community that they believe old-timers lend place. While many, if not most, people value the social characteristics of their neighborhood or town even as they celebrate its
physical features (e.g., the historic preservationist who celebrates buildings
and, at least implicitly, the gentrifiers who inhabit them), social preservationists place particular emphasis on those whom they deem old-timers. For
instance, a bearded writer explained the impetus for his move to Dresden
two decades ago: "I really loved the community . . . this real feeling of *real*
people and a *real* town." For him, "real" people are old-timers, and they lend
Dresden authenticity.

In this sense, social preservationists do not uniformly celebrate the social;
they narrow their focus to particular population groups. Indeed, many express distaste for the presence of others who share their economic and cultural traits. In this, they stand in stark contrast to the many homesteaders
and pioneers who turn to those who share their social traits for a sense of
community.

Preservationists frequently articulate the belief that affluent newcomers'
presence (their own included), like a bulldozer in the natural wilderness,

threatens the local social ecology. For instance, a middle-aged electrician suggested that newcomers endanger what he appreciates about Dresden: "The influx of flatlanders has increased, and their needs or wants have dampened the rural effect." His concern is not merely for landscape (threatened as it is by development) but for rural culture and small-town community as well: "The only person who'll be able to afford this will be somebody from out of state or somebody from in state who's just looking for a trophy house or a trophy property. It really does change dramatically the culture of the community."

Such criticism is rooted in preservationists' concern that gentrifiers' interest in profits and aesthetic improvements may contribute to old-timers' displacement and, therefore, to the declining authenticity of their place of residence. A Provincetown preservationist who works in nonprofit management complained: "Now every house in town has a construction truck in front of it." She expressed sorrow that "the fishing community is really gone"—and, with it, the impetus for her relocation—and concern that the wealthy cannot appreciate the town's true value as an ethnic enclave and fishing village: "People just want to buy a beautiful place without a sense of who they might be displacing." Another town resident said: "It's not a village anymore." Still another suggested that money "raped" the town. A Provincetown woman active in town governance referred to the death of "community character" and of a "sense of place": "Fishing plummeted, so money . . . became like a drug because people didn't know where it was going to come from, so, when they saw it, they grabbed it. . . . Materialism has basically rolled over the community in an irretrievable way. . . . You used to be able to sit in a restaurant without anyone having any idea whether you had a nickel or a million dollars. Now there are no people with nickels, . . . and to me that is just an absolute outrage." Another resident, a middle-aged teacher, likened displacement to an epidemic: "The second thing that right now is really in my face is housing. . . . I will not say it is as bad as AIDS because people rarely die with this, but it's really—how would I say—it has taken stamina out of the town when it comes to so many talents, so many exceptional people, people who have been here for years who had to leave just because they couldn't find housing. It's really sad. . . . A lot of artists had to leave, people who just needed houses."

Preservationists decry old-timers' displacement, not only out of empathy, but also, in the words of one Dresden man, because of their personal interest in maintaining the "sense of community" and "collective memory" that old-timers lend place. He spoke of the meaning old-timers bestow on the town:

> She's a fourth-generation Dresden person. . . . She has so much information, knows so much and cares so much about the town. It's fascinating: she's like a fourth-generation Dresden person in that same house. It's strange thinking back about how little I valued the sense of community at the beginning because now I really miss these older people. They're not related to me, but they really added something to the town. . . . They added their love for the town, I guess, and that's what they take away when they leave: this sense of appreciation, a sense of collective memory, maybe, that departs also.

Preservationists particularly celebrate the social networks of old-timers, identifying them as those with geographically rooted "friendship[s] that extend back in time" and as those with "forty thousand uncles and brothers and kids."[1] Social preservationists are particularly adverse to the disruption of such networks by gentrification. For instance, an Andersonville artist recalled witnessing a breakfast in Svea—a restaurant that serves Swedish fare in a room decorated in traditional blue and yellow shades—for a Swedish resident who had passed away. For him, the event indicated the finality of the neighborhood's transformation, a realization that brought him much sorrow: "[The man who died] was one of the big Swedish leaders, an old man who had been here forever. . . . His friends came in and had a wake for him. They were all talking about the time that they knew had passed. It's like all these people—the *real* Swedish community—the ones that got the [Swedish American] museum going and [who] used to have the Miss Andersonville Pageant and all the Swedish stuff. . . . I was just sitting back and watching it change."

Jason, a trim thirty-year-old man from an upper-middle-class family who cleans rooms at a Provincetown inn for board and a small salary, articulated similar concerns. With a degree from an elite journalism school, he regularly publishes articles—primarily on Provincetown—in regional and national publications.

While he is a gay man who often writes for a gay and lesbian readership, Jason reported that he did not move to Provincetown because it is an internationally renowned gay resort. Instead, seated in a small room under the inn's eaves, he said that he fell in love with aspects of Provincetown that remind him of what his hometown—a small New England industrial city—once was. Specifically, he appreciates the presence of Portuguese fishermen, who remind him of his hometown's tight-knit ethnic populations. He also confessed that he wanted to live in a town in which he could wear flip-flops to summer cocktail parties and go a few days without shaving.

Although he had lived in town for only a short time, Jason articulated a sense of loss related to the town's gentrification. He believes affluent newcom-

ers threaten Provincetown, just as they threaten his hometown and home-
towns across the world (he had recently traveled extensively in Asia and Europe). He said: "Just within the past few years, this really brash, obnoxious,
wealthy crowd has just come in. . . . They're just knocking down all these
homes, changing things so quickly." He believes that such newcomers have
"ruined town": "They brought in elitism. They brought in snobbery. They're
the ones—that type of gay person—who drove the real estate prices up. And
60% of real estate is owned by people who don't live here."

Jason worries this wealth threatens exactly what he moved to Provincetown
to find: an escape from a globalizing homogeneity. He said: "It's like the Wal-
Martization, and just the uniformity . . . I just don't want anything to do with
it. I want nothing to do with it. So I ran here more to get away from that. . . .
It feels like it's the '50s, this kind of cultural sterility. You can drive across
America, and every town looks exactly the same, and everybody's doing the
same thing, and everybody's listening to the same stuff. . . . I almost see this
desperation in people looking for something unique because it's so rare that
you can find that now." Specifically, he worries that affluent newcomers will
displace working-class Portuguese families: "If you look at the median wage
of P-town, it's really heartbreaking to see. Particularly people who have children. If you raise children here, there's no way, unless you pass on your home
to your kids, there's no way they're going to be able to live here. . . . You
look at the fishermen with their kids . . ." Despite the fact that his own position in Provincetown is tenuous—dependent as he is on his employer for
housing—he focuses his complaints on old-timers' displacement.

Jason's concern is apparent in his writing. While his articles cover a range
of topics, many celebrate Provincetown's working-class Portuguese residents,
highlight artwork that criticizes gentrification, or capture the words of others
who share his belief that wealth threatens what is unique about Provincetown. In fact, a weekly column that he wrote for a Boston publication often
read as a soapbox for such concerns and even as a love letter to the town's
recent past. Despite the fact that his sexual identity as a gay man informed his
relocation, his great love is for the town's Portuguese.

Some suggest that gentrifiers have already permanently altered the local
sense of community. For instance, a Provincetown preservationist, Jo—a
prominent business owner and civic leader—posits the past, before most
gentrifiers arrived, as the site of true community. Ironically, during our interview, she decried community decline while standing behind a table on a
busy sidewalk, selling raffle tickets to benefit a nonprofit that drives women
to Boston for medical treatment and provides other services for the critically ill. Running her hand through her short hair, and frequently pausing to

hawk tickets and chat with acquaintances, she said: "The difference between then and now is that people went out of their way to help you. If you were new in town—like I couldn't find a place to live. [A woman], she got me my first place to live. She went and found it for me. I didn't have to do anything. People, in general, were more friendly, I think. More community oriented." It was immediately clear to me—as an outsider and an observer—that Jo herself is very involved in Provincetown's community and that she goes out of her way to help others. In fact, in interviews, and during observation, a variety of residents—from newcomers to old-timers—praised her volunteer efforts. Yet she does not regard her own behavior as indicative of community participation. For her, community belongs to old-timers, and she does not identify herself with it. Indeed, she believes that newcomers like herself have stripped Provincetown of the community that attracted her to the town in the first place.

As Jo's words indicate, preservationists largely measure the authenticity of their neighborhood or town by the apparent strength of old-timers' community, rather than by the strength of their *own* communities. This concern for community shapes their residential choices. For instance, a twenty-three-year-old Chicago social preservationist explained why she moved from downtown: "[It's] not *neighborhoody* enough there." Yet, when asked how they evaluate community, social preservationists repeatedly refer to evidence of old-timers' social networks. For a sense of community, they turn to old-timers.

For instance, a Provincetown preservationist spoke with sadness of the displacement of the Portuguese, adding: "When you lose the [working-class] sector of people, you lose . . . the soul of the community." Another supports affordable housing "to actually keep a community because it's people that are moving out that are very much a part of the community. They're service people, they're people who have lived here all their lives, they're living with family. They're leaving." As such words indicate, preservationists celebrate what they regard as old-timers' privileged relationship to community and worry that gentrification threatens this: that money "erodes community" and "roll[s] over the community in an irretrievable way."

For this reason, preservationists regard newcomers like themselves as a threat to old-timers' authentic community, and, therefore, some *avoid* establishing a community with others who share their traits. Such preservationists lobby to draw the bridge up behind them, to prevent others like themselves from moving to their neighborhood or town. For instance, an Argyle resident wrote in opposition to proposed improvements to the neighborhood's main commercial strip: "The biggest reason that I like living in this area is the ethnic diversity and the range of incomes and social classes." Another wrote: "Try to keep Vietnam town a secret. Keep tourists and suburbanites away" (both quoted in Argyle Streetscape Task Force 2001).

Social preservationists value places that *lack* elements they associate with wealth. For instance, one Chicago preservationist described Andersonville's changes: "It's jogging strollers and Starbucks now, and it makes me sick." Another, a thirty-one-year-old lesbian, said: "I would probably get rid of some of the new yuppie types that have moved in and bring it back to its old neighborhood." A white, college-educated, politically active Argyle newcomer concurred: "I don't want to live somewhere where everybody walking down the street looks like me. I know people who look like me. A *lot* of people."[2]

Beyond the absence of newcomers, the *presence* of old-timers' families, particularly of multigenerational families, is of particular import to preservationists. At a Chicago protest demanding low-income housing, a middle-class speaker said: "We know that gentrification is horrible. It destroys families and communities." Another, a thirty-something white homeowner, spoke of his choice to live near Argyle: "We wanted to live in a community where families can afford to live. . . . We don't want to lose them." A Provincetown preservationist, a childless lesbian business owner, said: "If we lose the school, which is—that's a big one, that's *the* big one—if we lose the school, to me that puts the end of the town." As we have seen, in two recent years, Provincetown recorded only one birth, a fact that many preservationists mourn.

One preservationist complained that there are "more dogs in town than kids." This was a frequent complaint in all sites but Dresden. Indeed, as gentrification progressed in Andersonville and Provincetown, merchants began placing bowls of water and dog food beside their establishments, and Andersonville sports a dog bakery. On a warm Saturday in the late spring of 2006, I jogged along Andersonville's Glenwood Avenue from Uptown's northern border into Edgewater. Along the way, I found myself repeatedly stepping into the grass to make way for dogs and their owners. Almost always, the owners were a pair of well-heeled gay men. Recalling my Provincetown informant's comment—and the frequency with which I had to step off Provincetown sidewalks into Commercial Street to yield to similar processions—I began to keep a tally. After a dozen or so trios in three miles, I lost count, but the extent of the neighborhood's transformation was abundantly clear, as it was on a morning in Provincetown when I heard a man respond to a question about his dog by saying: "She'll be five years and three months on the second." Of course, I was also cognizant that my own Andersonville runs and long midday walks on Provincetown's streets and beaches were themselves evidence of—and likely a contribution to—community change.

In sum, preservationists often conflate families with those vulnerable to displacement; their complaints about dogs and demands for low-income housing are synonymous with a desire for a community populated by children. For instance, a Chicago preservationist said: "People who own condos and are worried about their property are not going to be tolerant of people who don't make very much money, have five kids living in a two bedroom house." For this reason, preservationists seek to prevent displacement by supporting affordable housing and schools: "If we could create five hundred affordable and moderately priced housing units, that would be the thing that would make the *town* a *community*. Families, artists, working people . . . particularly [housing] for families so we can maintain a school. . . . It's the bedrock of any community is kids. It's part of the important energy in a community."

As the evidence offered above demonstrates, social preservationists are oriented toward a place's social characteristics. While some share with homesteaders an appreciation for the built or natural environment, many are willing to sacrifice such concerns to maintain the presence of certain original residents. They do so because, for them, place character rests on old-timers, particularly on the sense of community that they provide. Specifically, many preservationists articulate the belief that old-timers have a unique claim to authentic community. For them, this authenticity has much to do with their perception that old-timers possess enduring, place- and lineage-based networks rooted in shared cultural traditions. In turn, they worry that affluent newcomers—among whom they count themselves—threaten those networks. This concern fundamentally shapes and even fuels their resistance to gentrification.

Resistance to Gentrification

Social preservationists decry gentrification and place high value on evidence that their neighborhood or town is ungentrified or at least less gentrified than surrounding neighborhoods or towns or other places where they have lived. For instance, a middle-aged administrative assistant who lives in Argyle identified the characteristics of her neighborhood that she most appreciates: "The food, the restaurants, the shops. The people, you know the different ethnicities, and then that *it is not so gentrified*." Social preservationists resist gentrification because they worry it threatens aesthetic, social, and cultural authenticity and will lead to old-timers' physical, cultural, or political displacement. Their resistance also emerges from general discomfort with economic and demographic change and concern that gentrification threatens the distinction between their place of residence and less authentic places.

Given their celebration of ungentrified space, preservationists are more likely than homesteaders to embrace the "background noise" of their neighborhood (Zukin 1987, 133), and some seek to befriend old-timers. A Chicago resident said that crime is necessary to prevent Argyle from becoming "too nice," and another, a middle-aged woman who lives alone in Argyle, prioritizes the preservation of old-timers over increased safety: "I don't like pushing anywhere from middle-class to poor people out. Where are they going to go? Because they seem to be pushed out of everywhere, you know. And that bothers me. But then I know that they say, 'The neighborhood is going to be safer if you get rid of them and blah blah blah.' But where are they going to go? And their kids?" Most preservationists are simply uninterested in the crime and safety concerns of other gentrifiers or express conflict between the desire to live without fear of personal safety and concern that increased anticrime measures will facilitate displacement. This understanding of the relationship between crime-reduction efforts and displacement may be a product of preservationists' exposure to other gentrifying neighborhoods as well as to academic and popular criticism of gentrification. In fact, preservationists tend to be well versed in the language of gentrification, and most are equipped with a sophisticated sociological vocabulary, using terms such as *urban pioneer, first wave,* and *social networks.*

Over coffee in a popular Andersonville bakery, Robert, a white, gay artist who has lived in Andersonville for more than a decade, articulated a general discomfort with gentrification and neighborhood succession common among preservationists: "I was walking around feeling like the neighborhood had moved, like I was still here but everything had changed. And then all of a sudden . . . I look down, and the building had little ground sprinklers that popped up and started. That building used to be a dump. . . . And now it was condos with landscaping. . . . Oh my God! It's like *you can move simply by standing still.*" Perhaps because he is an artist, Robert readily provided images of other moments that made him acutely aware of the neighborhood's transformation: the felling of a tree, the bronzed bodies of gay men overshadowing the presence of Swedes at the neighborhood's summer festival. Having endured more than a decade of such changes, he prepared to leave the neighborhood for a less gentrified alternative: Chicago's Devon neighborhood, which is characterized by Indian and Pakistani residents and their establishments.

This discomfort with change may originate in the fact that Robert has been a part of the gentrification of multiple neighborhoods. For instance, he recalled: "In the mid- to late '80s, I lived in Brooklyn, and, when I moved into my neighborhood, it was very inexpensive, nice, lots of old people, lots of old families. And then, one by one, the buildings turned over. And the same thing happened here, . . . and now it's filled with baby strollers and cell phones." Laughing, he added: "It's me, I guess. It's me." Watching Andersonville transform, he began to realize that he could no longer passively tolerate serving as a "seed for gentrification" (as another preservationist put it).

Robert's distaste for gentrification is rooted, partly in a concern for oldtimers, but also in a general discomfort with neighborhood change, particularly with population succession associated with gentrification. He watched as the place characteristics that drew him to Andersonville more than a decade ago disappeared. Despite the private and symbolic preservation efforts he engages in, Robert worried that he could not successfully preserve them. He recalled the moment he knew the neighborhood was gentrifying: "When my partner and I moved here [in the 1990s], . . . you hardly saw any gay men at all. We used to sit on our front steps, and, after living here for a few years, I remember one day seeing a gay man, a white gay man walk past our house, and I said, 'There's another one of us. What do you know?' And it just seems odd. [Now] it's a deluge of people moving into the neighborhood, and I've been here long enough to see it happen."

Robert recognized that he was a part of this transformation and that oldtimers' suspicion of him when he first moved to Andersonville was wellfounded: "[Our landlord] had been [here] forever, and you know people weren't necessarily happy to see us move in. And I don't blame them 'cause, once it happened, the turnover was amazing. . . . I remember when Starbucks moved in everybody said, 'Well, there goes the neighborhood.'"

Despite his virulent distaste for gentrification, Robert acknowledges that he benefits from some aspects of the process. For instance, he has displayed his art in Starbucks, and "a lot of the money that started coming in to the neighborhood started collecting my work, which was very nice, and I appreciate that." Similarly, he is grateful for an increased sense of safety that came with gentrification. On his neighborhood rounds, he now sometimes walks south into gentrifying Uptown, rather than limiting his strolls to Andersonville's borders. He now almost always encounters others on sidewalks late at night, which provides him with a sense of security.

However, these amenities come at the cost of authenticity, and, in the end, Robert prefers neighborhood characteristics that he believes gentrification has destroyed. For instance, he expressed nostalgia for the neighborhood's dingy past: "I remember the place that used to be there. . . . It was a piano bar and a restaurant, and it was *dirty*. I remember a friend of mine suggesting it, and I said, 'Is the food good?' and she goes, 'No.'" He laughed and expressed regret that such places have been replaced by upscale alternatives. He lamented:

"We're not really the Swedish neighborhood anymore. There are touches of Swedish, but we're more of a sushi neighborhood than anything." (Indeed, several sushi restaurants line Andersonville's Clark Street.) Robert believes that "gay men have taken over the neighborhood." Lest I should believe that he appreciates the presence of others who share his sexual identity, he added: "Sure, there's more gay men in the neighborhood, but they're like real estate broker gay men, you know."

Robert worries: "The quirkiness of the neighborhood, the thing that made it unique, could disappear completely unless something's done." Specifically, he worries about the neighborhood's atrophying Swedish character: "I know in five years that there's going to be some of these little Swedish places that just can't pay the rent anymore." As a result, the neighborhood will lose its spirit: "If the suits win it, you don't have a neighborhood anymore. You have a bunch of investments all together. . . . That's what could kill this neighborhood. . . . It stops having any spirit, you know." He likens the in-movement of gay men like himself to colonization: "I saw a thing about Hawaii on television last night, and this is the same thing Hawaiians went through when white people started showing up. It's like, wow, I was watching the show going, I know what they mean!"

Robert applies his artwork as a sort of triage to preserve what is left of what he terms neighborhood *spirit* and what other preservationists call the *soul of place*. With pride, he reported that he volunteered for Andersonville's prominent Swedish American Museum. He also displays figurines he crafted of women in traditional Swedish garb. Photographs of the figurines appeared on the glossy cover of a flier for an art tour sponsored by the Chamber of Commerce. More privately, Robert refuses to patronize a gym in Andersonville's center that he believes encourages affluent gay men to move to the area. Instead, he exercises at a much less upscale gym in an adjacent neighborhood (often—he reports—alongside Catholic nuns) and frequents local Swedish businesses whenever possible.

Despite such efforts, when I interviewed him, Robert had come to believe that the neighborhood was beyond repair. A few months later, I walked past his moving sale, his final preparation before leaving Andersonville for a less gentrified alternative. Robert is moving because he believes that he cannot reconcile contemporary Andersonville with the ungentrified, Swedish neighborhood he came to love in the early 1990s. He insightfully regards his personal reaction to gentrification as part of a larger pattern: "I remember hearing this thing about Key West. The fishing part of it was done. But the people, the people who came there in the '30s, would want it to be like it was in the '30s and, if you got there in the '50s, would say, 'Oh, it's a shame it's not like it was in the '50s.' Andersonville, it's a shame it's not the way it was in 1994." His words reveal two important components of social preservation: appreciation for what one imagines a neighborhood or town's past to have been and nostalgia for the character of one's place of residence on first arrival. Together, these sentiments encourage preservationists' discomfort with gentrification-related change.

While Robert differentiates between what his neighborhood was and what it has become, other preservationists differentiate their town or neighborhood from other places, particularly gentrified places. The stronger the distinction between the two, the more authentic they deem their current place. "Other" places, preservationists suggest, are characterized by the *absence* of old-timers and by the *presence* of affluent residents, aesthetic homogeneity, and retail chains.

For instance, a Dresden preservationist suggested that gentrification ruined the sense of community that she had experienced in a coastal Maine town: "In my previous house, when I first moved there, there was a real sense of neighborhood, and, as people changed, that sort of changed. Houses turned over. Folks moved in who weren't really into that." Similarly, a Provincetown social preservationist—a retired woman—worries that gentrification will destroy the town's character, just as she believes it has blemished other places: "Look what they did in Vermont. They should call it New York, not Vermont. I mean, look what happened there. I loved Vermont. I could've lived in Vermont." For her, gentrification specifically threatens the presence of those with a tradition of residence, which gives space its soul: "There's no soul [in Florida]. That's what I would say. . . . It's very beautiful when you talk about the ocean and the flowers and the trees. . . . But I don't find any feeling there. *Everybody's from somewhere.*"

Kyle, a white gay Argyle resident in his thirties, explained how a desire to live in a truly urban—and, therefore, in his mind ungentrified—place shaped his residential choices. We met at an independent coffee shop on a Chicago corner. As Kyle, a large man with curly hair, settled into a seat, he described his working-class childhood in rural New England. After studying creative writing at a Chicago college, Kyle decided that he wanted to move to the city: "I had been in a small town all of my life, and, when I decided I needed a big city, New York seemed scary, and Boston seemed small." For him, Chicago was the perfect compromise, and he cannot imagine returning to small-town life, despite the fact that his family remains in the rural Northeast.

However, it took Kyle time to find a Chicago neighborhood that felt like home. Alternately working as an administrative assistant and a bookstore clerk, he moved from one neighborhood to another, generally finding that they lacked the urban authenticity he craved. Of a neighborhood on the city's West Side, he said: "It was sort of a nonneighborhood for me. I was always leaving it to go to work, to go to a coffee shop, to meet people. It was just kind of purely residential."

It was when he was in his late twenties that Kyle discovered Argyle, which he appreciated—for practical, aesthetic, and ideological reasons—for being less gentrified than other neighborhoods in which he had lived. He recalled: "[It was] less polished. Also $400 bought me . . . a first floor wood-floored studio with big windows." He especially appreciated Argyle's urbanity, which for him is best embodied by the presence of those he regards as old-timers: "I like the idea that [the alderwoman] maintains a certain something about Uptown that makes it so poor people can continue to live here. I've read that Uptown has the highest percentage of homeless people in the city, and they've got to go somewhere. So I wouldn't want to see it all go condo."

Despite this appreciation for old-timers' sustained presence and distaste for gentrified neighborhoods, Kyle recognizes his own opposing impulse, which is to take advantage of new neighborhood amenities. He admitted: "When things go condo, I benefit because, like, Andersonville got a Starbucks." However, he did not suggest that we meet at Starbucks. Rather, he asked that I meet him at a nonchain coffee shop on a corner less gentrified than the upscale portion of Clark Street on which Starbucks sits. Furthermore, the presence of independent businesses like the one in which we met represents much of what he appreciates about the neighborhood. He particularly celebrates the fact that his favorite coffee shop is not "full of yuppies." Thus, it is a combination of the absence of certain characters—namely, yuppies—and the presence of other characters—old-timers—that lends Argyle the authenticity that Kyle sought in a neighborhood and indicates that the neighborhood is not fully gentrified or at least not as gentrified as other neighborhoods in which he has lived.

Social preservationists not only seek ungentrified space but also, as many of the individuals described above illustrate, work to prevent their neighborhood from gentrifying further. They evoke certain place names, such as Vermont, Nantucket, Vail, and Chicago's Lincoln Park neighborhood, as warnings of what their town or neighborhood might become should gentrification go unchecked. Specifically, they juxtapose their place of residence against the socially isolated suburb or affluent urban neighborhood.

In fact, social preservationists define themselves as much by where they do *not* live as by where they *do* live.[3] For instance, an Argyle preservationist said: "I would never live in Lincoln Park. . . . It's really crowded down there, just not diverse enough for my tastes." Meanwhile, an affluent, forty-five-year-

old woman, sitting in her smoke-filled office in Provincetown's center, suggested that the town is in danger of becoming like Nantucket, where wealthy
newcomers "liked watching the fishermen, they liked watching the hippies,
they liked watching the agrarian lifestyle—the remote island at the edge of
nowhere. Then they all bought into it, and what happens? They own it, and
the rest have to go." Similarly, a Dresden couple used the coastal town of
Boothbay to illustrate how gentrification erodes community: "Everything has
turned into hotels, and all the houses have turned into bed-and-breakfasts
and little shops and all that sort of thing. So there isn't any fishing side anymore. The fishing's gotten squeezed and squeezed and squeezed, and part of
what that's done is change the whole flavor of the community." In a more suc-
cinct manner, an Andersonville preservationist—a middle-aged researcher
who wears long hair and a beard—sourly drew a parallel between Andersonville's gentrification and that of San Francisco's Haight-Ashbury.

Such warnings and comparisons sometimes enter public discourse. For
instance, at a crowded November 2001 Andersonville meeting organized in
reaction to the proposed construction of a Borders in neighboring Uptown
(the chain did eventually move in to the space), the crowd—predominately
young and middle-aged white residents—voiced concern that Andersonville
and Uptown would become like other, less authentic places. In voicing his
opposition, one man noted: "I moved in because it was [a] culturally diverse
[neighborhood] . . . and away from the hustle and bustle of Lincoln Park [an
affluent Chicago neighborhood]."

Similarly, at a 2004 Provincetown selectmen's meeting, a board member
and retired attorney who often expressed preservationist concerns asked
whether anyone had seen a *Los Angeles Times* article on how Big Sur's "community is losing its vitality." Having seen the article earlier in the day, I listened carefully as he suggested that Big Sur serves as a warning. The article
reads: "For years, the phrase 'Save Big Sur' meant preservation of the timeless
forests and streams perched high above the Central California coast. Now
it means the people. 'A lot of people bemoan the loss of community,' said
Kirk Gafill, co-owner of the famed Nepenthe restaurant on Highway 1. 'That's
code for fear of the future of Big Sur.' Beset by sky-high real estate prices, rich
absentee landowners, restrictions on development and a shrinking, aging
population, many residents fear their community is losing its vitality" (Johnson 2004, 1). The selectman noted that Provincetown is also "beset by sky-
high real estate and an aging population" and that, as in Big Sur, the "school
is narrowly averting closure." Tellingly, his tone one of subtle reprimand, he
interjected his words into a conversation between other selectmen about how
best to market "P-town" as a tourist destination.

Thus, social preservationists universally express distaste for gentrification. While their collective concern with gentrification has several different sources, all report that they appreciate ungentrified space because it allows them to live alongside the old-timers they admire. Many appreciate the distinction that old-timers and their institutions and traditions lend a place, and they frequently rely on tropes about the inauthenticity of suburban and affluent neighborhoods. Social preservationists also commonly compare authentic community to the current state of their place of residence and rally to turn the clock back to a more authentic period in the local (recent) past and to prevent further change. For them, gentrification threatens the cultural, aesthetic, and social characteristics they value in their place of residence.

Assigning Accountability

While many homesteaders regard gentrification as an inevitable economic process or at least as a process larger than they themselves are—controlled by political and economic elites—social preservationists tend to hold affluent newcomers like themselves responsible for gentrification. For them, gentrification is not uncontrollable but, rather, fueled by certain actors, namely, gentrifiers, among whom social preservationists count themselves.[4]

Some preservationists emphasize gentrifiers' indifference to old-timers. For instance, at a meeting of the West Andersonville Neighbors Together block club, a resident complained: "I can think of a few [realtors] who have been predatory about development!" A Provincetown affordable housing advocate said of newcomers: "You see this intense consumerism. . . . People are just, 'I want that. I want that.' . . . And it just feels soulless." Similarly, at a table facing Provincetown's harbor in the ramshackle headquarters of the Beachcomber Club, a society for male artists, a middle-aged bearded poet said: "People stepped in that had a sense of making money, . . . and they chose to utilize the opportunity regardless of the motivations, regardless of the process or the priorities they chose to represent." At a Chicago protest against the construction of high-end condominiums, a white, middle-class man received cheers when he said: "Our gentrifying friends with their diversity . . . like Lincoln Park Zoo, where you can see a polar bear or a penguin. *They* want a neighborhood with three African American families, a few gays, a few Spanish-speaking people. *Ours* is a community that is as it is today with many people of different backgrounds."

Social preservationists generally worry that other gentrifiers' assessment of what is valuable about their place of residence is different from their own and even misguided. Gripping a cup of tea, Leslie fought tears as she described other newcomers' indifference:

> We came, and most of the gay people I knew at that time came [to Provincetown] not because it was a safe place to be gay. They came because they loved the place. They had a feel for the history of the town, for what makes it Provincetown, for the true diversity, for all the things we all yearn for here. So many of the people who are moving here now, they're moving here because it's a safe place, which means it could be *any* place. Many of the people moving here have no sense of the history, of what the town is. They have no idea that this was like the oldest continuous art colony in the country. They have no sense of the death of the fishing culture, and how the Portuguese came, and how the Yankees were here before, and how the Indians—the Native Americans—were here before. They don't have that sense. They don't care. They're here to make money. They're here to, you know, make money on their computers. . . . It's P-town, which is like fucktown. Let's get wild. It's like a theme park.

A middle-class, lesbian newcomer to Provincetown echoed this theme: "I sort of have a hatred for . . . the capitalist urge that happens here or . . . the rich people that move in . . . people who have a million dollars who [think] this is a great gay place to party and I'm just going to build a huge condo here so I can come here over the summer and party."

While such preservationists argue that other gentrifiers are more ruthless than they themselves are, preservationists generally are highly self-conscious about their own impact on old-timers, and none sought to identify themselves as anything other than gentrifiers. For instance, an Andersonville preservationist admitted: "We were part of the change process, unwittingly." Similarly, an Argyle preservationist acknowledged: "I'm a seed for [gentrification]." And an Ivy League–educated mother cried while acknowledging that her efforts to improve the local elementary school contributed to rising taxes and, therefore, to displacement. She criticized other gentrifiers for failing to recognize how their "basic needs" threaten to displace old-timers:

> You went down and bought land and bought the house. You had to drive down that dirt road. *You made a choice to live at the bottom of that road,* and now you're wanting to have it maintained to a level far beyond what it would have been if you hadn't moved. . . . That doesn't make sense. There's a way you could make a similar comparison to the school. . . . You have this quiet little country school, and you move to town with twenty or thirty other kids. . . . You're creating the need, but everybody's paying for it. So it feels like a mixed bag. It's still the new people moving in and creating a need . . . having the whole community pay for that new need.

In this way, preservationists engage in preservation work *because* they recognize the threat their own presence poses to old-timers.

When social preservationists' conflicting desires become self-apparent—to live in the central city while simultaneously preserving it or to maintain the presence of old-timers while simultaneously improving town infrastructure and, thereby, increasing taxes—they turn to practices to prevent or stall gentrification. Specifically, as the next chapter documents, their self-consciousness encourages efforts to limit or stall gentrification, primarily by preserving old-timers' presence. For social preservationists, old-timers distinguish their place of residence from inauthentic space, such as the suburb and the affluent urban neighborhood. They also differentiate the present state of their neighborhood or town from an imagined (or increasingly real) gentrified version of that space. For social preservationists, authentic people constitute authentic place and, therefore, valuable space. To uphold these claims of authenticity, social preservationists work to prevent the neighborhood from *becoming* inauthentic by resisting gentrification through symbolic, political, and private practices.

While this chapter focused on the social preservation ideology, the practices of social preservation are also apparent. We have encountered preservationists like Leslie, who advocates for affordable housing professionally and privately. We have met those like Sarah who use their personal pocketbook to ensure old-timers' sustained presence in town and many others who advocate against gentrification on radio programs, through artwork, and on the floor at town meetings. Others run organizations like the Provincetown Community Compact and an affordable housing company, while some work for a city alderwoman or serve on the board of selectmen. Preservationists almost universally report that their ideology shapes their everyday, private practices: from shopping choices to interactions with neighbors. Thus, as the next chapter will further explore, the social preservation ideology pairs with a set of careful practices.

Conclusion

This chapter introduced social preservation, previously unnamed and little explored. While, like all gentrifiers, social preservationists contribute to the transformation of their neighborhood or town, they differ from social homesteaders and pioneers in several ways. Social preservationists are characterized by an appreciation for the social character of their place of residence, specifically, for the sustained physical, cultural, and political presence of those they deem old-timers. In this sense, they invest economic, cultural, and social resources, not only in local property, but also in the nonmaterial value of places they deem authentic, places they believe old-timers' communities render more authentic than other neighborhoods or towns. As we have seen, many social preservationists relocate in search of the authentic community that they associate with old-timers, and, for this reason, they work to maintain the authenticity of their place of residence. Social preservationists are generally committed to limiting or preventing gentrification and, therefore, resist the in-movement of others who share their demographic and cultural traits, and they offer sophisticated criticisms of their own role in gentrification. Generally speaking, social preservationists celebrate old-timers and decry the risks that gentrification and its actors pose to old-timers and other place characteristics that they regard as authentic.

Social preservationists' sensibility mirrors that of those mid- and late-nineteenth-century scholars who worried that, with increasing urbanism, "the primary relationships of place and kin give way to rational, individualistic encounters typified by market transactions" (Hunter 1975, 538). According to preservationists, authentic community possesses extended families, shared values, culture, and identity, common labor practices, easily identifiable traditions, and regular interaction and belongs to those with a history of residence in the same locale. In this sense, the distaste for modern community forms implicit in their romanticization of old-timers' communities is not a new phenomenon, nor is it particular to them. In 1887, Tönnies suggested that, with increasing industrialization and urbanization, gesellschaft, "an interactional system characterized by self-interest, competition, and negotiated accommodation" (Christenson 1984, 160), would define community. Scholars such as Georg Simmel, Emile Durkheim, Max Weber, and many Chicago school sociologists echoed these concerns. In more recent years, but long before gentrification reached the height of its infamy, Albert Hunter noted among newcomers to a Rochester neighborhood "a very conscious rejection of suburbia, or rather a conscious rejection of the somewhat stereotyped 'image' of suburbia by residents in the area, and a correspondingly positive assertion of the values of 'urban living'" (1975, 546). We might think of social preservation as a continuation of this nostalgia for a "gold age of community" (Griswold 2000, 133), only, today, industrialization and suburbanization are no longer widely viewed as critical threats to real community. Today's specter is the widespread and popularly recognized return of the middle class to the central city and small town.

Having identified social preservation, we must now rethink our understanding of how gentrification happens, of how it is practiced and experienced by residents, particularly by those cognizant of their participation in the process. Social preservation draws our attention to gentrification as a *process*, expanding our conception beyond a set of predictable outcomes. It also pushes us to understand community change as a process fueled by a variety of actors who are characterized by *perspectival* positions as well as by a material hierarchy. This is not just a matter of documenting ideological variation; by neglecting social preservation, we have failed to acknowledge a set of norms and practices that, as the following chapters will reveal, influence local politics, local economies, local policies, and daily interactions in the central city and small town.

The Varying Strategies of Social Preservation

Despite some scholars' assumption that gentrifiers' concerns about old-timers are "no doubt practiced only 'in the mind'" (Rose 2004, 300), the beliefs and sentiments that social preservationists shared with me in their living rooms and kitchens are coupled with a set of practices through which they resist gentrification and seek to sustain markers of authenticity, most centrally old-timers' presence. They work to maintain the local social ecology to aid old-timers, preserve authenticity, and counterbalance their own presence. Most of their strategies are distinct from those the literature associates with gentrifiers, although homesteaders, pioneers, and preservationists sometimes rely on common venues, such as block clubs, to reach discrete aims.

Social preservationists' practices fall into three general categories: (1) symbolic, as in the use of festivals, streetscapes, and artworks that celebrate old-timers or theater productions that criticize gentrification; (2) political, from protests against upscale development to membership on an affordable-housing task force; and (3) private, such as the decision to support old-timers' businesses and to resist selling property for profit. Of course, the lines between the categories sometimes blur. Symbolic practices are often overtly political. Private efforts often arise from political concerns, and political choices can be very personal. However, the categories isolate the medium through which preservationists work. In other words, they distinguish between practices that rely on symbols, such as Swedish banners or a play criticizing gentrification, those that are exercised through overtly political organizations, and those that occur in the private sphere of everyday life. In this chapter, I first outline the dimensions of each type of practice. Then I describe the practices typical of each site and how context shapes preservationists' strategies.

Symbolic Practices

Through symbolic practices such as festivals, streetscapes, visual art, and theater, social preservationists seek to link place identity to old-timers.[1] In so doing, they underline local authenticity and sustain or encourage the popular association of place character with old-timers. Alternately, some use symbolic methods, such as plays, to speak against gentrification.

To buoy local authenticity, preservationists often create or support artworks and celebrations that embody old-timers. For instance, in honor of the 2005 Portuguese Festival and Blessing of the Fleet, the Provincetown Art Association displayed work inspired by the Portuguese. Argyle preservationists worked alongside other gentrifiers, politicians, and a few old-timers to plan an Asian-themed streetscape. In all sites, preservationists orchestrate or support celebrations that emphasize old-timers' ethnic, racial, or occupational identity.

When old-timers *are* displaced, preservationists rely on symbolic representation as a supplement to (though not a replacement for) the presence of remaining old-timers. For instance, as the number of fishermen in Provincetown dwindles, emblems of their labor gain significance. A worn wooden oar decorates the wall of a tourist cottage, while Dresden newcomers use rusted farm equipment as garden statuary. The space where old-timers gathered becomes significant: a pair of Dresden one-room schoolhouses, Provincetown's pier, a Swedish hardware store turned museum, and a conglomeration of Asian groceries and restaurants. Thus, old-timers' institutions offer the community a source of meaning, particularly when revitalization threatens old-timers themselves.

However, some—such as a Provincetown artist who argues that symbolic preservation amounts to "monuments to a dying culture"—resist commemoration, fearing that, like historic or landscape preservation, symbolic preservation may encourage gentrification.[2] As a result, some rely on symbolic means, not to commemorate old-timers, but to speak against gentrification and displacement. Such practices are distinct from those that embody or celebrate old-timers, but they share a reliance on representations, such as paintings, plays, and street banners.

Political Practices

Preservationists also employ political practices to prevent or stall gentrification. They represent their concerns on boards and committees and in civic organizations. Some also plan and attend protests. Through such forums, many engage in political discourse about gentrification. However, political practices also include conscious avoidance of forms of participation that might contribute to old-timers' political or cultural displacement.

Through a variety of political forums, preservationists work to prevent or stall gentrification. They represent what they regard as old-timers' interests (e.g., the preservation of affordable housing and blue-collar jobs). They serve on committees and boards and participate in block clubs alongside pioneers and homesteaders to block efforts that might encourage gentrification or otherwise abet displacement, such as urban safety initiatives and rural zoning restrictions. They frequently enter their voices in public discussions of gentrification and community change. Indeed, they often lobby against upscale development and for the preservation or construction of affordable housing. In Chicago, preservationists sometimes voiced such concerns through political rallies and protests.

However, social preservationists frequently maintain multiple political allegiances. As a result, some are caught between their political orientation and their desire to prevent old-timers' political displacement. For instance, a Provincetown preservationist is critical of newcomers like himself who have political influence: "A lot of the yuppies that moved into town, a lot of them came and the first thing they do is they got on boards. I don't mind politics if it was the same for everybody. But when it's not the same for everybody . . . I don't like it." Thus, preservationists sometimes face the quandary of honoring their political beliefs (often leftist) or voting against them to sustain old-timers' political power (many are comparatively conservative).

In this way, preservationists' political practices vacillate between, on the one hand, active participation and protest and, on the other, the careful avoidance of political activities that they hope will remain under old-timers' purview or that they believe contribute to gentrification.

Private Practices

Social preservationists express their appreciation for old-timers and desire to prevent gentrification through daily practices outside the political and symbolic realms in which their other practices take shape. Their ideology contributes to their consumption habits, interactional patterns, financial choices, and sometimes very personal emotional and professional choices.

Social preservationists often commit themselves to supporting old-timers' commercial establishments: patronizing a Swedish-owned bar rather than its gay or yuppie counterpart or shopping at farmers' produce stands rather than a grocery store. Such choices arise from appreciation for old-timers' culture and from a desire to sustain businesses threatened by gentrification and, therefore, to preserve old-timers and the establishments that mark their presence.

The social preservation ideology also influences some individuals' non-consumptive financial choices. Some make financial sacrifices to prevent displacement. For instance, some homeowners rent their property at a loss to provide affordable housing. Many choose not to "sell and run" as property values rise. Others donate money to old-timers' institutions, such as ethnic museums, or to nonprofits that provide housing for artists.

Preservationists' affection for old-timers and desire to prevent their displacement shapes their everyday interactions. We might find a Dresden preservationist who has pulled his car off the road to chat with a farmer or an Andersonville preservationist who has paused in a Swedish deli to speak with the owner. Preservationists often aid old-timers whose family members were displaced by gentrification. However, some resist such strategies. They consciously hover on the periphery of old-timers' lives, seeking to prevent the disruption of their practices and communities. For instance, one described sitting silently and self-consciously in a Swedish restaurant as old-timers reminisced about Andersonville's bygone Swedish heyday.

Social preservationists' ideology shapes their deeply personal choices. For example, an Argyle preservationist—a University of Chicago–educated social worker—ended a relationship with a boyfriend after an aldermanic election revealed their discordant beliefs about gentrification. His boyfriend supported crime reduction and property-value increases, while he himself sought to prevent old-timers' displacement. Social preservationists also report that their ideology shapes their professional choices. For example, one man established an affordable-housing company, and others choose to work for him. Similarly, several artists suggest that their aversion to gentrification and appreciation for old-timers influences the content of their work.

Social preservationists' private practices are not easily discerned from observation alone. However, during interviews, they reported the rationale behind their daily choices—their shopping habits, the neighbors they visit—and, in turn, these became apparent through observation. Of course, such practices are not without recompense. They fuel their hope for the sustenance of authenticity and ensure the pleasure they take from consuming old-timers' culture.

Place-Specific Practices

While social preservationists in every site engage in political, symbolic, and private practices, the forms that their practices take vary across the sites.[3] This pluralization of strategies is surprising given the constancy of the preservation ideology. Despite the variation of their social preservation practices, the diversity of their objects of preservation, and the fact that they have neither a name for their ideology nor organizations explicitly dedicated to their cause, preservationists maintain a remarkably consistent ideology across the sites. This is likely because they fashion beliefs and practices from popular ideologies accessible to their status group, including widespread criticism of gentrification. Thus, ideological variation cannot explain the diversification of preservation practices. Instead, three community-level factors shape the strategies that social preservationists employ: (1) the stage of gentrification within and in the area around the site, (2) gentrifiers' and longtime residents' demographic and cultural characteristics,[4] and (3) local activist groups, politicians, and business associations that advocate for or against gentrification.

In what follows, I detail the strategies that preservationists employ to prevent the displacement of old-timers and thwart gentrification as well as the factors that determine *which* strategies they employ (see table 7). My discussion secondarily reveals how the same factors affect the discourse and strategies of *all* gentrifiers, from pioneers to preservationists, and, in turn, how one group's discourse shapes others'.[5] Thus, in the following pages, we hear the voices and observe the practices of preservationists, homesteaders, pioneers, and longtime residents alike and learn how their discourse and practices are mutually constitutive. Finally, the conclusion considers the relationship between ideology and practice.

PROVINCETOWN

Provincetown social preservationists actively engage in public debate about gentrification and advocate for affordable housing. They also use symbolic means to speak against gentrification and the displacement of Portuguese residents and struggling artists and, secondarily, to emblematically wed place identity to old-timers. Finally, they employ private practices, such as cultivating relationships with old-timers and renting property affordably.

Many Provincetown social preservationists focus their political efforts on affordable housing. Attempts to create affordable housing began as early as the 1980s, when preservationists and artists, through an organization they named PALISS (Provincetown Artist Live-in Studio Space), advocated for

TABLE 7. Social preservation strategies by research site

	Political	Symbolic	Private
Dresden	Political abstinence	Supportive or appreciative relationship to markers of farmers' presence	Support farm stands and build relationships with old-timers
Provincetown	Affordable-housing advocacy, public policy work, participation in public debate about gentrification	Plays, art displays that mark old-timers' presence, and protests against gentrification	Good-neighbor policy with local old-timers; professional choices; consumption practices
Argyle	Affordable-housing advocacy; participation in public gentrification debate; paternalistic advocacy for old-timers; extralocal protests against gentrification	Proposed Asian-themed streetscape; Argyle Street mural; support of Vietnamese and Chinese cultural events	Support of local Asian-owned businesses (particularly Vietnamese); professional choices
Andersonville	Advocacy for neighborhood independent businesses; extralocal protests against gentrification and for affordable housing	(Subtly) Swedish-themed streetscape and artwork; support of Swedish cultural events	Support of local Swedish-owned businesses and institutions; good-neighbor policy with elderly Swedes whose relatives have left for suburbs; consumption practices; relationship choices

town-supported artists' housing.[6] Affordable housing is now a common theme at the annual town meeting as well as at monthly and weekly town board meetings. In 2004, preservationists, with other community residents, successfully lobbied for passage of the Community Preservation Act. The act places a 3% tax on real estate sales the proceeds of which go into a fund for affordable housing, conservation of open space, and historic preservation and matched by the state; in Provincetown, 80% of the funds are earmarked for affordable housing.[7] During the same period, residents frequently discussed other strategies for creating affordable housing. For instance, at a March 2004 planning board meeting, members discussed a proposal to provide amnesty to those who rent illegal accessory apartments in exchange for a pledge that they will rent them to residents affordably. Furthermore, in 1996, a preservationist established a company that aims to "preserve community through affordable housing." With a staff that includes several preservationists, in its nine years the company has, according to its Web site, built "57 year-round homes, and 23 artist studios" and has "an additional 85 year-round homes and 31 artist studios . . . currently under construction, in development, or in the planning stages."[8]

Even those Provincetown preservationists who do not work for the affordable-housing company or serve on town boards use political dialogue about the displacement of old-timers to win support for affordable housing. Utilizing a discourse about "community preservation," "living artists," and a "place where people do things with their hands," they appeal to town leaders and residents for support. At public forums, they often engage in debate on the topic with other gentrifiers. For instance, at a heated selectmen's (the town's governing board) meeting on affordable housing, a middle-aged gay preservationist angrily accused the town of "asking the fox to watch the chickens" by involving realtors in affordable-housing planning. Another repeatedly said that the limited number of proposed affordable units was an "outrage," and an elderly woman exclaimed: "It seems you have to be a millionaire to buy a house in town!"[9]

At the 2004 town meeting, a preservationist expressed support for the Community Preservation Act, saying: "It is about the preservation of *this* community. . . . [This vote] tests our real commitment to affordable housing and the fabric of our community." Also at the meeting a member of the housing partnership spoke from the town hall floor, saying: "There's no deadline [for the act], but there is a deadline for family members and coworkers who have to leave town now [because of affordability]." Another resident urged others to vote on behalf of those whom gentrification has displaced, asking: "How many aren't here because they can't afford to be?"

In contrast to such public declarations of social preservationist sentiment is the careful avoidance of the political participation that a few preservationists engage in to preserve old-timers' place in town politics. One commented: "Everyone understands that at one time there was a larger resident community here of native Portuguese, and they held sway in a way they don't now." A lesbian preservationist voiced concern about gay newcomers' active participation in town politics, acknowledging that their "credentials" are often more impressive than old-timers': "They've taken over most of the boards. . . . A lot of these people that come in, they're either retired or they're doing business on the computer, out of the house. . . . Lots of these have family money. . . . They had more education." Another preservationist—a professional affordable-housing advocate—implied that newcomers are morally obligated to ensure that there is space for old-timers in town politics:

> If you go to town board meetings, zoning, planning . . . the really important boards, a lot of what you have is business interests and primarily gay business interests. You go to the selectmen, and there's one straight person on it, . . . and the people who are working all these jobs trying to keep their homes or keep an apartment, they don't have time [for the boards]. [People on the boards] talk to so-and-so, "You're complaining about housing. Why don't you go get on a board?" Well, so-and-so puts all their energy into just figuring out how they're going to make ends meet. . . . How would I feel, what would I do if I were in their position? Well, you know what? I wouldn't have energy to worry about what the hell is going on in town hall.

Those with this perspective compose a small minority of Provincetown social preservationists. Most do not hesitate to engage in local politics to shelter old-timers from gentrification's effects.

In addition to political practices, Provincetown preservationists readily employ symbolic strategies. Some emphasize the presence of the town's Portuguese population: for example, an organization with preservationist aims placed five huge portraits of elderly Portuguese women on a town pier "as a tribute to the Portuguese community and its fishing heritage." Their weathered faces loom against a pier building, their eyes seemingly watching the harbor, which leisure boats now primarily populate. The Web site iamprovincetown.com detailed the impetus for the installation: "They came from a long line of hard-working people. . . . Their families fished the waters off Cape Cod for over 200 years, built a major fish packing and distribution industry. . . . Economic uncertainty combined with the pressures of an inflated real estate market has pushed much of the native population to other towns and job markets. This shift in the demographics is a threat to the rich Portuguese heritage and its contribution to the history and culture of Provincetown. . . . [The] installation was designed to help keep the spirit and the presence of this culture alive."[10] Social preservationists also support other symbolic representations of the Portuguese population, such as the annual Portuguese Festival and Blessing of the Fleet.

More frequently, Provincetown social preservationists engage in symbolic practices that directly criticize gentrification. One fondly recalled a play that emphasized wealthy lesbians' exploitation of Provincetown's Portuguese. Another play depicted "seagulls attacking Provincetown [because they were] angry over the way humans have ruined the town" (Desroches 2004). Newcomers' disregard for the town's history and culture enraged the gulls: "The last straw for the attacking seagulls is when developers build condos on stilts over the town cemetery. The fear that prompted this play is that people won't realize what is happening to Provincetown until it is too late" (Desroches 2004). An article in the *Advocate*, a gay periodical, suggests that the playwright, Ryan Landry—a popular town figure—intended the play to speak against gentrification: "When he first visited P-town almost 20 years ago, hippies still strummed guitars on street corners and drag queens chatted with Portuguese fishermen in front of rustic art galleries. Now condominiums outnumber sand dunes and Commercial Street is clogged with circuit boys and yuppies in SUVs. 'It's like watching a parent die,' says Landry. 'You're watching the cancer grow'" (Desroches 2004).

In addition to plays, some use art installations and performances pieces to criticize gentrification. For instance: "[A] Venezuelan-born artist . . . created a statement on the gentrification of Provincetown. With black spray paint and clippings from the Provincetown Banner's real estate section, [he] presents 'Provincetown for Sale.' . . . [The artist] points out, 'Provincetown's image as a quaint fishing village and an artists' community is being used to sell real estate.' The very thing that drives the inflated real estate market could well drive out the fishermen and artists" (Desroches 2003b). Likewise, Tom, a slender, middle-aged gay man who has lived in Provincetown for two decades, used art to comment on Provincetown's transformation from a Portuguese fishing village to a gay resort by conceiving a fictional "gayted theme park":

> Only gay, lesbian and bisexual survivalists are allowed at [the] camp, which pokes fun at what [the artist] calls the attitude of the growing gay majority [in town]. "I think some people would like to see [it] just strictly as a gay town and I think that's what I play with in this survivalist campground," he says. . . . The map shows the camp resort, a "gayted theme park," called "P-town, Inc.—Formerly Provincetown," and all of its assorted attractions . . . includ[ing] the Center for Fecal Studies at the Heritage Museum, the P-Town, Inc. Gayted Corporate Compound at Telegraph Hill, and Frequent Flyer's Boat Yard for after-hours male bonding. (Butler 1999)

The same artist draped over an abandoned house that he anticipated would soon be demolished or renovated a banner that read: "The Outermost Alms Museum." He hoped that it would encourage other residents to think about "the character, the nature of the alms, the generosity of the community being eroded by all the money and all of it coming to town. It was a lament really." The property owner charged him with trespassing and defacement (charges were later dismissed) and demolished the building (during the winter of 2004, I rented a studio overlooking the building's cellar hole). The artist's interest in the building was driven, in part, by the fact that "'it was, literally, one of two dilapidated houses in town'": "I knew it was being torn down and felt the passage of the building needed to be marked not just for the building, but for what it symbolized"' (Harrison 2002a, 35). Previously, in 1986, at the town's Year-Rounders' Festival, he performed the song "Sold Cape Cod": "If you're fond of shopping without a care, quaint condominiums everywhere, you're sure to fall in love with *sold* Cape Cod."

The town paper also prints cartoons that comment on displacement and gentrification. In one, two characters mused about Provincetown's past: "Remember when Ptown was full of all kinds of people from all walks of life n' we'd have deep discussions into the wee hours. . . . 'Course this was back when people like us could afford to live here" (Andert 2004b, 25). In the same month, a character complains: "Nobody lives here anymore. Just weekenders and summer people." He and his friend list "more interesting" places to which they might move, such as New York, Amsterdam, and Vegas. Finally, one suggests that they move to Worcester (a city in central Massachusetts rarely celebrated for its cultural attractions), to which the other responds: "Yeah . . . That *would* be more interesting" (Andert 2004a, 24). In this way, Provincetowners use symbolic strategies to criticize gentrification and old-timers' displacement and only secondarily to associate town identity with old-timers and their cultural traditions.

Outside the theater or the town hall, social preservationists engage in private practices to prevent old-timers' displacement. These practices range from selecting a job that professionalizes preservationist concerns, as those who work for the affordable-housing company have done and as the artists discussed above do, to renting property at a loss to provide year-round affordable housing. For instance, a preservationist celebrated her sister's decision to rent a house year-round, rather than using it as a vacation home, and another his friends' similar decision: "They live in the two-bedroom [cottage], they rent it out in the summer at market rate, and then the one bedroom they rent affordably. They don't have to, but they do." Similarly, a landlord explained: "I'm really one of the lucky ones. We have four units that we rent year-round [rather than for a high profit to summer tourists]. We are trying to keep it that way because it's vanishing."

More subtle practices include building friendships with elderly Portuguese whose children live "away." For instance, on a snowy morning, I arrived for an interview at the home of an elderly Provincetown old-timer and offered to shovel her walk. No need, she told me. A gay neighbor takes care of that. Later, when she mentioned that she had been sick, she assured me that a neighbor (another newcomer) would bring her soup. Others employ struggling artists and frequent Portuguese-owned businesses. Other practices include the decision to remain in town, rather than to sell for profit, which would further contribute to rising home values and taxes. For instance, at the 2004 town meeting, a thirty-something mother explained why she supports affordable housing: "I'm sitting on more property value—I purchased before 2000—than I ever imagined. Thank you for the opportunity to give back." This reveals that social preservation predicates itself on the sense that one has resources to give back.

Why do Provincetown preservationists' practices take the shape they do? The significant gap between housing costs and residents' income is largely responsible for their attention to affordable housing. The past decade has seen housing values skyrocket; 2006 asking prices ranged from $139,900 for a 246-square-foot condo to $4,999,000 for a three-bedroom waterfront home.[11] While some newcomers purchase such houses, displacement threatens many longtimers. In 2000, 15.4% of Provincetown residents lived in poverty (compared to 9.3% of state residents), and 55.1% qualified as low or moderate income.[12] Despite the disparity between housing costs and income, at last count only 6.55% of town dwelling units were certified as low- and moderate-income housing by the Massachusetts Department of Housing and Community Development, and the number of year-round housing units decreased by 324 between 1990 and 2000.[13] According to nearly all informants, this has forced many artists and Portuguese families to leave town. These factors—products of the town's advanced gentrification and tourist economy—focus preservationists' attention on affordable housing.

A different but related dynamic explains Provincetown preservationists' engagement in public debate about displacement and gentrification. Their participation arises from a broader dispute between newcomers and longtime residents about change, particularly about the town's popular identity as an upscale gay and lesbian enclave, a debate many preservationists feel they must intervene in. For their part, pioneers often celebrate transformation. Indeed, such sentiments regularly color discourse at town meetings, where gentrifiers argue that condos have saved the town, affluent retirees, such as a retired professor who enjoys extended foreign travel, ask who is thinking about relief for property owners, and gay business owners worry that the town is not as widely recognized as a gay resort destination as it ought to be.

A lesbian social preservationist reported that gay pioneers have told her: "'We just want to get rid of all, I can't stand these townies,' or this type of thing." She said that she argues with them, for she believes that gays "invaded" the town: "They're wrong. I don't know why they think as they do. This works two ways. We talked about discrimination against gays. The gays do a lot of discriminating." Specifically, she believes that this "reverse discrimination" has significantly altered the composition of town governing boards.

The fact that many gay gentrifiers value Provincetown as one of a few places where gays have achieved cultural, economic, and political power buttresses their ideology and discourse. Many Provincetown gentrifiers experience a sense of entitlement, predicated on the dual beliefs that they will improve the town (a belief common to gentrifiers everywhere) and that they are caretakers of a gay "promised land" (a belief specific to Provincetown), and preservationists often react against this.[14]

Pioneers' celebration of Provincetown's transformation may also be read as a reaction to Portuguese longtime residents' vocal opposition to the town's demographic shifts. On an overcast winter day in the harbormaster's office, fishermen—dressed in work shirts, jeans, and boots—offered an assessment of how the town has changed and who has changed it: "The way I look at it, if [gays are] born that way, they're born with a defect. . . . They think defected and everything, you know? . . . They took this town over." They recounted an instance when old and new Provincetown clashed: "We came in with a load of fish and couldn't get them onto shore because the pier was full of gay men on ecstasy bumping and grinding!" Old-timers frequently speak—with disgust—of witnessing sexual encounters between men and report that they are no longer comfortable bringing their children to the town's commercial center.

Over beer in one of two Provincetown bars popular with old-timers, a manual laborer admitted that, as a teenager in the 1980s, he participated in a violent attack on a gay man, a practice known as *rolling*. Sitting in a corner of the dark bar, he sheepishly explained:

> I'm not really proud of this, but . . . at the time the big thing to do, as far as town kids were concerned, was to go and "roll faggots." So I knew what that meant, I knew what it entailed. . . . The person going down to the beach would be the decoy. He'd be standing on the beach, waiting. And [we would target] a guy who wanted a little sex, and just because you were there indicates that [you also wanted to have sex]. . . . There were four guys waiting for this to happen. And they'd all come out. . . . Once the clay pigeon, so to speak, came out, all the other guys would jump out of the shadows and beat the living shit out of this guy and take his wallet. And that was that.

A man two decades his senior similarly confessed: "This [gay] kid used to come down, and we'd beat him up. You know that kind of stuff." And a middle-aged female old-timer suggested that gays' "inappropriate" public behavior invited violence: "There was gay bashing and stuff like that . . . there's no question. And I don't think anybody . . . I mean nobody thought it was right. You know what I mean? Just 'the way it is.' . . . It was becoming so right in front of your face. I think that's what started it. And I'm not saying it's right, but I think that's what started it."

These are not merely private sentiments or clandestine acts of homophobia. At public meetings, longtime residents, using more subtle language than that reproduced above, articulate the belief that gays "took" Provincetown. At a meeting about the potential privatization of the town-owned nursing home, a Portuguese lifelong resident complained that the town cares more about gay tourism than elderly residents. He evoked a decades-old discourse about year-rounders versus tourists: "I was born here and will be buried here, and my parents, too! We spend more on naked shave your armpit weekend [than on elders]. I don't care about debt. The elderly, these are the people who gave us life, who gave us this town, who fought to keep it this way." Similarly, at the 2004 town meeting, another Portuguese man urged support for schools, saying: "This community owes it to itself and to the families who have it hard enough existing in such a multidimensional community." In general, longtimers suggest that gay gentrifiers pose a threat to families and their heritage. For instance, one wrote a letter to the editor of the local paper: "If you take away character and heritage, this town will have nothing. Everything that once kept people here will be a distant memory, leaving a fading fishing fleet, commercialized dune shacks, and absolutely no children" (*Provincetown Banner*, November 25, 2004, 8).

This debate about whether Provincetown is a gay enclave or a Portuguese fishing village is infused with class tension, for the town's moneyed elite is primarily gay while old-timers are generally working- or middle-class.[15] The infusion of homophobia in old-timers' protests against their cultural, social, and economic displacement only intensifies gentrifiers' defensive response. In this way, the debates about sexual identity and gentrification become conflated, and gay pioneers and homesteaders respond to complaints about change and displacement with a defense of their sexual identity. This, in turn, fuels preservationists' discourse as they aim to defend old-timers during public conflict with other gentrifiers, such as by supporting their stance on school issues or by backing affordable housing. Thus, gentrifiers' characteristics entangle with old-timers' homophobia and outspoken resistance to change, consequently fueling an intrepid social preservation discourse.

While newcomers' and old-timers' collective characteristics contribute to Provincetown preservationists' bold public discourse, social preservationists' own traits explain their reliance on symbolic strategies. In 1899, proclaiming Provincetown's light spectacular for painting, Charles Hawthorn opened the world renowned Cape Cod School of Art in the town's East End, and "the mostly Portuguese population . . . made room for a heady mix of painting students, anarchists, and writers from Greenwich Village, and expatriate artists escaping Europe and World War I" (Shorr 2002, 52). This legacy, as well as today's Fine Arts Work Center, continues to attract artists and writers who continue the "heady" tradition by readily enlisting their skills in support of their preservation concerns. Their infusion of art with politics is rooted in a combination of Provincetown's reputation as a haven for liberals and for gays and lesbians, many of whom were politicized by the gay rights and feminist movements and by AIDS activism.[16] These factors, combined with the fact that some artist-preservationists once narrowly averted displacement, encourage the use of art to achieve preservation aims.

Thus, Provincetown social preservationists engage in public debate about gentrification and old-timers' displacement and advocate for affordable housing. A few consciously abstain from town politics to preserve what little remains of Portuguese old-timers' hegemony. These political strategies are a result of the town's advanced stage of gentrification, limited affordable housing, and public discourse about gentrification fueled by conflict between gay gentrifiers and Portuguese Catholic longtime residents. Finally, the fact that many Provincetown preservationists are artists or writers contributes to their use of symbolic strategies.

ARGYLE

Despite the fact that Argyle is at an earlier stage of gentrification, the practices of neighborhood preservationists are similar to those of their Provincetown counterparts. They rally for affordable housing and speak publicly against gentrification. However, Argyle preservationists uniquely emphasize *paternalistic advocacy.* That is, rather than carving space for old-timers' voices, they speak on their behalf. Through partnerships with local politicians and gentrifiers, they use symbolic strategies to link neighborhood identity to Asian, particularly Vietnamese, old-timers. Privately, they consciously support old-timers' restaurants and stores.

Argyle preservationists regularly advocate for old-timers and articulate their concerns about gentrification and displacement at meetings at which progentrification sentiment prevails. For instance, at a tense block club meeting about aesthetically displeasing burglar bars on Asian-owned businesses, Vietnamese and Chinese merchants sat on one side of the room, while a crowd of predominately white gentrifiers filled the primary seating area. A preservationist reminded other gentrifiers: "[The fact that you] want things better . . . scares some [Asian] business owners because they'll be displaced by Starbucks." Another interrupted complaints about Asian-owned businesses to say that the Asian community is the "goose that laid the golden egg that made this a neighborhood you wanted to move into." Yet another referred to his decision to move to the neighborhood: "I moved here because I think it is incredible that there's this Asian community on Argyle. I think they're so solidly set here in this community."

A social worker used elderly residents' claims to old-timer status to encourage support for the construction of an assisted-living facility: "Many of them are selling homes because of [rising] taxes. Their families are leaving because they can't afford it. These people aren't used to moving all over the city. They have a right to stay in the neighborhood. Seniors are the backbone of our neighborhood." At other meetings at which they were very much in the minority, social preservationists said that "the other part of the story is tons of gentrification" and that "[there are] some who are very antipoverty."

At another meeting, a preservationist—a religious leader—argued that Asian and Jewish senior citizens should be able to remain in the neighborhood "where they are comfortable," surrounded by their food, language, and culture. He asked: "How do we ensure that our neighbors who lived here before all of you, when the neighborhood was *really* bad, will be able to stay?" On an Internet board, he also wrote: "I know . . . the pain and suffering of the elderly who can no longer live on their own and must leave their friends and neighbors to reside in assisted-living facilities in other parts of the city. We have many seniors in our area representing many ethnicities. We should view them as our collective responsibility."[17] In this way, Argyle preservationists engage in public discussion about gentrification and old-timers' displacement. Specifically, they advocate against the disruption of Asian businesses and for affordable housing.

However, Argyle preservationists extend their advocacy for affordable housing beyond their immediate neighborhood to the official Uptown neighborhood in which Argyle sits. They do so through membership in several community organizations that advocate for affordable housing and against gentrification and by participating in protests and community meetings. Such organizations include the Organization of the North East, a nonprofit United Way organization dedicated to "building a successful multi-ethnic, mixed-economic community on the northeast side of Chicago."[18] Another organization, the Community of Uptown Residents for Affordability and Justice, aims to preserve existing and create new affordable housing in the neighborhood. Queer to the Left (Q2L), established in 1999 and primarily composed of gay, lesbian, bisexual, and transgender individuals, devotes much of its work to maintaining low-cost housing in Uptown. Q2L is largely concerned with the

FIGURE 4.01. Uptown Rally (Chicago)

role of gays and lesbians in the gentrification process and advocates for those facing displacement.

Argyle social preservationists attend public hearings and protests—often held just outside Argyle—organized by such groups. At a rally against a proposed Borders, a twenty-six-year-old graduate of a Seven Sisters college spoke: "Here is a neighborhood that is vital and diverse and beautiful already. If the city keeps sponsoring gentrification, it's gonna reverse history. Its gonna *re-segregate* us! . . . This gorgeous fertile mixture of people will disappear. People come from all over to go to cities because cities offer a place where people can be fundamentally different and still share space. It is essential that we protect that vision. It is imperative that we save what we have built from homogenization, the corporate appetite, and spiritual death." The protestors, many of them highly educated, young and middle-aged white adults, chanted: "We don't want Starbucks or the Gap; low-cost housing is where it's at!" A rain-drenched speaker said: "The rich are *not* enriching! [We want] vibrant racial and sexual diversity."

Preservationists also attended a raucous meeting to ensure that the city include nearly two hundred affordable units in the development of a vacant lot. Affordable-housing advocates and prodevelopment activists appeared wearing distinctly colored T-shirts (each provided by different organizations), and debate grew so vehement that a representative of the college hosting the event threatened to disband the meeting, and police took to the stage. In this sense, Argyle preservationists rely on local organizations to help stall Uptown's gentrification.

The leaders of such groups serve as models for social preservationists and shape their participation in public dialogue about gentrification. For instance, at a September 2004 hearing on affordable housing, an administrator of one such group expressed concern for the preservation of particular populations: "Many working people are leaving the neighborhood who bring vitality [to Uptown]. . . . People will tell you to be afraid of your neighbors, but Uptown has a history of being full of generous, helping people." Such words are a discursive resource for Argyle social preservationists, and, in turn, such organizations use listservs, posters, and word of mouth to recruit social preservationists and other volunteers.

In contrast to these far-reaching (i.e., Uptown-wide) affordable-housing efforts, Argyle preservationists focus symbolic and private efforts on the preservation of residents on and around Argyle Street. Some attend the neighborhood's Chinese New Year celebrations. Indeed, the New Year Parade is attended by at least as many well-educated whites as Asian residents and tourists. In 2002 and 2003, dozens of white couples stood with strollers watching drummers and dancers. Young white men gathered with cameras as smoke wafted from firecrackers and later walked home through streets strewn with cabbage and firecrackers. At meetings frequently held at the Chinese Mutual Aid Association, preservationists speak with affection of markers of old-timers' presence, such as the pagoda that sits atop Argyle's el stop, and insist that streetscape plans include symbolic representations of the pan-Asian community.

FIGURE 4.02. Argyle Commercial District during Chinese New Year

Some try to emphasize or complement the presence of Vietnamese and Chinese residents and businesses. For instance, the owner of a Mexican restaurant painted his establishment's door red because, as he put it, "I know in the Asian culture red doors . . . is good luck . . . I said, 'You know, I'm their neighbor.'" Between 2001 and 2005, preservationists worked, alongside homesteaders, pioneers, politicians, business leaders, and a few old-timers, to establish an Asian-themed streetscape. Responding to a survey about the proposed streetscape, many expressed appreciation for old-timers and concern about displacement. Among the comments were the following:

I love the diversity and authenticity of Argyle Street.

Do everything to improve the small indigenous business owners. No national chains. No displacement.

Argyle is a beautiful part of Chicago—culturally rich, and the diversity should be preserved.

The social preservationists on the committee beamed as an aldermanic aide read such quotes and tallied them on a board.

Argyle social preservationists are committed to supporting local Vietnamese-, Chinese-, and Thai-owned restaurants, shops, and grocery stores. They do so to sustain old-timers' businesses but also out of appreciation for the culture and practices of their old-timer neighbors. For instance, a chef said: "I use the Vietnamese markets. For me to purchase Asian ingredients there, as opposed to going to, say, a Whole Foods, it is just amazing." In fact, several informants corrected themselves after beginning to say that they frequent old-timers' businesses in an effort to sustain them. They acknowledge that Asian customers—from Chicago as well as from surrounding states—primarily sustain Argyle's Asian businesses and that they personally support the businesses because they enjoy doing so (this is likely true in all the sites).

What accounts for the contours of social preservation in Argyle? For one, the strong presence in Argyle of frontier and salvation rhetoric and practices encourages preservationists' emphasis on affordable-housing advocacy as well as their public debate about community change. In recent years, Argyle has witnessed a significant escalation of condominium conversions and the (re)establishment of block clubs that seek to improve the neighborhood through crime reduction and beautification and by preventing the establishment of additional affordable-housing units as well as homes for the elderly or mentally ill. Many members of such block clubs readily deploy a discourse justifying neighborhood change.

Social preservationists react against pioneers' vociferous public complaints about homelessness, panhandling, and neighborhood cleanliness, such as claims that "our buildings still look like slums" and that low-income housing is "unhealthy" for the neighborhood. For instance, at a 2004 meeting, pioneers told a densely packed room of newcomers that "40% of the population in Uptown lives below the poverty line," that the neighborhood has reached "social service saturation," and that there should be "drug testing and criminal background checks" for low-income residents. Earlier, at a March 2001 Argyle Streetscape Meeting, newcomers complained about frequent problems with individuals cutting the power to streetlights. Richard, the pioneer we met in chapter 2, said: "They've been so brazen as to say, 'Can't you see that we want the streets dark? We're going to keep doing this.'" In response, a preservationist asked: "Who is *they*?" Richard responded: "Thugs."[19] Another asked: "Do we know who they are?" And another expressed concern for those cutting wires, saying: "They could get killed cutting wires!" Richard retorted: "Then we'd know who they were. Then we'd be rid of them."

At another meeting of the same group, residents gathered in the lounge of a neighborhood single-residency-occupancy (SRO) building for a conversation with the social workers who operate the institution about the "nuisances" newcomers associate with SRO residents, such as panhandling, public drunkenness, and loitering. Implicit in the conversation was concern—on the part of pioneers and some homesteaders—that the public presence of the mentally ill would stall gentrification. For instance, one argued: "[The clients create an] image that [the neighborhood's] not safe. This image problem is a big thing." Another implied that gentrification is a natural process interrupted only by institutions like SROs: "With gentrification [the SRO's] suppressed tax rate keeps things here unnaturally." Three years later, at meeting on a proposed assisted-living facility for the elderly, a pioneer said: "I'm relieved that it will be for the elderly [rather than others], but this is a social services Mecca. It is very much saturated. I don't see my property values going up with that monstrosity in the neighborhood."

Others engage in boosterism by not referring to the neighborhood as Argyle (the main commercial strip, primarily composed of Asian-owned businesses) or Uptown (the official community area of which Argyle is a part). Instead, they refer to it as "SOFA: South of Foster Avenue" (a bar with a similar name recently opened in the space where Andersonville and Argyle meet), "Andersonville Gardens," "Edgewater," "the bastard child of [the alderwoman]," or "Margate Park."[20] A new condominium just south of Argyle Street—with one-bedroom units starting at $240,000—identifies the area as "Sheridan Grande." An Argyle masseuse reported that she avoids referring to the area as Argyle or Uptown: "I usually say . . . Edgewater because when I say Uptown people get a weird look. . . . So it is Edgewater, just east of Andersonville."

As we have seen, social preservationists interject their voices in such debates. The public and volatile nature of pioneers' discourse encourages preservationists to take to the streets in protest and to attend meetings at which revitalization strategies are discussed. If pioneers' discourse were more muted, Argyle social preservationists might turn their attention from protests and block club debates about gentrification and concentrate on symbolic or private practices.

But, of course, this is not the case. In fact, there rarely is a middle ground in Argyle debates. Comparatively few Argyle gentrifiers engage in efforts to preserve neighborhood character so common among homesteaders. At meetings, pioneers sometimes contrast themselves with Andersonville new-comers, noting the discrepancy between their focus on upscaling Argyle and Andersonville preservationists' and homesteaders' efforts to preserve their neighborhood's character. They imply that the preservation of character is a luxury of advanced gentrification. For instance, at a meeting of Uptown block club leaders, one said: "There's a movement coming out of Anderson-ville to block big national [chains]." Another implied that this was acceptable in Andersonville because it is more "charming" than Uptown: "Research on independent businesses isn't a bad thing. Andersonville has a charm all its own." Similarly, at another meeting, a resident expressed support for An-dersonville's preservation: "I don't want Clark Street to turn into Southport or Halsted. I want to keep the character." In response, another newcomer carefully distinguished between Andersonville's character and that of Argyle, suggesting that, to preserve Argyle's character, one would have to revert to its economically depleted past: "We could sell our six-flat, and they could put in a church or a halfway house!" Similarly, another—a white man in professional attire—suggested that Uptown has little worth preserving: "The problem is in our neighborhood mom-and-pops are dollar stores and cell phones. It would be good if they were driven out. It would bring something new in." In this sense, because of its early stage of gentrification, there may be less room for homesteading discourse and practices in Argyle (although they are present) than in the other sites. Tellingly, the conditions that discourage homesteaders' public expressions support social preservation.

Social preservationists are not the only Argyle actors whose discourses and practices are shaped by those of other gentrifiers. Indeed, social pres-ervationists influence homesteaders and pioneers. For instance, some social preservationists have joined block clubs to dissuade others from disrupting the neighborhood. In this context, they engage in public discussion about displacement and gentrification, and preservationists' influence on other gentrifiers' discourse is apparent. For instance, at a block club meeting discussion of strategies for curbing litter, a pioneer—a man who frequently sweeps neighborhood streets of refuse—defended improvement strategies by disassociating them from gentrification: "People are scared to talk about a neighborhood becoming healthy. I refuse to use *gentrification*. It doesn't matter how much money you have. There's healthy and unhealthy. . . . We've won because there are less and less buildings that *haven't* been gutted, changed into condos, . . . [and there are] fewer drug dealers."

Such defensiveness often arises in response to social preservationists' words. For instance, at an Argyle block club meeting, a preservationist said: "There are two kinds of people in this neighborhood: ones who don't want change and people who turn a quick profit and leave." A pioneer retorted: "I'm not sure who you mean. Some people move here for jobs. It is not just profit." Thus, while pioneers' discourse acknowledges social preservationists' concerns, most pioneers remain rooted in their own ideology.

This adherence to a strict frontier and salvation ideology (Spain 1993) is encouraged by the articulation of overtly progentrification rhetoric by an Uptown organization that formed to resist plans to construct low-income housing on a tract of vacant property. Calling for artists' housing instead of low-income housing, they argued: "We live here, too! We have rights!" At a public meeting, a white woman affiliated with the group who appeared to be in her thirties explained that, when she moved to the neighborhood in 1997, it was significantly "different than it is today": "The neighborhood has changed for the better. . . . Change is not a bad thing. I like to feel safe and want to feel that I can raise children here." A man suggested: "Any real estate should be for purchase, *not* rental." Someone else argued: "Affordable housing can lead to crime." Argyle pioneers borrow from the rhetoric and strategies of this organization—by promoting, for instance, drug testing for affordable-housing residents—and, when social preservationists call for affordable housing, pioneers turn to the organization for data on the number of existing affordable units to buoy their resistance efforts.

In the context of pioneers' vociferous rhetoric, Argyle social preservationists, more than those in any other site, aim to speak *for* old-timers. Many engage in *paternalistic advocacy*.[21] That is, out of the belief that they better articulate old-timers' concerns than do old-timers themselves, they seek to speak on old-timers' behalf and represent their perspectives in negotiations with other power holders (e.g., other gentrifiers and politicians).

Argyle preservationists rely on paternalistic advocacy, in part because language barriers prevent many Asian old-timers from having an equal voice at meetings where gentrifiers promote change. At most such meetings, Asian business owners compose a small minority and defer debate to white gentrifiers and social preservationists. In recent years, a few of the Vietnamese and Chinese business owners' adult children, who are highly educated and fluent in English, have become increasingly vocal at such meetings, and gentrifiers of every ilk seek relationships with them. The recent appointment of two highly educated, American-raised young women, one Vietnamese and one Chinese, to ethnic business associations has increased dialogue among Asian merchants and between Vietnamese business owners and non-Asian members of Argyle's largest block club. Still, for the most part, Argyle old-timers are much less a part of gentrification debate than are old-timers in the other sites, and language differences largely explain this.

As a leader of a Vietnamese business association suggests, the presence of overlapping service and advocacy organizations for the various facets of the Argyle Asian community also contributes to old-timers' muted participation in local debate. Leaders sometimes mistakenly assume that another organization is responsible for a given issue: "It's hard because there are so many groups. . . . There's like all these associations who aren't in communication with each other. How many meetings can we have, too? I mean, we just started getting everyone's e-mails and everyone's phone numbers, who's who and who's part of what. So I think there's just a big lack of communication with everyone in this area. There's just so many associations in one little community, and I think because there is that lack of communication no one knows what's going on." Furthermore, some Vietnamese business owners do not believe that it is necessary to develop relationships with newcomers, for, despite boosters' claims that an aesthetically improved commercial district will build business, they believe that their primary client base will remain Vietnamese Americans from the Chicago area and surrounding states. For instance, a Vietnamese small business owner reported: "All over are my clients . . . out of state too. Yeah, a lot. Michigan, Indiana, Milwaukee, Ohio." The leader of an Argyle ethnic business association concurred, saying that rising commercial rents "aren't an issue" for merchants "because Argyle is so busy all the time. We attract customers from the suburbs and . . . out of state." For these reasons, many old-timers seem—at least for now—content to let preservationists battle for them.

Finally, one of Argyle's alderwomen (the neighborhood is split between two districts), with the support of city policy, is partly responsible for Argyle social preservationists' use of symbolic strategies. Chicago's Neighborhoods Alive! program funds streetscapes that often celebrate a local group's ethnic identity (Reed 2003). This alderwoman used city funds to create a streetscape in the adjacent Andersonville neighborhood and sponsored an Argyle streetscape planning process. While her office, pioneers, and homesteaders speak of the proposed streetscape, as well as a mural, as a revitalization mechanism, it also provides preservationists with a forum for the symbolic demarcation of the neighborhood as belonging to old-timers. Without such preexisting efforts, preservationists may not have initiated the symbolic conservation of old-timers and may instead have focused on preventing physical displacement.

In summary, Argyle preservationists engage in heated debate about gentrification and displacement, both within their neighborhood and in the official neighborhood in which it falls. A rise in condominium conversions and several proposals for the construction of affordable housing, as well as pioneers' frequent and publicly visible transformation efforts, encourage debate. Local activist groups on either side of the debate over affordable housing and gentrification advance a vigorous discourse, and this encourages preservationists' affordable-housing advocacy. Organizational confusion among Argyle's Asian populations and language barriers between old-timers and most newcomers contribute to social preservationists' paternalistic advocacy on behalf of old-timers. Finally, politicians' attempts to elevate the neighborhood's appeal as a destination (and, in so doing, to encourage gentrification) through a planned streetscape and mural encourage preservationists' use of symbolic strategies.

ANDERSONVILLE

More than those in any other site, Andersonville social preservationists rely on and value symbolic strategies. They also actively employ private practices, particularly through consumption and daily neighborhood interaction. Their political strategies are oriented in two directions: to the preservation of independent businesses, particularly those owned by Swedish Americans, and to the preservation of affordable housing in the adjacent Uptown neighborhood (the same housing that Argyle preservationists advocate for).

In contrast to the sharp ideological and practical divisions among Argyle gentrifiers, in Andersonville many gentrifiers articulate concern about the disruption of local character by further gentrification. Homesteaders and preservationists alike organize and attend meetings about how to retain the independent businesses that dominate Clark Street, such as Swedish delis, Women and Children First Bookstore, and two Swedish restaurants. At one meeting, held in the Swedish American Museum, social preservationists talked about the negative impact of chains on the local *community*, while homesteaders spoke of the implications for the *built environment*, namely, its historical integrity and aesthetic appeal. One homesteader said: "The bummer is that everything just looks the same. It's homogenized." At another meeting, social preservationists noted that the neighborhood's history is "fundamentally about a small business community that began with farming and became a thriving Swedish community with a bell ringer" and that it is an "amazing community." They proposed a "sustainable Andersonville fair" and asked whether one could use an ordinance to prohibit chains from moving in. Social preservationists were among supporters of the Chamber of Commerce's study of the comparative local economic contribution of independent businesses and chains.[22] For social preservationists, it is paramount that Andersonville retain, not only independent businesses, but also small business owners and Swedes.

In subtle contrast, a homesteader—a young gay man who once owned an Andersonville gallery—articulated reservations about chains coming to the neighborhood: "I'm like, 'Not another Gap.' It turns the neighborhood into just any other neighborhood, nothing unique. You might have all of these *lovely little upscale restaurants or shops* or whatever, but [with chains] it does lose something, and this just becomes like Lincoln Park North or something" (emphasis added). His concern is for preserving, not people, but "lovely little upscale" independent businesses. Nonetheless, regardless of the impetus for his position, he shares preservationists' concern about chains and allies with them in the preservation of local businesses.

When developers disclosed plans to build condominiums and a Borders bookstore in the building that once housed Goldblatt's Department Store in Uptown, Andersonville social preservationists and their allies crowded a feminist bookstore to strategize about methods to block the development. They feared that the store would threaten independent businesses, raise property values, and contribute to the displacement of poor and working-class neighborhood residents. Thus, for many, the issue was not limited to the preservation of businesses. Their concomitant concerns were the preservation of place character embodied by independent shops, Swedish business owners, and working-class residents. For instance, at the meeting, a resident said: "I saw a poster against low-income housing . . . this morning. I don't ever want to see this again!"

As a result, some Andersonville preservationists, alongside their Argyle counterparts, joined Uptown activist groups in protests against the proposed Borders and advocated for the construction of affordable housing. In an interview, a twenty-something Andersonville preservationist explained that she helps plan such events as part of a general resistance to gentrification: "[The] gentrification you see taking place . . . is a threat to city culture. The demands increase homogeneity: homogeneous households, homogeneous retail outlets, homogeneity on the streets and in public places." In this way, like those in Argyle, Andersonville preservationists take part in protests against gentrification *outside* their immediate neighborhood. However, in contrast to that in Argyle, such participation here translates, not into affordable-housing advocacy within Andersonville, but, rather, into efforts to prevent the displacement of independent businesses and their owners.

In fact, social preservationists often direct their efforts toward the preservation of Andersonville's Swedish character through symbolic means. For instance, some helped plan the city-sponsored streetscape on busy Clark Street, which incorporated markers of the neighborhood's Swedish population, such as Swedish flags, bells imprinted on sidewalks, and blue and yellow lettering on street signs. Others attend and contribute to festivals and Swedish cultural events, such as the Santa Lucia Procession and the Midsommar Festival (which some complain is decreasingly Swedish). Neighborhood business leaders (some of whom are social preservationists and some of whom are not—few are Swedish) primarily orchestrate the festival. Regardless of its accuracy, it contributes to the neighborhood's Swedish identity, as do notices of street construction from "Sven," a fictional neighborhood correspondent depicted with a Viking helmet and a seven-foot-tall sculpture of a blue horse shipped from Sweden that stands on a busy corner. Individual preservationists help plan or participate in all these events. Some attend events or emphasize the community's Swedish heritage through artwork, such as sculptured figurines of women in traditional Swedish garb.

Andersonville social preservationists consciously support local Swedish businesses, such as a family-owned bar, two Swedish breakfast restaurants, and a pair of delis that sell Swedish cheese, meats, beverages, and candy.[23] As do their counterparts in Argyle, they do so to sustain local businesses and because they enjoy sampling Swedish culture. For similar reasons, they befriend elderly Swedish residents whose relatives relocated to the suburbs.

However, over the last two years of fieldwork, the frequency with which newcomers spoke of Swedes diminished. Recent interviewees (interviewed in 2004 and 2005, as opposed to between 2001 and 2003) even included a preservationist—a prominent attorney—who worried about the displacement of mentally ill residents whose facility closed to make way for redevelopment. She felt that they brought "color" to Andersonville and expressed sorrow about their displacement:

I think it is a really bad thing. . . . To have [that building] in the neighborhood was kind of a guarantee that the neighborhood would never be, like, the crème-de-la-crème. 'Cause you had a lot of people wandering around who were obviously troubled. And you're gonna have some group of people who says, "I don't wanna live there." And, in addition, one of the reasons I wanted to live here was for the diversity, and that's part of the diversity. I can say to my sons, you know, "Let me tell ya why they look that way," or "Let me tell ya why they act that way." . . . You know what I'm saying? It's another way to say, you know, this is, this is life. . . . They've been gone for weeks. And that's sad to me too because part of that neighborhood, for me, was them. So that's a big change.

Others refer to conflicts about affordable housing in Uptown as if they directly affect Andersonville residents despite the fact that some of the *same* informants consider the neighborhoods to be separate and distinct. In this way, for reasons discussed below, social preservation seems to be shifting, away from Swedish old-timers toward either independent business owners or populations outside Andersonville.

Part of the explanation for this shift is that, arguably, the time to fight for affordable housing and against condominium development in Andersonville has passed. In 1999, 13.6% of neighborhood households made $100,000 or more, while only 11.5% earned less than $14,999. In addition, in the same year, the median owner-occupied home value was $300,167. Today, realtors list one- and two-bedroom condominiums for $150,000 and $350,000, respectively, and single-family homes for between $500,000 and $1.5 million.[24] That preservationists' traditional object of preservation, Swedish residents, tends in this case to be middle-class homeowners who are not in great peril of displacement (certainly some were displaced, but many suggest that displacement is largely already over) compounds with Andersonville's advanced stage of gentrification to diminish the likelihood of local affordable-housing advocacy and gentrification protests.

Furthermore, vigorous debates about affordable housing in adjacent Uptown distract Andersonville social preservationists. As with Argyle preservationists, Uptown antigentrification groups draw Andersonville preservationists to their protests and meetings, but, unlike their Argyle counterparts, who bring these debates home, Andersonville preservationists seem to think of them as distinct from their own neighborhood's changes.

Why, then, do they devote attention to the Uptown debates? In recent years, the gentrification of central Andersonville has expanded into southern Edgewater and northern Uptown. In all four directions, newcomers attracted to Andersonville have purchased condominiums or rented apartments in areas bordering Andersonville proper. Simultaneously, the Andersonville Chamber of Commerce extended the Andersonville streetscape several blocks south of Foster Avenue and north of Bryn Mawr, which many once thought of as the neighborhood's southern and northern borders. In so doing, they symbolically (and, subsequently, economically) extended the neighborhood into Uptown, where social preservation activity increasingly concentrates. Finally, several businesses catering to middle and high-end clientele opened in the areas south of Foster and north of Bryn Mawr Avenue, such as a popular breakfast place, an upscale cookware store, and a busy bar and grill that attracts many lesbians. In a sense, this expansion of residential, commercial, and government-sponsored gentrification outward from Andersonville encourages a sense of responsibility among Andersonville preservationists as they hold their own neighborhood accountable for the increasing gentrification of adjacent neighborhoods. This may be because preservationists have encountered evidence of the *intentional* expansion of gentrification, as I did in a 2005 newsletter: "The Andersonville portion of Clark is hot, full of boutique retailers, and the North Clark section is not. It lacks a cohesive character and suffers from a high vacancy rate. . . . A group of community organizations and the 48th Ward have decided to undertake a vision process" (Edgewater Development Corporation 2005, 1). While Argyle preservationists wish to prevent the gentrification of Uptown to block the gentrification of their *own* neighborhood, Andersonville residents seek to prevent Uptown's gentrification *by* their neighborhood. They have insights about gentrification to share with Uptown, while Uptown has lessons for Argyle.

Finally, some preservationists who live in the area bordering Andersonville and Argyle consciously choose to identify with Uptown rather than with Andersonville out of resistance to Andersonville's expansion south (via gentrification). As this suggests, there is some discord among informants about Andersonville's boundaries. Residents of the same block differ over the neighborhood's proper name (Andersonville vs. Uptown vs. Argyle). For instance, one preservationist—an affluent professional and longtime homeowner—said: "I don't like that it's becoming so hard to rent an apartment. . . . It's gonna push everybody out. I don't like that. I don't like that it's become 'Andersonville.' . . . It's a push created by developers to show that we are more upscale." Thus, for such residents, it would seem superfluous to focus protest efforts on Andersonville, and they turn instead to Uptown.

In contrast, pioneers and homesteaders offer a range of responses to Andersonville's incursion on Uptown and Edgewater. For instance, a homesteader and realtor who often lists houses in the neighborhood suggested that his street—which technically falls in Uptown—shouldn't "really count as

Uptown" but acknowledged that, when he purchased his home several years ago, "there was a big [symbolic and economic] difference between South Foster and being in North Foster." Thus, he benefited from this symbolic distinction because, in his estimation, the association of his property with Uptown lowered its value. However, he now works discursively to minimize that difference, not because he anticipates selling soon, but because he enjoys identifying as an Andersonville resident. At a May 2004 meeting of a block club whose territory extends into Uptown, a member spoke appreciatively of the distinction between Andersonville and less-gentrified Uptown: "We are the furthest north in Uptown. Other people have many more problems, like crime." A gray-haired homesteader responded: "Yes, the only thing is that Uptown is becoming gentrified and raising our taxes." To this another member said: "Yeah, that's the kicker with gentrification. It's a nicer place to live but more expensive." Like social preservationists, these newcomers recognize gentrification's encroachment on Uptown, but they do not associate this with a loss of social authenticity.

We can best understand the contours of Andersonville preservationists' strategies, particularly their emphasis on symbolic and private practices, in relationship to old-timers' characteristics. Andersonville Swedes are arguably in a better position to avoid physical displacement than those in the other sites. After all, they remained in the neighborhood after many were displaced in the 1990s. Several Swedish old-timers I interviewed own (or reside in family-owned) two- or three-flats that provide rental income, which helps them meet rising property taxes.[25] Others depend on protective tax policies for low-income elderly. Thus, they do not warrant the types of strategies (namely, affordable-housing advocacy, protests, and public debate about displacement) that we see in Argyle and Provincetown. Finally, old-timers who do leave often do so because of old age. They move in with their children, to a retirement community or nursing home, or even die. For this reason, despite the fact that less than 4% of Andersonville's 2000 population was Swedish, the slowly declining number of elderly Swedish residents *appears* to be a natural phenomenon. As a result, Andersonville social preservationists focus their attention on that which gentrification most obviously threatens: neighboring Uptown and markers of Swedish culture.

Widespread popular concern for Andersonville's independent businesses and institutional support (via the Chamber of Commerce) for such enterprises draw social preservationists' attention to independent business owners. Indeed, many neighborhood actors share a concern for local independent businesses—key symbols of Andersonville's Swedish identity and character—and the Chamber of Commerce (whose director frequently espouses social preservation rhetoric) has spearheaded efforts to prevent their displacement. Andersonville preservationists often join Chamber efforts, despite some old-timers' resistance. At a meeting about a proposal to prevent chains from moving to Andersonville's commercial district, the owner of a local home-supply store said that his business was started in "1886 when my grandfather came over" but added that he was "worried about the protectionist language," about barring chains. Andersonville preservationists largely disregard such concerns, perhaps because their appreciation for Swedes does not rest on admiration for their fierce independence, as it might if their object of preservation were a New England farmer or an Idaho rancher.

Cultural events sponsored by the Swedish American Museum, combined with the Chamber's interest in marketing the neighborhood as Swedish, reinforce Andersonville preservationists' appreciation for symbolic manifestations of old-timers' culture. These institutions often cosponsor events that highlight Swedish culture and were at the forefront of the establishment of the Swedish streetscape. Such efforts not only direct preservation practices to Swedes but also provide venues for symbolic preservation.

Finally, because the neighborhood is so resolutely gentrified, and because Swedish old-timers rarely publicly contest gentrification, many homesteaders share preservationists' concerns about the disruption of the neighborhood by zoning changes and retail chains, and this helps produce Andersonville social preservationists' comparatively muted public rhetoric. While homesteaders focus on preserving the neighborhood's physical integrity (in contrast to social preservationists' concern for its social integrity), they share with social preservationists an interest in maintaining neighborhood character. At a community meeting about proposed zoning changes, one homesteader said: "I want the *look* of Andersonville only. I advise you to look at other neighborhoods around here, and [you'll see] there's nothing good happening there. . . . Developers will screw up the essence of a neighborhood." Another said: "There are many historically and architecturally charming buildings. [Would zoning] impact decisions about demolition?" Similarly, when rent increases displaced a shopping center that housed several independent businesses, "Save Andersonville" signs appeared throughout the neighborhood, and the displacement disturbed social preservationists and homesteaders alike.

Despite their distinct logics for why preservation should occur (to save people or to save buildings), many Andersonville homesteaders and social preservationists are on the same side of preservation debates. They share an appreciation for neighborhood history and a concern that encroaching capital threatens local character. While it is unclear whether strong support for preservation is the cause or the result of earlier preservation work (i.e., whether social preservationists' ideology affected homesteaders' discourse, which in turn shapes social preservationists' strategies), such as the creation of the Swedish American Museum and the Swedish streetscape, public discourse is so weighted in favor of preservation that it leaves little space for the more virulent social preservation discourse of Argyle and Provincetown. At this time, Andersonville homesteaders' dedication to preservation discourages the ideological contestation that supports public social preservation discourse in Argyle and Provincetown.

In summary, Andersonville social preservationists use private practices to ensure the continuation of independent businesses and to build relationships with Swedish old-timers. However, they focus political practices on neighboring Uptown, where they attend protests and public forums. This is a result of several interrelated factors: Andersonville's advanced stage of gentrification, its expansion into surrounding neighborhoods (which produces both social preservationists' guilt and their disassociation from Andersonville), the middle-class character of and high homeownership rates among Andersonville old-timers, and the presence of Uptown activist groups. Given these factors and the preservation efforts of homesteaders, Andersonville social preservationists primarily focus on the preservation of independent business owners and symbolic manifestations of local Swedish culture. The Chamber of Commerce, the Swedish American Museum, and local and city politicians' boosterism (particularly in the form of the streetscape) encourage their symbolic practices.

Andersonville poses important questions about social preservation and gentrification. First, will social preservation continue if the Swedish population continues to diminish and the neighborhood further gentrifies? Might it perish altogether, or is there early evidence of a shifting object of preservation, from Swedish old-timers to independent business owners or Uptown residents? To what extent is the extensive *symbolic* preservation of Swedes through a streetscape, artwork, and the Swedish American Museum responsible for social preservationists' subdued political activities within their neighborhood? In some instances, might, despite preservationists' insistence otherwise, a critical mass of symbols stand in for people, especially in a context in which the old-timers whom preservationists admire are few in number and not readily (visually) identifiable apart from such symbols? Finally, did independent shops and Swedish history draw homesteaders to Andersonville (as opposed to, say, Argyle, where one might better anticipate encountering a rugged social authenticity or the challenge of the urban frontier), or do advanced gentrification, a small population of old-timers, and old-timers who primarily welcome neighborhood change encourage expression of the homesteading orientation?

DRESDEN

At first glance, the political practices of social preservation do not seem to exist in Dresden. Social preservationists have not established affordable housing for old-timers, and they rarely engage in public debate about gentrification. In fact, they are notably quiet at town meetings—held either in the town's drafty meetinghouse or in the modern elementary school cafeteria—and are relatively absent from regulatory committees, such as the conservation and planning boards, boards otherwise popular with newcomers, many of whom moved to town out of appreciation for its pristine forests and open slopes along the Eastern and Kennebec rivers. In the early days of my fieldwork, the disjuncture between the quiet demeanor of social preservationists at town meetings and the passionate concern for old-timers that they expressed in the kitchens and living rooms of their restored farmhouses and solar-paneled homes puzzled me.

However, I soon realized that this quiet approach is not rooted in apathy. Rather, Dresden preservationists actively and purposely engage in *political abstinence*. That is, they consciously abstain from certain political acts. In contrast to political avoidance, which Nina Eliasoph (1997) finds to be rooted in a sense of political powerlessness, political abstinence requires a sense of agency or influence. It also requires an awareness of one's personal competing political or ideological commitments and a recognition of the political saliency of strategic abstinence. Specifically, interviews and observations reveal that, out of deference to old-timers and a keen awareness of their own ability to alter local politics, many Dresden preservationists actively and purposely abstain from certain votes, committees, and commentary, for fear that they will contribute to old-timers' political or physical displacement. In so doing, like some Provincetown preservationists, they seek to leave old-timers' political structure and community life relatively undisturbed.

Specifically, Dresden social preservationists articulate an awareness that some newcomers "pushed their weight around" and seek to avoid replicating this. One woman, an articulate but soft-spoken small business owner, recalled: "We participated a lot, but I never got any flack about 'those new people coming in here and telling us what to do.' I never got any. . . . *I'd like to think that it was because I was very low-key and didn't tell anybody what to do*" (emphasis added). This dedication to remaining "low-key" often requires that preservationists abandon their other ideological commitments. For instance, respect for old-timers' resistance to regulations leads Dresden social preservationists to forsake other forms of preservation (e.g., of the town's rural nature, which they might preserve through zoning regulations) and their environmentalist tendencies. As one social preservationist said: "Through ordinances, you can say, 'I'm sorry you can't develop your land because we all want to look at it,' but philosophically that's not fair. It's like, all of these nice wonderful farms that we have here in Dresden, you know it is easy for us to say they need to keep farming so we can look at it, but the fact is they have to make a living." This is a somewhat surprising position, given that Dresden social preservationists universally express adoration for the town's natural beauty. Despite this appreciation for landscape, social preservationists support old-timers' right to use their land as they see fit. They complain, as one middle-aged preservationist did, that *other* newcomers "want their chunk of land": "They don't want the farmers or whatever to have the use of their land the way they have always done it. They come up from Boston to get the rural identity, and they want to make it like Boston." They do this by restricting old-timers' traditional but "unsightly practices," such as operating gravel pits on their property and spreading manure. In contrast, Dresden preservationists are committed to *abstaining* from complaints about even those practices of old-timers that they deem reprehensible, such as illegal dumping, as well as those that they find aesthetically jarring, such as dilapidated trailers and large trucks traveling to gravel pits. For instance, on an impromptu driving tour of Dresden, an elderly social preservationist—in between expressions of appreciation for the landscape and frequent cautions that I drive well below the speed limit—voiced distaste for yards littered with scrap metal and gravel pits in view of the road, yet she resists efforts to restrict such practices.

In contrast, a pair of *homesteaders*—a leftist academic couple—are critical of old-timers' practices and, therefore, of their character. The wife said: "I'm not sure how much a part of the community I can become unless they share my beliefs, and I'm not sure they do. . . . We have guys driving shooting out of their trucks. . . . [I don't want to] sound too snobby, but, with some of these people, it'd be impossible to find community. They're like grizzlies. Some of them are Grizzly Adams, I mean rough. These are rough people." Sitting in their book-lined living room, they specified particular concern about old-timers' relationship to nature: "I guess the hunting thing scares me . . . the littering, the stuff that they bring into the woods, like old stuffed chairs, batteries, mattresses, and stoves." These are exactly the concerns that

FIGURE 4.03. Rusting Cars, Dresden

Dresden social preservationists forsake through political abstinence. In so doing, despite their environmental concerns, they demonstrate a preference for the preservation of local characters over landscape.

If Dresden social preservationists' political practices are characterized by turning a blind eye to the less pastoral elements of rural life, then their symbolic and private practices celebrate the opposite. For instance, many take part in the annual Dresden Harvest Supper, which honors the work of Dresden farmers with a traditional New England meal served at the elementary school. At the 2004 supper, elderly residents clustered together, passing plates of cranberry sauce and rolls, entertaining newcomers with recollections of "old" Dresden, while a selectman and other community leaders scrubbed pots in the kitchen. In the 1980s, newcomers sponsored the revitalization of old-timers' baked-bean-supper tradition, a tradition common in many Maine small towns. A drive through Maine reveals hand-lettered signs for bean suppers benefiting volunteer fire departments and schools. Today, preservationists applaud the symbolic recognition of old-timers through the Dresden Farmlands campaign—organized by farmers and funded by a state agency—which placed signs advertising local farms on the town's state highways. For instance, one preservationist said: "The whole issue of appreciation, of valuing agricultural activity . . . the Dresden Farmlands Project identifies the agricultural as something that has tourist benefit, and I think it is very positive. Otherwise, people could just sort of drive up and down . . . but not really have any idea of the sort of gem that's out here." Thus, like Dresden farmers, she recognizes tourism's import for the sustenance of farms. However, preservationists are seldom at the forefront of symbolic practices. The Dresden Historical Society organizes the Harvest Supper, and farmers planned the Farmlands campaign.

Dresden preservationists' private practices include shopping at farm stands to support old-timers. For instance, a pair of informants—young parents who peppered me with questions about goats after learning that my family had pastured one—explained that they shop at stands because "we're about community," and another informant described them as the "heart" of the community. In the same vein, they participate in a community-supported agriculture program, and one middle-aged couple served only local produce at their recent wedding (held on the riverbank behind their home), listing the farm that produced each product in their program. The bride said: "It was time to walk the walk." Social preservationists also volunteer at the town's thrift shop, where they work alongside old-timers, whose rural New England frugality they admire: "The thrift shop is also just so wonderfully Yankee. The

FIGURE 4.04. Dresden Farmlands Sign

'use it up' or 'do without' [sensibility], you know. It's just kind of fun to [go there] and to have that as a common thread of community."

They especially enjoy the presence of the working-class old-timers who operate the transfer station (where residents bring refuse). At a Dresden Conservation Committee meeting, homesteaders and preservationists joked that Nate (one of the operators), whom many identify as an iconic old-timer, might be evoked to encourage recycling. One man said that he does not mind recycling because "I enjoy seeing Nate," to which a woman responded: "My husband seems to enjoy it, too." Laughing, the committee chair said: "We just need a poster that says, 'Meet Nate.'" In this way, newcomers use such venues to interact with old-timers.

Dresden preservationists' use of political abstinence is rooted in an appreciation for old-timers' character, a central component of which is a libertarian worldview, an ideology popular in northern New England. Maine demonstrated the breadth of this ideology in 1992 when Ross Perot "won 30.4 percent" of the state's vote, "more than then-President Bush" (McLaughlin 1995, 2), and again in 1994 with the election of an independent governor. Today, more Maine voters register as independents (technically "unenrolled") than as Democrats or Republicans. In fact, in 2004, unenrolled voters outnumbered Democrats by more than 68,000 and Republicans by more than 90,000. Dresden's voters mirror the statewide picture: in 2004, out of 1,217 voters, unenrolled voters outnumbered Democrats by 259 and Republicans by 183.[26]

As part of this libertarian ideology, old-timers and some newcomers who moved to Dresden from other Maine towns view the regulation of town life—from farming practices to building codes—as reprehensible and work to thwart their own displacement by *preventing* policy changes, rather than advocating for protective measures (e.g., affordable housing). For instance, an old-timer of much influence said: "I just don't like all the restrictions. . . . It seems like they get more stringent every year. People can't do nothing without permits and all that kind of stuff." Another, who serves on a town board believes that newcomers have increased regulation: "The people that used to live in this town are typically Maine people [who] mind their own damn business, but these people who are moving in from Massachusetts—the biggest part of them—once they get what they want, they want to change everybody else's freakin' regulations. . . . [They've] got it so you can't have a horse near a brook because the horse is going to eat the vegetation, and that's disturbing the wetland. Well, how damn ridiculous can you get?" An old-timer whose family has lived in town since its permanent settlement echoed this sentiment. She complained: "Some people want it to be a conservation town."

This distaste for town regulations affects town policy. At the 2004 town meeting, citizens narrowly voted to *remove* wetland protection from an aquifer that is the water source for several towns. At a 2003 meeting in Dresden's small meetinghouse, citizens debated whether to institute a committee to assist the planning board with ordinance evaluation. As the planning board chair penned in the town newsletter, "the tide of development has arrived in Dresden," and advocates argued that the board was overwhelmed with permit applications as many sought to build along the town's protected riverbanks. However, several old-timers worried that the new board would add unnecessary government oversight. One objector angrily said: "[I'm] not in favor of this in any way, shape, or form. . . . This is like a black dog over [the board's] shoulder." (In contrast, a newcomer suggested that the committee would serve as a "helping hand.") Despite early, enthusiastic support for the proposed board, old-timers, with the aid of social preservationists' abstinence, won a vote to floor the topic for a year, an outcome that puzzled and even infuriated some social homesteaders.

One old-timer's report of a planning board conflict suggests that at the root of this ideology is a concern for economic survival and a tradition of dependence on one's land:

> [A newcomer] said we better make an ordinance that a trailer has to have a cement slab under it. . . . People buy house trailers because they can't afford houses or foundations. When they finally get money, especially a young couple, they move the trailer and build a house where the trailer was. They don't want a cement slab to move. And what's it hurting? . . . He didn't have a good answer, so it didn't pass. . . . I got quite upset and told him . . . "What the hell. What do you care if a guy's got cement under his trailer? It's none of your freakin' business. You just sit here and want to pass something and make some poor guy who can't afford it do something he don't need."

In this way, old-timers evoke a libertarian ideology to ensure their continued presence in Dresden, and their aversion to regulation is evident at most town meetings.

Social preservationists side with old-timers, in part, because their appreciation for them rests on a sense of their character, an important component of which, in Dresden, is a libertarian sensibility: a resistance to government regulation and a commitment to personal freedom, particularly that which they relate to land use. One social preservationist said: "I just have always loved the . . . down-to-earth, pragmatic, no-nonsense, salt-of-the-earth personalities that I see as very Yankee." She recalled how old-timers' libertarianism explicitly shaped her stance on regulation, saying: "I remember vividly when someone . . . want[ed] to put in one-acre zoning. [A newcomer said], 'We've got to put in one-acre zoning, we've got to preserve Dresden.' And [an old-timer] said, 'Now let me get this straight, which Dresden are you trying to preserve: the one from 1830, when it was all farms and there wasn't a tree to be seen. Or was it the one when we had a dance hall? Or is it this one here when there was no people?' *He was definitely an old-timer.*" In this way, Dresden social preservationists define old-timers, in part, by their libertarian ideology, and their political abstinence mirrors this ideology. Despite the fact that some Dresden social preservationists worry about, for example, overdevelopment and runoff from illegal dumps, they consistently avoid voting against old-timers' interests and abstain from homesteaders' efforts to preserve the town's rural character through the conservation and comprehensive plan committees as well as the planning board.[27] In so doing, Dresden social preservationists demonstrate a preference for the preservation of rural *characters*, rather than of the rural landscape. If Dresden old-timers were to fight gentrification by *proposing* town ordinances and policy, social preservationists would likely be at the forefront of proactive strategies to prevent old-timers' displacement. In other words, Dresden social preservationists' strategy of political abstinence is tactically parallel to old-timers' ideological tradition and stands in opposition to homesteaders' use of regulations to conserve landscape.

It is possible that Dresden social preservationists will exchange their strategy of political abstinence for one of political intervention, should the character or pace of Dresden's gentrification shift. For now, despite rising property and homeownership costs and a significant reduction in the number of farms, the displacement of old-timers is rare. Only one of more than forty Dresden informants could, when asked, name anyone displaced by rising costs: "A few people were displaced—but by their own doing. Drugs, alcohol, etc." The most common response was: "If they can't afford it here, I can't think of where they *could* afford it." This partially explains preservationists' and old-timers' strategy of working to *avoid* town change, rather than utilizing preventative measures to halt displacement.

The belief that Dresden is as affordable as anywhere else may be rooted in the fact that its gentrification is at an earlier stage than is that of Lincoln County (in which it is located) overall. Between 1990 and 2000, both Dresden and Lincoln County experienced population growth and an increase in the number of residents with a bachelor's degree or higher. However, as of 1999, Lincoln County contained a greater proportion of households earning $100,000 or more than Dresden: 7.8%, compared to 4.2% in Dresden.[28] Similarly, in 2000, Dresden's median home value was $97,900, while Lincoln County's was $119,900. Today, Dresden properties listed over $250,000 are rare, while homes frequently sell for twice that in the county's coastal towns. Next door to Dresden, the rural town of Alna and the coastal village of Wiscasset have each broached the topic of affordable housing. In short, residents' belief that Dresden is affordable for old-timers is rooted, in part, in the fact that the town's level of gentrification is slightly below that of its county.

Several internal political-economic factors also discourage displacement and, therefore, interventionist social preservation practices. Often in communities with high rates of homeownership, gentrification hits the young, adult children of longtime residents hardest, while older residents may be sheltered from rising housing costs because they owned property prior to gentrification, as is the case in Provincetown (see Faiman-Silva 2004).[29] For several reasons, this is *not* true of Dresden. Dresden preservationists prefer residents who farm, are descended from farmers, or otherwise work the land. By definition, such residents are likely to own land that supports or supported their livelihood. This slows old-timers' physical displacement in two ways. First, land sales are an income source for those for whom rising property taxes and limited employment opportunities are financial hardships. Second, landowners may give (or sell affordably) land to family, allowing their children to remain in town despite rising homeownership costs.[30] For instance, when I asked an old-timer where her adult children live, she told me her son lives "in back": "We gave him a piece of property in back."

Finally, unlike many gentrified rural communities, Dresden does not restrict individual trailers.[31] Trailers, when combined with gifted land or inexpensive family property purchases, provide an affordable alternative to traditional homes.[32] In the words of a town official, in Dresden "affordable housing equals trailers," and trailers compose 24.4% of Dresden housing stock. Throughout town, single and double-wide trailers, in varying states of repair, stand beside antique Cape Cods and Victorian farmhouses.

Town and state tax policy also shapes Dresden social preservationists' strategies. A town official explained that "no one is forced out of town" for failure to pay property taxes, for "the town never takes property because of back taxes if there is a dwelling on it." However, the town reserves this policy for residents: "If there is property that no one's paid on, and they aren't around and can't be found, the town sells it for back taxes." Furthermore, the state of Maine has a "circuit breaker program" that provides tax relief for individuals or households under a certain income threshold.[33] Similarly, the town office maintains a small charity fund that provides small amounts of money and heating oil for those in need. Moreover, a Dresden foundation has a similar fund that distributes money to impoverished residents. While these programs do not alleviate the financial stress faced by many families—

after all, in 1999, 15.4% of the town's households made less than $14,999 —
they diminish the frequency of displacement.

Longtime residents' resistance to restrictions on trailers and their means
of livelihood (e.g., gravel pits), social preservationists' political abstinence,
and Dresden's relative affordability help prevent displacement and, therefore,
mitigate the need for proactive political practices. Longtime residents' commitment to self-sufficiency, arguably rooted in their libertarianism, also as-
suages preservationists' symbolic practices. As we have seen, Dresden farmers
organized and won support for the Dresden Farmlands designation without
preservationists' aid. In this and other ways, Dresden longtimers resist forg-
ing partnerships with newcomers. This, combined with the current stage and
character of gentrification within Dresden and homesteaders' use of zoning
changes to protect landscape, shapes social preservationists' strategies, focusing their efforts on the careful work of political abstinence.

Conclusion

As we have seen, social preservation takes a distinct form in each site. Specifically, preservationists employ unique political, symbolic, and private strategies to prevent gentrification and displacement. However, my findings reveal
that preservation strategies are not particular to rural or urban sites or to a
given stage of gentrification.

Instead, several interrelated factors influence such strategies: the stage
of gentrification within and around the site, the demographic and cultural
characteristics of gentrifiers and longtime residents, and local activist groups,
politicians, government agencies, and business associations. Social preservationists fashion strategies to assist with the application of their ideology to the
particular characteristics of their place of residence. In this way, their ideology
interacts with "place character" (Paulsen 2004), or what we might think of as
their "community toolkit," to shape strategies of action particular to their
site.[34] That is, actors have both a "cultural toolkit" (Swidler 1986) and a community toolkit with which to devise strategies of action. Each kit, in a sense,
determines which tools from the other kit one can use: certain cultural skills
enable a preservationist to draw upon particular community resources, and
vice versa. This serves as a reminder of both the power of ideology and the
"contexts that govern action" (Swidler 2001, 160; see also Wuthnow 1989).

Inadvertently, the examination of social preservation's variability has
uncovered how homesteaders' and pioneers' discourse and practices also
vary, ranging from Argyle pioneers' antipoverty rhetoric to Andersonville
homesteaders' devotion to local character. The stage of gentrification, in

part, determines the character of discourse and practices. However, this does not alone account for homesteaders' and pioneers' discourse or practices. Provincetown is as gentrified as Andersonville, yet discourse is nearly as virulent there as in Argyle, which is at a much earlier stage of gentrification. Long-time residents' ideology—both after and before gentrification—contributes to the frontier and salvation discourse, muting it in some places, and elevating it in others. Demographic variables, such as the proportion of old-timers to newcomers and the level of gentrification relative to a larger area, are also contributing factors.

This chapter demonstrates the importance of comparative sites for identifying causal mechanisms. Without the diversity of sites included herein, our image of social preservation might be limited to political abstinence or paternalistic advocacy. It also reveals that preservationists' diverse strategies are a product, not of ideological variation, but, rather, of differences in local context. This is not to suggest that ideology or culture is less powerful than previously argued. After all, the chapter reminds us of the constancy of the preservation ideology and of the practices and sacrifices that it engenders. In so doing, it underlines the coherence of social preservationists' ideology *and* encourages consideration of how place characteristics mediate *expressions* of belief.[35]

While we have seen that social preservationists' practices vary by place of residence, the question of whether social preservation exists in all locales has not been directly addressed. While its presence in four very distinct sites suggests that the process is widespread—existing in small towns and urban neighborhoods, in two different regions of the United States, and in communities at different stages of gentrification—there may be contexts in which it is less likely to occur.

For instance, a recent *New York Times* article profiled Bentonville, Arkansas, a Bible Belt town. Bentonville, the home of Wal-Mart's headquarters, has recently experienced a substantial increase in highly educated, affluent families. Among newcomers are Jewish Wal-Mart employees, who have established a synagogue and Jewish-support networks, educated their neighbors about their faith, and changed school policies that prioritized Christianity. Rather than preserve the culture and presence of Bentonville's old-timers, Jewish newcomers labor to preserve their *own* culture. The author of the article writes: "The Jews of Benton County say they have become more observant in—and protective of—their faith than ever before" (Barbaro 2006, 1).

This anecdote underlines the fact that social preservation is predicated on the sense that one's own culture is threaten*ing* rather than threaten*ed*. It also suggests that residence in a place relatively inhospitable to one's social group may encourage self-preservation, rather than social preservation. For instance, Bentonville's Jewish newcomers have "become eager spokesmen and women for" their religion and culture: "[The Rosens] . . . moved . . . from Chicago in 2000 after Mr. Rosen was offered a job in Wal-Mart's technology department. The family did not attend a synagogue in Chicago because, Mrs. Rosen said, 'you didn't need a synagogue to have a Jewish identity.' There were Jewish neighbors, Jewish friends, Jewish family. But not in Bentonville, where her daughter brought home from day care a picture of Jesus to color in. Suddenly, *a synagogue did not seem like a luxury anymore, but a necessity to preserve her family's Jewish heritage*" (Barbaro 2006, 1 [emphasis added]). Thus, context shapes newcomers' relationship to their own identity and, in turn, to others' cultures and identities as well. In a context in which a newcomer must elbow his or her way into a community, social preservation may be less likely to occur. This suggests that context and newcomers' identity conspire to shape, not only preservation strategies, but perhaps also the likelihood of social preservation itself.

The Real People:
Selecting the Authentic Old-Timer

On a summer afternoon in 2005, I entered a small grocery in southern Andersonville owned by an Iranian couple.[1] This was not my first time in the
store. Once or twice a week, I would take a break from writing and walk to
the store for a soda. I sometimes heard local gossip as I walked between racks
of candy and chips to the beverage cooler or while standing in line beside a
bin of pistachios, my eyes tracing the covers of Iranian cassettes behind the
counter. On this hot July day, I opened the shop door to the trembling voice
of one of the owners as she spoke to a customer about relocating the store.
She explained that she and her husband would have to close their business
because, without warning, the building owner sold the space and the new
owner raised their monthly rent "from $2,200 to $6,000." She decried the
increase with much feeling, saying, "I have to pay my mortgage!" and, "He
wants us out!"

After the other customer left, I handed the owner money for a soda and
asked whether she was OK. Without pause, she said: "I have had three, four
lives. Because of the war with Iran and Iraq, I lost everything. I had to leave. I
never saw my house again. But I was OK. I was strong. I came here. But this
is worse because where are my human rights? My human rights!" She asked:
"What will I do without my people? . . . They are cutting my root! My root!"

While she believes that she and her husband helped improve the neighborhood by opening a business and supporting troubled young people when
Andersonville was—in her estimation—dangerous, she also felt that some
improvements were enacted *against* her. Andersonville's streetscape, the
lengthy construction of which was detrimental to some businesses, incensed
her. She expressed bitterness that, having absorbed losses during construction, she would not benefit from the finished product and other beautifica-

tions: "They come, they put in flowers [for the streetscape] and condominiums, and then I have to go!" While she did not say so, the Swedish theme of
the streetscape may have perturbed her and other Middle Easterners whose
establishments have closed. Certainly, it may explain why her store, rather
than a Swedish equivalent, closed. A few newcomers told me that they were
sorry to see the store and its owners leave. They also expressed dismay when a
high-end frame shop (the neighborhood's fourth) replaced it. However, none
suggested that their loss threatened Andersonville's character. For most Andersonville social preservationists, that character is Swedish, and the Iranian
store fell outside the scope of their primary efforts. While the store may have
closed even if preservationists had rallied behind the owners, the fact that
they were not Swedish was likely of consequence.

In the store's final month, I often spoke with the owners. While taking
my dollar for a sundry item, they provided updates, but the wife had become
shy in the manner of one who fears she shared too much with a stranger. I
was thus unprepared for their closing: one day the store was there, shelves
stocked, and the next it was gone. Two blocks away, tables were full at Swedish restaurants, and customers waited in line at Wikstrom's Deli and the
Swedish Bakery.[2]

The Iranian couple provides a miniature of a larger pattern: the displacement of residents whom preservationists do not regard as old-timers and
of whose dislocation they are relatively unaware. These include young gays
and lesbians unable to secure affordable housing in Provincetown, Argyle's
working-class and poor African Americans, residents of Chicago halfway
houses, and middle-class Russians who only a few decades ago disbanded
their Dresden settlement. Preservationists are not entirely indifferent to such
displacement. Occasionally, they publicly defend the presence of those they
do not regard as old-timers. For instance, they counter pioneers' calls to rid
Argyle of gangbangers and the mentally ill. However, social preservationists
do not regard such groups as essential to local character or as having equal
claims to authentic community and, therefore, rarely advocate on their behalf
unless prompted. They work to preserve some groups and not others. Thus,
even under social preservationists' watchful gaze, not only are some longtime
residents displaced, but they also leave without the benefit of public outcry
that old-timers' moving vans and empty storefronts generate.

Sometimes preservationists overlook those we might imagine they would
celebrate, like descendants of Provincetown's WASP whaling captains. For
instance, in the 1990s, a successful Provincetown artist and a school administrator met to establish a "cultural sanctuary" to preserve "a distinct mix
of people and place, a peculiar cultural ecology." They wished to prevent

the displacement of residents whom gentrification made vulnerable and the related "destruction" of local culture. For them, the town's culture, like its dunes, was a threatened resource. However, implicit in their words about "a distinct mix of people" was concern for specific groups, namely, Portuguese fishermen and struggling artists; to preserve culture was to preserve *certain* residents.

Despite their heightened self-reflexivity about gentrification, social preservationists are chiefly unaware that they select "preservable" old-timers from a pool of residents. They are also largely unconscious of the fact that *old-timer* is not a fixed category and that they engage in the construction of old-timers by emphasizing specific groups and certain of their traits. For instance, they celebrate a fisherman or a farmer when he fishes or farms, *not* when he watches television, shops at a mall, or engages in other activities that preservationists also engage in.

This chapter examines four questions about preservationists' selection process. First, which longtime residents do they seek to preserve? Specifically, whom do they believe has the greatest claim to authentic community and, therefore, to a place? Second, what beliefs, stereotypes, and political-economic factors influence their association of a place with a particular group? Third, of what consequence are notions of authenticity for authenticators? For preservationists, what are the effects of associating community with traits they do not share? Finally, if definitions of authenticity affect preservationists—those with the privilege to select, not only their own identity, but also, to an extent, that of their place of residence—what does this say about the power of authenticity and of ideology?[3]

In answering the last two questions, we learn more about preservationists and how they think of themselves than about old-timers. As a result, this chapter departs from most studies of authenticity, which often attend to the consequences of definitions of *authenticity* for the authenticated, rather than for authenticators. Later in the book, I discuss the costs of social preservationists' construction of authenticity for those they do and those they do not aim to preserve, but this chapter primarily explores their costs for authenticators. By documenting the qualities that social preservationists admire in old-timers and that they believe they do not themselves possess, it lays bare how they think about themselves, especially about their relationship to community.[4] In this sense, it calls us to recognize that how people define the authentic both reflects and shapes their ability to recognize such qualities in themselves.

The chapter also departs from traditional ways of thinking about the social construction of authenticity. While social preservationists construct the authentic, they do so by emphasizing certain groups and traits—for instance, by turning their attention to Portuguese fishermen rather than to WASP whaling captains' descendants and, secondarily, by celebrating an Andersonville Swede's ethnic pride and not his homophobia—rather than through the strict process of "fabrication" (Peterson 1997) or "invention" (Hobsbawm and Ranger 1983), to which contemporary authenticity theorists are so attentive. They do not construct the authentic old-timer and her community from scratch but instead whittle existing characteristics of people and place until they complement their preconceived notions of authentic people, place, and community. Put differently, like all social actors, they construct authenticity by evaluating and assigning meaning to people, places, and objects, but—perhaps because they are so attentive to old-timers' perspectives—they do not simply impose meaning onto silent or passive old-timers, nor do they ignore preexisting local definitions of authenticity.[5] They engage in "talk, interaction, and negotiation" (Holstein and Miller 1993, 139) and borrow from local and translocal signifiers to define the "real" old-timer and her community. This muddies the distinction between that which they imagine and that which others have already imagined or constructed for them. For instance, they do not seek to present Dresden old-timers as fishermen, despite many Maine preservationists' admiration for New England's seafaring tradition. Instead, they celebrate potato and strawberry farmers but describe their lives as somewhat more insular than they in fact are. Similarly, Provincetown preservationists celebrate those old-timers who are fluent in Portuguese or who fish but do not present or even imagine all Provincetown old-timers as such. Rather, they engage in the collective and individual work of "interpreting" characteristics of place and person, and this chapter highlights their "patterned interpretations" within and across my sites (Holstein and Miller 1993, 133).[6]

Having outlined these key departures, in the next section I discuss preservationists' selection of those they regard as the "real people" or the true tenants of a place, those whom they believe embody independence, tradition, and a close relationship to place—traits that they believe are beyond their own purview and that they admire in others. Next, I document what I term social preservationists' *commitment to virtuous marginality* and examine the consequences of associating community with those unlike oneself. I conclude by discussing the implications of social preservation for understandings of authenticity and community.

The Real People

On the first Saturday in March, Provincetown residents gather at their town hall, a white clapboard building in the town center, for the Year-Rounders'

FIGURE 5.01. Provincetown Year-Rounders' Festival

Festival. The event includes informational booths, dinner, a variety show, and a navy band. In 2002, during the variety show, an amply sized drag queen, Isadora with More and More-ah, took to the stage. The audience was notably different than Isadora's summer audiences at a local bar. Children ran between chairs. Grandparents held babies. Neighbors, dressed in work shirts and jeans, sat beside each other.

Isadora, wearing a blonde wig and a flowing pink gown, prefaced her performance by saying: "This is my seventeenth year in Provincetown." The audience met her announcement with much applause. "Some of us," Isadora said, "come across the bridge [onto Cape Cod] and never leave." The crowd remained quiet. "So . . . seventeen years," Isadora spoke carefully, "am I a townie yet?" The auditorium resounded with silence. Finally, a few residents replied yes, but a collective no countered their affirmation. "I'm not?" Isadora asked, her tone unsurprised. "Well, then, how long do I have to be here to be a townie?" This time the audience agreed, a chorus of "Forever!" rang throughout the hall. "Well, you better get used to it," Isadora said before breaking into song, "because I'm going to be."

While it is not altogether surprising that Provincetown's Portuguese-Catholic lifelong residents do not consider Isadora a true townie, it may surprise some that social preservationists—many of whom are gay or lesbian—concur and would never bestow old-timer status on lifelong residents who share their traits. For instance, seated at a patio table, T-shirt sleeves rolled to catch the July sun, Amy—a young lesbian social preservationist—was eager to share her observations of the rules that govern who can be considered a townie: "I meet these old lesbians that [a]re like in their sixties. . . . They've lived [here] for years and years and years and years, but there's always this thing about who's a townie, who's a native, who lives here, and who doesn't. . . . [They've] lived here for twenty years, they're still not a native." Amy recognizes that she abides by unspoken rules about "who's a townie, who's a native," and that ethnicity, class, and occupation figure into her identification of residents. In fact, in her mind, these rules are so strict that they shape her ability to accurately recognize residents' traits:

> I always thought [this guy] was this old Portuguese fisherman because he has really brown and leathery skin and he's always lived in P-town . . . but he's not. He's black. I mean, he's African American. . . . Somebody said, "He doesn't want anyone to know he's black." And I'm like, "He's not black, he's *Portuguese*." And then [the man we were discussing] looked at me like I fell off the turnip truck. I'm like, "Wait, you're black?" He's like, everybody's like, "Look at him! Look at him! How can you not know?" And I'm like, "I just thought he was brown from the sun. I just thought he was old and weathered." And they were like, "What?!" Because it's just your assumptions. You don't assume that any African American person lives in Provincetown or is a *true* townie because *everybody* here is Portuguese.

Thus, Amy realized that she did not have a place for an African American fisherman in her conception of a townie, nor could she, despite her own sexual identity, regard a longtime lesbian resident as an old-timer. And, while she acknowledges that she excludes such individuals from the *old-timer* category, she remains unaware that the Portuguese compose only a minority of the town's population (roughly 23% at the 2000 Census). That is, for Amy, Provincetown's place identity is so linked to the Portuguese that she dramatically overestimates their presence, even misidentifying an African American longtime resident as Portuguese.

Alternately, some preservationists recognize the presence of non-old-timers but believe that one group has an indisputable claim to place. For instance, over dinner in a Mexican restaurant, a social worker acknowledged that Argyle contains many groups but suggested that Asians are its bedrock: "On the corner, you'll meet people from Croatia, from Russia, from Pakistan, from anyplace in the world . . . , [but] the underlying thing is the Asian community."

In one way or another, in each site preservationists select old-timers from a pool of actors, but they do not always work to preserve those who constitute the majority or those with the longest residential history. In Andersonville, they overlook Middle Easterners and lesbians in favor of Swedes despite the fact that they compose less than 4% of the local population. In Argyle, they seek to preserve the Vietnamese but neglect longtime African American residents who constitute, 22% of Argyle's population.

Social preservationists are not alone in this. A variety of residents collectively engage in the public definition of the *real* resident. Certain events, such as public festivals and celebrations, provide opportunity for the construction or affirmation of such definitions. On the evening of Isadora's performance at the Provincetown Year-Rounders' Festival, a social preservationist proudly affirmed Portuguese performers' old-timer status, introducing one as "a graduate of Provincetown High School" and another as "a native son." The audience met such performances with warm applause. Later, a graying, bearded resident—a newcomer wearing a tapestry knotted under his neck and a matching cap—performed his composition "Prelude to the Butterfly." In the course of his performance, a middle-aged Portuguese man turned to the woman beside him to say: "I can't believe this is our town. There's some guy up there acting like a butterfly and blowing on a flute! Can you believe this is our town?" Speaking loudly enough for others to hear, the man discursively separated the flutist from "our town." Thus, social preservationists are not the only residents who differentiate between old-timers and others.

Notably, not all longtime residents call for their own preservation. For instance, at an Argyle meeting, residents sought to decide which local groups a new mural should represent. Bob, a white, middle-aged resident whose Swedish family has lived in Argyle for generations, argued for the inclusion of Appalachians and Asians. He said that the "diversity in [Argyle] was amazing" when Appalachians lived there and as Chinese and Vietnamese arrived. He did not call for the inclusion of his own group, associating them instead with neighboring Andersonville.

However, when pressed, he expressed strong views about which longtime residents the mural should *not* represent. The leader of a nonprofit funding the mural (a woman who espouses leftist beliefs but *not* components of the social preservation ideology) said: "I want to get back to ethnicity because that's what the mural is about." Referring to the neighborhood's single-residency-occupancy hotels, she asked: "When is the heaviest time when they declassified the mentally ill?" The longtime resident responded incredulously:

FIGURE 5.02. Argyle Mural

"That's an ethnic group?" And it is likely that some of the social preservation-ists present concurred with his view. The administrator retorted: "Ethnic or not, they've had a big impact." Later, she pressed him to consider another group for inclusion, saying: "Another group is gangs in the '60s, '70s, '80s." This seemed to offend the longtime resident even more than including the mentally ill: "We don't want them in the mural!" The woman pressed ahead: "But [they] helped the affordability." In this sense, even those longtime resi-dents whom preservationists do not consider old-timers (e.g., Argyle Swedes) participate in the selection of groups for preservation and do not necessarily advocate for themselves.

These anecdotes reveal that preservationists borrow from a local logic about the real old-timer (in large part because their definitions are place based) that old-timers and other longtime residents also share. This may par-tially explain why it is rare for those not included in the *old-timer* category to complain about their exclusion. After all, the Iranian store owner we met in this chapter's introduction did not suggest that Andersonville's streetscape should have acknowledged Middle Easterners' presence and legacy, nor did a young, gay Provincetown artist—on the afternoon of his moving sale—call for his own preservation.

While preservationists borrow from local definitions in their selection of the old-timer, they also draw from two sets of beliefs. First, they express nostalgia for "traditional" community rooted in fear of the constant evolution of space, particularly, its transformation by gentrification.[7] While reaction to change also fuels historic preservation (see Lowenthal 1999, 394–95; and Francaviglia 2000, 68), social preservation is not an effort to return to an earlier era.[8] Rather, it is an effort to prevent change, to freeze a place before gentrification alters it. For preservationists, this longing arises, in part, from nostalgia for their lived experience of community (Davis 1979, 8) as well as for a form of community that none has experienced because it never existed (see Coontz 1992; Griswold 2000, 133; and Suttles 1972, 187). This opposition to change encourages the preservation of those they associate with ungentrified space.

Second, preservationists associate authentic community with certain others to whom they assign social value (see Walton 2001).[9] In analyzing their association of community with certain old-timers, this chapter builds on Herman Schmalenbach's theory of community recognition, which suggests that, rather than noticing our own attachments, we are more likely to see *others'* communities: "Only through contrasts and disturbances does a community become an object of attention" (Schmalenbach 1922/1961, 334; see also Erikson 1976, 187). This explains why preservationists notice, seek proximity to, and aim to preserve communities of greatest contrast to their own. Their association of authenticity with others contributes to their belief that real community is that of which they are *not* a part. However, Schmalenbach does not help us explain why they do not associate community with *all* others. Consequently, we must identify how beliefs about racial, ethnic, economic, and sexual identity groups' claims to real community influence the preservation of *certain* residents.

Despite the unifying ideologies that I have noted (nostalgia for real community and an association of such with certain others), preservationists' selection of old-timers is place specific, rooted in each site's traits and local history. However, there are notable patterns that extend from Chicago to Maine. They seek old-timers with qualities they associate with real community and that they believe they do not personally possess: independence, tradition, and a place-based way of life. Table 8 lists those whom preservationists aim to preserve and those they do not.

In the rest of this section, I identify the specific traits that preservationists associate with strong community and believe are absent from their own biographies and networks. I also demonstrate how they construct old-timers as actors worthy of preservation by framing such traits in positive terms. After all, they do not, as some might, regard Yankee independence as mean or detrimental to the environment, nor do they treat tradition as an indication

TABLE 8. Social preservationists' selection of old-timers from among longtime residents

Andersonville	Argyle	Dresden	Provincetown
Middle Easterners	African Americans	Russian Americans	WASPs
Lesbians	Appalachians	Factory workers	Descendants of Pilgrims
Middle and upper-middle-class longtime residents	**Asian immigrants** (Chinese, Cambodian, Laotian, Korean, Thai, **Vietnamese**)	State employees **Farmers and their families**	Gays with history of residence Non-Portuguese working class
Swedish-Americans		**Small-business owners**	
Independent business owners	SRO residents	(who perform manual labor)	**Portuguese fishermen**
	Ukrainian Jews		**Struggling artists**
SRO residents (who were displaced during period of study)	Descendants of theater district and movie studio past		
	A handful of Swedish longtime residents		
	Latino residents		

Note: Selected populations are given in boldface.

of backwardness. Finally, I note how institutions and other actors shape preservationists' choices.

INDEPENDENCE

Social preservationists appreciate those who appear autonomous: those who seem to live a simple life and depend on their own labor, land, bodies, families, and local networks. They believe that, in the city, merchants best embody this. In the small town, they turn to those who depend on their body or the land for labor: fishermen, farmers, and struggling artists. In this, social preservationists borrow from the long-standing popular association of self-sufficiency with the farmer, entrepreneur, and artisan and, in turn, with an "idealized ethos of the township" (Bellah et al. 1985, 40), departing from this only in their distaste for those whose labor has led to wealth.[10] Most preservationists negotiate networks that extend far beyond their place of residence and read old-timers' permanency as evidence of an admirable independence and self-reliance as an indication of rootedness.

Chicago preservationists celebrate old-timers who own small, moderately successful independent firms. The prototypical old-timer, according to Argyle preservationists (who are themselves a fairly diverse lot, including, among others, professors, administrative assistants, and social workers), owns an ethnic restaurant, bakery, or butchery: "struggling new citizens in America [who] have their own little ethnic businesses." A preservationist explained his resistance to gentrifiers' efforts to "improve" Argyle, arguing that Vietnamese merchants are the heart of the community: "It's really Vietnamese. . . . Those merchants [are] the people who saved this community, without a question."[11] Fighting with other gentrifiers about the risks of upscaling for old-timers, a silver-haired social service director described Asian proprietors as the "goose that laid [Argyle's] golden egg." Preservationists admire concerns that employ family members: where a daughter waitresses or brothers cut hair side by side. At a dinner party, a leftist twenty-five-year-old praised an Argyle eatery: "The best part was that we were served by an eight-year-old girl, the owner's daughter. I mean, I'm all against child labor, but there's something about being served by an eight-year-old girl."[12]

At a meeting organized to block a high-end development, a thirty-something Andersonville leader said: "The thing that strikes me the most [about Andersonville's history] is how fundamentally [it's] about the small business community." She described its evolution from a place of farms to a Swedish neighborhood marked by family businesses, signifying how business owners' independence provides Andersonville with its own autonomy or uniqueness. Another resident said: "After the [older owners] leave, who knows what's going to happen? . . . The landlords are probably licking their chops. . . . What has always been really unique and wonderful . . . is the mom-and-pops, the individuality. . . . Now, with Starbucks, Einstein's, and Mail Boxes Etc., it's starting to look like Main Street anywhere" (Larson 1998).[13] Another preservationist agrees: "[The] key element that makes the neighborhood is small, independent businesses." In this sense, preservationists simultaneously appreciate old-timers' self-sufficiency and the sense of independence and distinction from chain stores that they lend space.

Small-town preservationists also venerate small business owners, such as those who own Dresden farm stands, which are considered a locus of community. However, they prefer those whose independence seems to originate in *dependence* on their body or land. As Sarah—a Dresden Republican (briefly mentioned in the previous chapter) who is married to a lawyer and who relies on others to mow her fields—said: "I have a real fondness in my heart for the people who live off the land." This fondness is rooted in the belief that certain life attitudes—in this case libertarianism, tied as it is to a sense of independence—affix to such work. Sarah explained: "I've always loved the down-to-earth, pragmatic, no-nonsense, salt-of-the-earth personalities I see as very Yankee . . . people who live their own lives and don't try to live [others'] lives for them. I'm really very fond of that, and it's always very endearing. They can be frustrating . . . because they're very opinionated, but they're just salt of the earth. Just very *live and let live.*"[14]

Preservationists extend "salt-of-the-earth" status to some who do not live off their land per se. For instance, in one breath, a young Dresden stay-at-home mom and ardent churchgoer who recently employed an old-timer to dig a pond in her yard said: "There are a lot of families who had this land in their family for three hundred or two hundred years, and they're making a living off of their land, cutting wood, plowing driveways." In this mode, preservationists categorize fishing, worm digging, plowing, and even painting with farming. They admire manual laborers: carpenters, volunteer firefighters, and day laborers, self-employed men who depend on their bodies and who, in contrast to nearly all preservationists, do not appear to rely on others' manual labor.[15] For instance, the stay-at-home mom admires a resident who backhoed a pond in her yard, and a Provincetown preservationist speaks fondly of "the people who do the excavating and the people who are the septic people . . . *you know, the septic people,* it's been in their family for a million years."

Similarly, Provincetown preservationists see parallels between artists and laborers primarily because they work with their hands and face financial uncertainty, two measures of independence and qualities that few preservationists themselves possess.[16] They express dual appreciation for artists and working-class Portuguese and believe that both are endangered: "Artists and working people. I don't want this to be a place of just retirees or vacationers or people involved in the tourism industry." An artist worries that "'the inflated real estate market could well drive out the fishermen and artists'" (quoted in Desroches 2003b) and uses his art to "call attention to the transformation of the community from a laid-back fishing village and art community to a high-priced resort" ("Jay Critchley" 2002). Similarly, at a heated affordable-housing hearing in the town hall, a Provincetown woman argued that, by supporting affordable housing, she feels "not so much that I am doing something for someone else, but for myself. I want a diverse community. . . . I want poets, writers, carpenters—*people who do things with their hands.* Not a chichi community where people just come down for the weekend."

In 2004, members of Provincetown's arts council reported their recent activities to the board of selectmen. The conversation revealed the logic that associates artists with fishermen because of their common financial struggle. A council member explained: "The poet laureate idea came up to help make poets and artists more visible in town. They're a cultural resource that is integral." In response, someone asked: "How do we define what a Provincetown artist is? What about people who live in [neighboring towns] because they can't afford to live here but still consider themselves Provincetown artists?" A selectwoman called out: "Like fishermen!" And to this the town manager responded: "Right." In this way, preservationists not only separate artists and anglers from others but also engage in the construction of what an artist or a fisherman *is*, deciding that, given Provincetown's advanced stage of gentrification, both categories may include those who have relocated to neighboring towns in search of affordable housing and that artists and fishermen are more alike than different because financial struggle and a related independence mark both livelihoods.

Of course, very few social preservationists work with their hands. Implicit in an appreciation for the fisherman, the farmer, the artist, and the entrepreneur is the suggestion of independence *from* the very resources on which preservationists rely: global financial and information networks, institutional support, economic and cultural capital, and even old-timers. The form of independence varies across the sites but is salient in each. In all, preservations borrow from a long-standing appreciation for the self-made, self-reliant individual, departing from this only in their *dis*taste for financial success.[17]

Farmers, fishermen, the owners of Vietnamese restaurants, and other self-employed old-timers embody a locally based form of self-employment, seemingly detached from global networks of capital and information. For social preservationists, who are by definition from "away" and reliant on "commuter friendships" and often on national professional networks, old-timers' (seeming) freedom from such networks is indicative of a particularly attractive form of independence, one they view as more organic than her own lifestyle.[18] For instance, an Ivy League–educated preservationist worries that gentrification leads old-timers to lose an independent lifestyle that she views as more organic than her own. In fact, preservationists frequently highlight the distinction between their privilege, particularly that experienced as children, and old-timers' economic struggle. A lawyer whose father was an insurance executive said: "I never really had to want for much. It just created a sense of responsibility for me, being fortunate." A Provincetown businessman who lives in a home with water views says that the contrast between his affluent suburban hometown and the cities in which he lived as a young adult drew him to old-timers: "[Housing issues] were something I came to care about the more I was exposed to what the problems were outside my suburban upbringing." A woman from an upper-middle-class family argued that recognizing the contrast between one's own "good fortune" and the plight of the displaced encourages criticism of gentrification. Similarly, a language teacher who co-owns several Provincetown properties said: "I'm one of the lucky ones." She recalled that this sense of luck drove her to rent her property affordably.

Preservationists criticize old-timers whose economic lives resemble their own. An Ivy League graduate complained that gentrification leads old-timers to become reliant on luxuries that she herself enjoys: "This was a hardy lot, and increasingly, if they don't have their jacuzzi and their central heating, they're not here." Provincetown preservationists do not honor employees of large, commercial fishing vessels, preferring family-owned and -operated boats. Dresden newcomers celebrate old-timers who farm rather than employees of Bath Iron Works, a producer of military ships. They appreciate those who seem to embody the rural life, a life distinct from the city ways many preservationists know well. In this, they borrow from the venerable Jeffersonian romanticization of the rural and those who work the land. Fiscal struggle and local networks affirm preservationists' sense of old-timers' lives as those of spiritual rather than economic importance (Schmitt 1990, xix).

In fact, economic success and the extralocal networks to which it is tied can remove one from the *old-timer* category. For instance, Argyle preservationists express distaste for capital flows that stretch beyond the neighborhood, even those internal to Chicago, preferring Vietnamese to Chinese shops opened as part of an entrepreneur's effort to create a "Chinatown North." Preservationists eschew this marketing by referring to the neighborhood as *Vietnamtown* or *Argyle* rather than *Chinatown North* or the city-designated *Asian Marketplace*. One said: "Chinese businessmen are the biggest capitalists around. . . . Gentrifiers are probably something [they] aspire to. They would really like to be gentrifiers themselves." In Dresden, that a Russian immigrant's real estate venture led to the in-migration of other Russians may explain preservationists' disinterest in Russian roots. Their association with a profitmaking venture suggests a relationship to capital more like preservationists' than that of the "self-made" farmer. In this mode, preservationists rebuke those who use gentrification to realize mobility and celebrate the Portuguese fisherman or Dresden farmer when farming or fishing, not when sitting in a jacuzzi or driving a high-end car.[19] Again, this suggests that they not only choose groups for preservation but also reward individuals for maintaining certain traits.

Likewise, Provincetown preservationists prefer artists detached from institutional support or financial success. Leslie, an affordable-housing advocate, reminisced that, when she moved to town, "everybody was a marginalized person . . . that was a wonderful thing" and criticized successful artists who spoke against affordable housing: "It was just so appalling. Some of the people who got up and spoke against us were artists, longtime artists who had gotten theirs." Another distinguished between art institutions and "real artists": "'The irony is that the art institutions in town are doing fantastically. . . . But the number of artists living and working here has declined since few can afford it'" ("Jay Critchley" 2002). He explained: "If you succeed as an artist, you are almost selling out. . . . There's the whole thing about being a struggling artist. . . . Ambition is looked at critically." A language teacher argued: "People who have money in their pockets [replace artists]. . . . It's like cutting down a tree and cutting so many branches that the trunk is . . . exposed." Fueled by such sentiment, these Provincetown preservationists support "living artists" through residencies, grants, and affordable housing and by purchasing their work.

Even the suggestion of a history of privilege detracts from a resident's independence. Following a dominant American theme, preservationists value hard work over the status that goes with inheritance.[20] Provincetown preservationists ignore those one might consider the prototypical custodians of New England ports: the descendants of Pilgrims and WASP whaling captains. Even though some such descendants struggle financially, in preservationists' eyes the suggestion of a legacy of privilege (imagined or real) detracts from perceived independence. This is telling given that many preservationists are themselves associated with such privilege.

For preservationists, economic struggle is a marker of independence, despite concern that it will lead to displacement. Workers' financial struggle seems to indicate a life apart from the preservationists' realm. For instance, an alumna of an elite college defines old-timers as those who "don't eat at all those fancy restaurants": "They're living in these side streets, not a brand new condo. . . . [They] spent a lot of time in unemployment." Similarly, a Dresden man who owns a large riverbank home described old-timers as "pretty self-sufficient": "They all have their own ideas. They have several jobs instead of just one. . . . They're very *real,* I guess."

In this vein, old-timers' economic struggle and their independence from newfound wealth and economic institutions that might shelter them simultaneously suggest independence and a mutual, in-group *dependence* that preservationists regard as indicative of strong community. By selecting those *without* outside ties, preservationists assure themselves that old-timers are at once independent (from the outside world) and mutually *dependent* (within their group) and, for these reasons, possess a strong community. Embedded in this admiration for independence is displeasure with their own reliance on extralocal, and even global, financial and social ties. Perhaps Wordsworth put it best in "Resolution and Independence," in which he pairs fondness for "Him who walked in glory and in joy / Following his plough, along the mountain-side" and for the leech digger's "Employment hazardous and wearisome! / And he had many hardships to endure" with criticism of his own lifestyle:

As if life's business were a summer mood:
As if all needful things would come unsought
To genial faith, still rich in genial good:
But how can He expect that others should
Build for him, sow for him, and at his call
Love him, who for himself will take no heed at all?
(WORDSWORTH 2000, 261–63)

In their taste for independence—for those who they believe depend on their own labor, land, bodies, families, and local community for survival—preservationists borrow and depart from popular veneration of those who achieve success through independence, ingenuity, and hard work.[21] Like many, they romanticize the struggling merchant, the undiscovered artist, and the yeoman farmer. However, they dismiss those who achieve success because their appreciation for independence is rooted in esteem for those *unlike* them and for the mutual in-group *dependence* required of those who face financial struggle, which they take as an indicator of strong ties. In fact, their appreciation for the independent old-timer comes with the implicit stipulation that they balance this with other claims to community, namely, tradition and a strong relationship to place.

TRADITION

For social preservationists, tradition weds the independent old-timer to a particular place, a particular culture, and enduring ties. Preservationists are attracted to those who have traditionally resided in a locale, who have publicly visible traditions, and rely on customary labor practices and extended family. Essentially, they value old-timers' membership in a subculture of those who "share a defining trait, associate with one another, are members of institutions associated with their defining trait, adhere to a distinct set of values, share a set of cultural tools . . . and take part in a common way of life" (Fischer 1995, 544; see also Parsons 1951). Yet they implicitly appreciate and recognize *certain* traditions: distinct food and language, community centers, symbols, and public celebrations. For this reason, local institutions and old-timers' self-preservation efforts combine with social preservationists' taste for traditions they associate with real community to shape preservationists' admiration for certain groups.

Preservationists prefer those who seem to have a special claim to local history and culture. As an Argyle preservationist said of himself: "[I] feel more comfortable in a place that has a sense of history." Preservationists derive this "sense of history" from living people and their continuation of traditional practices, such as speaking a non-English language and eating traditional foods. For instance, over bagels, a rabbi who, as we have seen, argues that a kosher deli does *not* belong in Argyle advocated for low-income housing for Argyle's elderly Asian residents so that they would be near their language and food: "so they can enjoy the community where they have the local flavor and shopping and speak their own language." He elaborated: "I've heard people say, 'There's too many Asians here.' That's ridiculous. . . . If you're Asian and you go to the grocer, all the vegetables and products you need for your style of food are readily available, and the owner is attentive to what you need. There's nothing wrong with that. I enjoy that very much. I think it's a beautiful street." Similarly, a young journalist wrote: "The food at 'the Pot' isn't just seafood, it's the result of the hardworking fishermen that keep P'town grounded in its history" (Desroches 2003a).

In contrast, social preservationists rarely express a connection to the heritage of their place of origin. For instance, a man from a Boston suburb characterized his hometown as "suburban imitation." For this reason, preservationists enjoy witnessing the continuation of old-timers' traditions, which help connect them to the local past (see Griswold and Wright 2004, 1444). Thus, at a Provincetown Historical Commission meeting, a gay newcomer said: "[I'm] in support of fishermen, in support of a valuable part of *our heritage*. [We should avoid] making it unaffordable for a valuable part of *our heritage* to be here."

Of course, no group is solely responsible for local heritage. The institutionalization of particular groups' presence promotes and preserves a particular rendering of local history and shapes preservationists' sense of which group has the greatest claim to local tradition. For instance, Andersonville's Swedish American Museum serves as an unofficial community center. Other ethnic groups do not have such a museum, and the presence of the museum and its public events partially account for preservationists' understanding of Andersonville as Swedish. Similarly, in Dresden, the only historical sites are that of the Brick School House Museum and a colonial courthouse, each pointing to a frontier history. Besides a rusted sign for a Slavophil Society, no other ethnic groups have institutionalized historical presence. In Provincetown, the former Heritage Museum focused on the Portuguese, containing fishing artifacts. However, the town is also home to the Provincetown Art Association and the Pilgrim Monument. These museums symbolize three pillars of the local past, but they do not alone explain preservationists' attention to artists and fishermen and lack of interest in Pilgrims' descendants, gays and lesbians, and Jamaican laborers.

Why do social preservationists align with groups associated with some

FIGURE 5.03. Procession to the Harbor, Provincetown

institutions and not others? Two cultural factors—the popular association of certain groups with particular places and nostalgia for certain historic moments—explain what institutions do not. For instance, attention to Andersonville Swedes is consistent with Chicago's popular image as a city of Western European immigrants. Their move to Andersonville after the Great Fire embodies a vivid moment in Chicago history. Similarly, Argyle's Vietnamese residents reflect recent historical shifts and Chicago's increasing cosmopolitanism, as Asian immigrants moved to the city in the late twentieth century and the early twenty-first. Dresden's farmers affirm a popular image of New England as a place of farming villages. Provincetown's Portuguese fishermen likewise represent the East Coast's heyday as a maritime center and the opening of America to Southern Europeans, a "safe" other for preservationists to venerate.

Finally, while such groups affirm popular renderings of local history, in some instances they also embody a unique tradition. For instance, a preservationist boasted that the presence of Asians on Chicago's North Side, rather than in Chinatown, adds a novel dimension: "It adds a flavor Chinatown never had. Chinatown is Chinese. This is not. So it is a very unusual situation. I think it is a wonderful [example] of what we should be doing in the inner city." Provincetown preservationists point to the town's unique blend of Portuguese village and avant-garde artist's enclave. Sitting in his book-cluttered kitchen (quite distinct, in this sense, from most old-timers' homes), a masseuse said that one of his favorite aspects of the town is that "you go into old [Portuguese] people's houses and they've got one or two original oil paintings." The distinctive partnerships that placed the paintings in working-class homes captivate him: "Artists would rent rooms and donate and exchange art [to pay rent], and [it's] one of the unique things about town."

Preservationists also appreciate old-timers' connection to an other place or time, verified by ethnic, religious, and labor traditions. Specifically, they value cultural continuity: "They're Vietnamese or Thai; it is their culture. That's how they lived in their country. And . . . they were here before we moved here, and maybe we moved here because of Argyle, because we love Thai food or because we could tell the street was charming and [wanted to be] around it." (Of course, this is how preservationists *imagine* old-timers lived in another country. In Andersonville, for instance, some Swedes object to the Chamber of Commerce's representation of their traditions.) They appreciate space marked by references to other places and festivals that celebrate old-timers' (foreign or historic) heritage and cultural traditions. These include Andersonville's Midsommar Festival, replicating a traditional Swedish rite, and mid-December's Santa Lucia Procession. Each year, Argyle holds at least two Chinese New Year celebrations, complete with dragons and drums. Provincetown hosts a Portuguese Festival and Blessing of the Fleet, during which fishermen process with a statue of Saint Peter. In Dresden, many celebrate farmers at the annual Harvest Festival. In three sites, the ethnic groups mentioned above are the *only* ones with festivals, and Dresden's Harvest Supper is the *sole* townwide celebration. The only groups in symbolic (festive) competition with Andersonville's Swedes and Provincetown's Portuguese are those who take part in Andersonville's Dyke March or in Provincetown's predominately gay Carnival. By emphasizing certain cultural events and not others, social preservationists engage in the construction of local tradition as well as of what constitutes the local community.

FIGURE 5.04. Dyke March, Andersonville

FIGURE 5.05. Andersonville Midsommar Festival

Those who might compete for preservationists' attention are at times complicit in this construction by celebrating old-timers' traditions themselves, often to attract business. Andersonville Middle Eastern merchants sell Swedish glögg, and, during the Santa Lucia Procession, gay and lesbian business owners stand beside their shops, candles in hand. Gay Provincetown merchants sell Portuguese crafts and, thus, shape or affirm preservationists' prioritization of Portuguese traditions.

When it comes to the individual old-timer, preservationists appreciate those who engage in activities that they imagine are traditional to their group. For instance, a Dresden political consultant who frequently travels for work and relies on farm stands for her produce explained that she values farming because "it is a continuation of what is important in terms of the culture and the heritage of this area." She and her husband worry that *other* newcomers "don't want the farmer to have the use of their land the way they have *always* done it." They supported legislation to "require realtors to identify active farmland," to protect farmers' traditional practices from newcomers who complain about "noise early in the morning, insects, [and] smells." The wife, who wears tailored but simple clothing, said: "It's when reality meets life [that] folks with no understanding of agricultural policy really go nuts. . . . We talk a lot about farm *land* preservation, and we need to talk more about *farming* preservation."

Similarly, a lesbian who has an administrative position at a university said that "mom-and-pop shops that were undiscovered" attracted her to Andersonville and worried that Bank One would displace a bank "that's probably been here forever. . . . [But] the advantage of having an ATM across the street is that I won't be paying my ATM fees, so I can see the benefit. I look beyond my personal gain and think, 'What's it doing to the neighborhood?'" A man also lamented: "All those quirky little neighborhood stories—they will just disappear because that just doesn't happen at Starbucks." Provincetown preservationists grieve for the fishing industry. One said: "'The town has been in mourning . . . because of the loss of the traditional fishing industry. . . . That's been a major cultural and economic loss.'" He explained: "You go through the whole thing of loss: anger, fear, sadness" (quoted in "Jay Critchley" 2002).

Notably, social preservationists value traditions that involve publicly visible practices that shape the land, sea, or streetscape. In each site, they prefer those whose labor traditions are visible from streets or sidewalks, evidenced by farms, shop awnings, an artist's easel, and fishing boats. They do not celebrate those whose traditional practices are inaccessible to them, such as Dresden's factory workers or Argyle's Chinese American lawyers. Similarly, they recognize those whose celebrations fill public space, such as Andersonville's Swedish Midsommar Festival and Santa Lucia Procession. The neighborhood's Middle Easterners never fill its main street, and its lesbians take to the street only once per year for the Dyke March, which, with its bare-breasted participants, is less traditional than the Santa Lucia Procession, whose participants, dressed in white, sing as they process. And, aside from a few gays and lesbians who take part in the Dyke March or Provincetown's Carnival, preservationists are observers of, not participants in, such rituals.

Preservationists are attracted, not only to traditional practices, relationship to local history, and cultural customs, but also to families that are traditional in two senses: those that have traditionally resided in a locale and those that affirm popular images of the working-class family, marked by mutual dependence and cohabitation.[22] They often recognize old-timers by their local blood relationships. A young lesbian Provincetown preservationist who lives apart from her own family described old-timers as those "with relationships that extend back in time," and another said that they have "forty thousand uncles and brothers and kids." Such place-based family ties stand in stark contrast to preservationists' residential mobility. In fact, preservationists conflate blood relationships with length of residence: they describe old-timers as having lived in a place for "a million years," as having "been here forever," as "multigenerational families," or as families that "go way back." They refer to some as having lived in the community for "multiple generations," as if community membership extends beyond a lifetime.

Social preservationists' sense of old-timers' longevity is derived from an acute awareness that "*they* were here before *we* were," as an Argyle woman said of her Asian neighbors. They view old-timers' longevity as the most fundamental justification for their preservation efforts: "A local person that's lived here for *many, many* years can't afford the taxes on their property, even if it's paid for. And then someone will come along and offer them an outrageous price for their property, and what do you do?" Dresden's well-coifed political consultant said: "It has to do with the ability of people to remain in their communities, which I think is really important. . . . [Some said] I was very naive to suggest that people had any right to raise kids in the community where they grew up. To me, *that kind of continuity is what community's about.*" Of course, very few social preservationists raise their children in their own natal community.

Importantly, old-timers are not necessarily old; social preservationists extend old-timer status to even the very young descendants of old-timers, whom they believe inherit a local legacy. For instance, sitting against a bank of windows overlooking her rolling green fields, an affluent Dresden preservationist described a variety show performance by a pair of working-class old-timers: "These were the old-timers. They'd done this in the old days." However, the "old-timers" to which she referred were a young couple, and the "old days" she referenced was their childhood and adolescence, before Dresden gentrified. Preservationists wish to reside alongside old-timers and their children because the young embody the continuity of local tradition. For instance, a young Provincetown artist expressed concern that "local families who've been here for a long time . . . people who have actually been living here for decades and decades actually can't afford it anymore." He was concerned about the displacement of "local families," despite the fact that *he* was moving out of town because he could not find an affordable rental. In fact, the interview took place during his moving sale. Similarly, a Yale-educated writer said: "You look at the fishermen with their kids, and I'm like: that would . . . be so upsetting to be here for so many generations and to know that this is it." Provincetown preservationists worry about the potential closure of both schools and the town-owned nursing home: "You take a school out of the town, and you don't have a community. You take the school out, you take the Manor [the nursing home] out, what's to stay for?" In Argyle, a single, thirty-something social worker suggested that intergenerational interactions among Vietnamese residents are central to her appreciation for residence in the neighborhood: "There's a sense of community, especially in the Asian community. . . . You get the sense that the guy across the street is watching the kids play on the corner. I think part of that is the generational thing, so it's not only the little kids, it is the old Vietnamese men sitting in the bakery every morning. . . . It's the intergenerational thing that goes on; there's a sense of family." This "intergenerational thing" contrasts with her own family, which is dispersed across the United States. Similarly, a Provincetowner said: "I envy [old-timers] in a lot of ways because they have the familial connections of being here. . . . You know, I miss seeing my family all of the time. They come [visit], they love it here, but you know they're still far away."

For social preservationists, old-timers' families embody permanent community composed of those with a tradition of residence in the locale.[23] One preservationist suggested that those with a tradition of residence lend place character: "You have some very beautiful houses that've been totally renovated, and next door you have a house that somebody's lived in for fifty, sixty years—*that's* the charm of it, *that's* my neighbor." At a Provincetown meeting, a woman spoke in favor of affordable housing. She referred to the privilege she has of living in "a beautiful house on the water" and of her daughter "set[ting] up a lemonade stand for a friend's grandmother whose apartment was turned into a condo." Similarly, an Andersonville researcher is nostalgic for his childhood neighborhood, dominated by extended families: "It is hard as you get older? It is hard, the constant transformation and lack of stability. . . . It doesn't give you anything to hold on to. This is not an intergenerational community. When I grew up, people's grandparents were there and parents. When Mrs. —— yelled at me, I knew it was so-and-so's cousin. Grandpa was still speaking Hungarian every other word, and my other grandfather Swedish. . . . All that's gone. There's a total break in continuity." This nostalgia has transformed into a desire to reside alongside those bound by the qualities he appreciated in his childhood home. For him, as for many others, community suggests "a sense of familiarity and safety, mutual concern and support, continuous loyalties" (Brint 2001, 2).

Preservationists' appreciation for old-timers' traditions and for traditional old-timers is a product of broader resistance to change (see Kefalas 2003, 62; and Logan and Molotch 1987, 19), particularly to gentrification, as well as to a sense that contemporary American society (particularly, the segment of which they are a part) is without tradition (di Leonardo 1998, 30). The presence of those whose ancestors settled a place, who embody a traditional culture, are part of a multigenerational local family, and perform ancestral trades, lends preservationists a sense of stability and tradition. By preserving old-timers, they seek to maintain the historically enduring character of their place of residence, to live in a place where tradition and its practitioners define that character, where tradition balances independence, and where traditional families and their traditions are sheltered from gentrification.

PEOPLED PLACE AND PLACED PEOPLE

Just as (perceived) separation from one's traditions promotes attention to others' traditions, the more detached we feel from our place of origin, the more place matters (Griswold and Wright 2004). Preservationists admire those who have a demonstrable relationship to place, whose traditions, practices, and even independence are rooted in their current locale or an other place. This section explores which residents they regard as having a close bond to place and how they come to think of some as a natural" part of the social and physical landscape.

Preservationists believe that connection to place strengthens community; for them, old-timers' rootedness in a place is a vital measure of community viability. As an Argyle man said: "I like the idea of somebody making their living and living in the neighborhood where their store or business is because then they're also more committed to . . . the community." Here again, social preservationists appreciate traits distinct from their own; they prefer place-based relationships to their—in the words of one preservationist—"commuter" friendships.[24] A bearded Provincetown poet who spends part of each year aboard a houseboat said of his own relocation: "It's nice to have an opportunity to explore the newness of the moment, which you don't have where you've [lived before] because where you've been you're still who you used to be. . . . There's an opportunity [in a new town] to redefine yourself." Indeed, it is this sense of personal *im*permanence that draws preservationists to old-timers, whom they imagine are rooted in place. For instance, a Dresden man suggested that an affluent, summer community he belonged to as a child was inauthentic precisely because it was fleeting: "We were Massachusetts people who came up [to Vermont] for the summers and a couple of weeks during the winter to go skiing." For preservationists, *community* refers to relations that take place over time in common space.

Small-town preservationists appreciate old-timers who literally entangle with the landscape. In Dresden, they devote special attention to those who work the land (and whose work marks the landscape), and, in Provincetown, similar attention is devoted to those who labor on the sea (fishermen, boat captains) or who embody appreciation for the landscape (artists). At a historic district meeting, officers debated a young artist's proposal to build a house next to his grandfather's historic home, facing the harbor. Some worried the house would destroy one of the few remaining vistas of the ocean in an area in which houses now crowd the beach. A few argued that the houses would destroy a traditional "seascape" and "streetscape," elements essential to their sense of place: "I lament that we're closing off all these sightlines to the water. The goal of historic preservation is not just about structures but about landscape too. . . . Sometimes you can't even tell that you live on the waterfront!" The artist's architect reminded the board that he wished to build on his grandfather's property and promised that the new house would not "be a cookie cutter, not a Nantucket shed. . . . He is an artist and wants it to reflect his individuality and creativity." The artist added: "I want to live here the rest of my life." Because he was armed with these claims to old-timer status— family legacy, intended longevity of residence, and his identity as an artist— the committee approved his proposal. However, it was an articulation of the belief that the artist himself would contribute to the landscape that closed the debate. An officer exclaimed: "I am so glad that a *real* artist is going to live on the waterfront." The board's decision united a symbol of the town's social landscape with its natural landscape.[25] Thus, not all eschew the protection of the built environment in favor of people; some seek to preserve both.

In Argyle, and to a lesser degree in Andersonville and Provincetown, old-timers' status rests on ties, not only to their locale, but also to an ancestral home. An Argyle man said that it is "amazing" for old-timers to "still have a piece of what they've left from their country." Indeed, urban preservationists largely measure old-timers' ties to other places through commerce: storefront names and products sold within that reference another place.

As a Jewish Argyle preservationist we met in the introduction demonstrates, preservationists seek to prevent the disruption of these meaningful streetscapes, often at the expense of their *own* cultural tradition:

> This community is an old community . . . built upon various cultures and ethnicities that has made it what it is, good and bad. And to deny it through pressure . . . against certain ethnicities, even in terms of the stores, is absurd. [We] want to have a kosher delicatessen here—and, God, we need a kosher delicatessen—but it doesn't have to be on Argyle. I mean that's the point. Putting a kosher delicatessen on that street would be an eyesore. Let's be perfectly honest; it doesn't belong there. So what's wrong with having an Asian street? There's nothing wrong with it.

In this way, preservationists highly value cultural and aesthetic *congruity* (Burgess 1967, 150). They seek unification of a place's social and physical traits, especially those tied to old-timers.

Old-timers themselves sometimes serve as boosters. Their churches, agencies, and clubs influence preservationists' choices. In addition to old-timers and social preservationists, other gentrifiers, longtime residents, associations, and politicians all have an interest in and influence over the association of people with place (and vice versa). While local power holders often fight attempts to resist gentrification, preservationists are not alone in their attention to particular residents and certain of those residents' traits. In this sense, the construction of the old-timer is an interactional process. Some old-timers engage in it by emphasizing certain of their traits, such as their ethnicity rather than their wealth or local knowledge rather than a college degree.

Andersonville demonstrates the role of multiple interests in linking place to old-timers. In the 1960s, Swedish business owners, with local politicians' support, dubbed the area *Andersonville,* after a Norwegian minister who purportedly aided ill Swedish immigrants (Lane 2003). While this is not an official name, it weds the neighborhood in Chicagoans' minds to a time when Swedes were omnipresent there. In 2000, an alderwoman and the Chamber of Commerce began planning a streetscape for the shopping district. After months of meetings, the city allotted $6.4 million for widened sidewalks, trees, flowers, faux brick crosswalks, and banners (Rado 2002). The intent was to "create and maintain the unique sense of place which celebrates [Andersonville's] Swedish roots, its architectural styles, its multi-cultural businesses, unique shops, neighbors and friends. All culminating into a quaint, walkable, timeless village" (Andersonville Streetscape Committee 2002). The streetscape emphasizes Swedish roots, embodied in the colors of the Swedish flag and Andersonville's name itself—the design's focal point—despite debate about whether the area is Swedish, multicultural, or lesbian.

Many factors shaped the predilection of Swedish identity. Swedish institutions—the Swedish American Museum, a Lutheran congregation, and a hospital and college founded by Swedes (both west of Andersonville)—may have influenced the decision. Decisionmakers included the Chamber, block clubs, politicians, preservationists, Swedish old-timers, and those who christened the area Andersonville. As a result, an artist identified real Swedes as those linked to such institutions: "The *real* Swedish community—the ones that got the museum going . . . they used to have the Miss Andersonville Pageant and all the Swedish stuff."

As Andersonville's streetscape suggests, multiple players contribute to the definition of place character; such definitions are a product of interaction (Hunter 1974, 194). Specifically, institutions are mediating factors in each site (see Breton 1964; Kornblum 1974; and Suttles 1968). For instance, as we have seen, a state-funded marketing campaign for Dresden farms placed signs on highways advertising the "Dresden Farmlands." The impetus for such choices varies by site and does not always align with preservationists' other political commitments. Provincetown's Chamber may emphasize the town's Portuguese heritage to avoid addressing its gay and lesbian population (as some argue), and the same may be true of the Andersonville Chamber's attention

FIGURE 5.06. Provincetown Art Association

to Swedes. Regardless of the impetus, by signifying that a place belongs to a group, boosters influence preservationists' selection process. As a man said of a pagoda a booster placed on the Argyle el stop in 1991: "At that time, the neighborhood could have gone either way. The pagoda stamped it as Asian." Another added: "[The pagoda] convinced me to live here." And a woman said: "I thought it was interesting when I bought. It gave [Argyle] character." Preservationists often turn to symbolic commemoration when old-timers are hard to find, such as in Andersonville with its tiny Swedish population.

This is not to suggest that landscape or symbols are all that preservationists desire. While they vary in their concern, many concur with a preservationist who said that institutions cannot stand in for people: "We are going to have a new theater, a new art association, [the] Fine Arts Work Center. . . . These are all monuments to a lost culture. To a dying culture. . . . Not lost. To a dying, living culture. It is still alive, but a lot of artists have left town, families, working people. . . . So you know it is ironic that all of these institutions are flourishing but the community is dying."

Despite such reservations, many preservationists work to emphasize the relationship they perceive between old-timers and the place they live. A Provincetown artist, funded by an organization with social preservationist aims, hung photographs of elderly Portuguese women on a pier "to honor the women who've made Provincetown what it is and the women behind the fishermen." Argyle preservationists (alongside other gentrifiers) lobbied for an Asian-themed streetscape. In this way, their predilection for certain groups interacts with the landscape; that is, symbolic markers of a group's presence shape preservation choices, and their choices mark the landscape (see Small 2004, 122; and Bearman 2005). This is but one way in which preservationists shape the space where they live and contribute to placemaking, or the construction of their neighborhood or town's identity.

However, landscape itself informs social preservationists' choices. For instance, a Dresden preservationist—an author active in the local Democratic Party—complained that fields are lost to overgrowth because there are few farmers to clear them. Driving his old Saab through town, he spoke with a conservationist about methods for preserving fields, asking: "Is there any way to encourage farmers to move to Dresden [to keep fields clear]?" His words suggest that some appreciate old-timers because they are custodians of a particular landscape and/or that appreciation for open fields facilitates esteem for those who clear them. Of course, pastures are no more indigenous to Maine than the gentrifier is to Dresden; they are the product of labor performed by particular people (see Alanen and Melnick 2000, 3; Hardesty 2000, 171, 200; and Barthel 1989, 89). In this sense, for some, to preserve a Dresden old-timer is at the same time to preserve a rural landscape, and, given an orientation to view landscape as social, to preserve a landscape is to maintain a particular social meaning associated with it. Better than most, preservationists recognize that farmers need fields and that fields require farmers, that place is peopled and that people are also placed.[26]

We have seen that, for social preservationists, those whose traditions, history, and culture most mark public space are a place's real tenants. Independence and tradition are inconsequential if they are not bound to a place. This reveals the import of place in their definition of *community* and their implicit concern that their transience limits their own claims to community. However, their appreciation for place marked by people and for people marked by place—a harbor dotted with fishing boats and the faces of fishermen darkened and weathered by the sea—also suggests that, while people lend meaning to place, place also gives meaning to people. In other words, the important insight that social characteristics influence understanding of landscape clouds the related fact that landscape, however socially constructed, shapes people—or, in this case, appraisals of people and of their authenticity.

Virtuous Marginality

As we have seen, social preservationists appreciate characteristics of community—independence, tradition, and a close relationship to place—

not popularly associated with those of their social class. I propose that this is not happenstance: their admiration for old-timers' community reveals aspects of their own identities and networks that they regard as *in*authentic. That is, they associate authentic community with, and highly value, traits they do not share.

This is readily apparent in the contrasts they draw between themselves and old-timers and their desire to live *apart* from those like them. They criticize newcomers' transience, secondary ties, and wealth. One, having identified himself as a newcomer, said: "[Newcomers are] here for a couple of years, they've got friends elsewhere, and they're just not as . . . community involved." A homeowner said that old-timers "were replaced by people who weren't interested in getting to know their neighbors . . . [people] here as speculators." Similarly, after commenting on the "richness" of old-timers' place-based community, one newcomer said that she engages in "commuter friendships." Another confessed: "At some levels I'm a yuppie." Yet she moved to Argyle to escape yuppies: "There's a certain polish people have . . . that diminishes as you proceed north [to Argyle]. So this area, especially when I moved here seven years ago, was a lot more Asian. . . . It's not as polished. This coffee shop, if it were ten blocks [south], would be full of yuppies." When asked why he does not wish to live near yuppies she said: "[I don't like their] sense of entitlement, which I share. [Their] certain economic standard, which I also share." Similarly, an Argyle University of Chicago graduate is dissatisfied with other English speakers' presence in his building: "I liked being one of the few. I liked being the minority." Another preservationist said: "I'd say that living here is alienating—but nice." Many gay preservationists express a desire to live apart from those who share their sexual identity. A white, gay Argyle man reports that he "couldn't" live where gay men were the majority because it would be too "homogeneous" yet enjoys what he regards as the homogeneity of the Vietnamese community. A gay Andersonville man complained: "There was an old marketing slogan for gay people called, 'We are everywhere.' And it's really true and unfortunate. . . . When they came in, I was like, this is not the people I want in the neighborhood. And then our little [Swedish] Midsommarfest became a sea of topless men. And now it's all condos with landscaping." This desire to live apart from those like them uncovers an implicit self-critique: by seeking to preserve those unlike them, preservationists reveal their sense that networks of highly educated, mobile persons are *in*authentic.

In fact, preservationists worry that their own presence and that of others like them disrupts old-timers' communities and, therefore, maintain distance to preserve its authenticity by policing the borders of old-timers'

networks and consciously balancing the participation that preservation requires with distance from old-timers. For instance, a preservationist shared
her favorite line from a book: "You'll never be a native if you're not." Another
said: "Would I ever feel like a townie even though I was married to a townie?
No, never. *It's a line you don't cross.*" To prevent the disruption of old-timers'
community, preservationists sometimes consciously abstain from certain political acts. As we have seen, despite distaste for the trailers that dot the countryside and concern about illegal dumping, Dresden preservationists support
old-timers' libertarian aversion to regulations and remain passive at town
meetings.[27] In Provincetown, some preservationists refuse to join boards for
fear of disrupting old-timers' tradition of governance by "lineage."[28] More
generally, in their daily rounds, preservationists seek to leave the social wilderness untrammeled, to minimize evidence of their presence.

Given this orientation, preservationists hold gentrifiers morally liable:
"It's the old idea: I'm the new guy, I've got the fancy building, and I'm going
to call the punches. And that's wrong. Absolutely wrong!" One said: "Real
estate brings out the worst in people. . . . It makes their heart cloud. . . .
When people are driven by greed and power, they do all kinds of things that
go against what . . . they know they ought to [do]." Another said: "[Gentri-
fiers] just want to take. They don't want to give back to the community. . . .
They don't care who gets hurt." Similarly, a Provincetown poet said of gentrifiers: "People stepped in that had a sense of making money . . . , and they
chose to utilize the opportunity regardless of the motivations, regardless of
the process or the priorities." Another preservationist said: "Older people
who made the town what it is are pushed off into low-income elderly housing
because they can't afford to stay in their houses anymore. They're replaced by
people who just moved in who—I mean, they're not bad people, necessarily,
but . . . " Another suggested that many gentrifiers approach old-timers with
an attitude he finds reprehensible: "'I don't know you. I don't really care how
you live. I don't know you. I don't care if you're poor or not.'" Yet another
echoed this: "It has been sold up the river by a faction of the community
wthat . . . put profits first." More generally, social preservationists view the
decision to move to gentrifying space as indefensible and hold themselves accountable for this. They regard themselves as culpable gentrifiers.

Consequently, preservationists assign moral value to (their own and others') efforts to prevent the disruption of old-timers' community. For instance,
the man who suggested that Provincetown has been "sold up the river" argued: "Maybe we shouldn't be selling these places for that much money. You
know maybe the moral thing to do . . . is to keep it affordable or sustainable."
Similarly, a woman said of an affordable-housing advocate: "You don't lead a

bad life by doing the right thing . . . what a good heart!" Generally, preservationists view participation (their own included) in gentrification as unforgivable and assign value to resistance efforts. For example, one said of his preservation work: "It makes me feel righteous, like I care." Thus, preservationists are committed to *virtuous marginality,* which exists when people associate authenticity with and highly value traits they do not share and, consequently, out of a desire to preserve the authentic come to regard their own distance from it—their marginality—as virtuous (Brown-Saracino 2007).[29]

What are the implications of virtuous marginality, of defining authenticity as that which one is not? Preservationists' definition of *authentic* community is associated with a lack of recognition of their own community membership, despite their engagement in the very measures of community that scholars use to assess its vitality.[30] As we have already seen, most engage in civic and social life. They attend political meetings and civic events and volunteer. All have a strong attachment to place. Few reported a paucity of friendship, professional, or identity-based networks.[31] Without intending to be or acknowledging that they are, preservationists are part of an "ideological community" (Hunter 1975) composed of others who resist gentrification. Yet they contend that old-timers' communities are more real than their own. In this, they ignore their "elective communities" and identity subcultures. They regard community as a birthright rather than a practice and associate it with certain others.[32]

In 1975, Albert Hunter found that the "sense of community" that Rochester residents reported was inconsistent with customary measures of community. Residents believed that they were a part of community when scholars would have concluded otherwise. Hunter argued that ideological loyalty to community strengthened his informants' sense of membership. Social preservation reveals the reverse: how those engaged in community come to believe that they have *limited potential* for membership in community. Preservationists' sense of community (or lack thereof) is rooted in ideology, in the association of community with particular old-timers in particular places.

In a sense, preservationists write themselves out of community. Because they associate authentic community with old-timers and believe themselves incapable of belonging to real community, they fail to recognize their own networks, participation, and ideological community.[33] For instance, a year-round Provincetown resident, a writer, said: "I almost want to say [to old-timers], 'I know this is more your community than mine.'" He added: "I'm just a visitor. I may even own a home here, but it's not my community." Similarly, when asked, "Do you have a sense of community?" or, "Tell me about your community," preservationists responded, "*My* community? I don't have a community," and "There is *no* sense that there will be an ongoing community here." These words are puzzling given the dense networks of which these individuals are a part. For instance, the man who said, "I don't have a community," also described partnerships to prevent gentrification, local friendships, and neighboring practices. The other respondent—a former stockbroker—described local households, referring to several as "friends" and fellow activists. Greetings from his friends interrupted our interview, held in a high-end beer pub. Moreover, the note of anticipated community disruption in his words does not likely refer to his friends' displacement, for they are professionals, and almost all are homeowners. Finally, since he deems his family's relations with old-timers estranged, we might expect him to welcome others like him. Rather, despite having lived in Andersonville for two decades, like the other man he views himself as a visitor, associating real community with old-timers.

While this disavowal of membership in place-based community may be a particular problem for those dedicated to virtuous marginality, the common sentiments that it borrows from—the belief that certain others are more likely than those with high cultural and economic capital to be a part of real community—may contribute to a general sense of community decline. This affirms the notion that community is imagined (Anderson 1991) and underlines how ideology shapes perceptions of community viability (Hunter 1975; Ryle and Robinson 2006).[34] It also points to the implications of definitions of *authenticity,* indicating that they are of consequence, not just for those defined by the term, but for those with the power to define it, for authenticators define themselves in the process and risk essentializing, not only the other, but also themselves and, in this case, their communities as well.

Conclusion

Social preservation is predicated on a search for authentic community. While others have noted similar quests for authentic people, place, and community, most regard them as isolated, market-oriented practices that enhance the seeker's authenticity.[35] By contrast, preservationists' search is neither fleeting nor market driven: it informs their residential choices and is predicated on a sense of the *in*authenticity of their own identity and networks.[36] This sense of personal inauthenticity is not without benefit, for, when balanced with the practices of social preservation—particularly intentional distance from old-timers' community—it affirms or repairs their moral identities (Kleinman 1996) and, they hope, preserves the authenticity they admire in others.

In their departure from a temporally limited quest for authenticity, social preservationists align themselves with those who seek to preserve the built or natural environment. Like material preservationists, social preservationists select objects of preservation from a pool of candidates. They do not automatically work to preserve those who constitute the numerical majority or those with the longest local legacy. While beliefs about racial and ethnic groups' claims to community inform their choices, traditional racial and ethnic hierarchies (or a precise inversion thereof) do not explain why they prefer the Portuguese to WASPs or the Vietnamese to Chinese merchants. Instead, they seek old-timers who embody three mutually supporting traits. These choices, like those of all preservationists—from historic to cultural—are rooted in ideology. Social preservationists' choices take root in the belief that authentic community belongs to certain independent, traditional others with a close relationship to place—because membership in community is a birthright, rather than a practice—and in a nostalgia for a "gold age of community" (Griswold 2000, 133) based in resistance to population succession. They wish to freeze old-timers' lives in the present because they fear community dissolution. They worry that Dresden farms will close, causing farmers to lose their independence and connection to the land, and that Argyle's Vietnamese will "Americanize," losing insularity and ties to an other place.

While this chapter identifies the breadth of preservationists' ideology and patterns of old-timer selection, it also demonstrates that such choices are place based.[37] There is no universal old-timer: a Portuguese fisherman would not be preserved in Andersonville, nor would a Vietnamese merchant in Dresden. Preservationists draw from local history, demographics, and landscape to identify old-timers, and boosters influence their choices. Those they do not preserve in my sites may elsewhere be an object of preservation. As Mary Pattillo (2007) has found, African American professionals in a gentrifying African American neighborhood may work to preserve poor or working-class African Americans. It is also possible that the object of preservation will change *within* a site. This is not to dismiss the import of independence, tradition, and rootedness in a place. It instead suggests that the articulation and arrangement of those qualities varies by context.

How do preservationists reconcile concern for the displaced with advocacy for only some? They are largely unaware of alternate candidates for old-timer status. As we have seen, they do not celebrate old-timers by actively demeaning or opposing other longtime residents. Instead, they rarely mention those they do not regard as old-timers unless pressed to defend them in the context of pioneers' public evocations of revitalization strategies. Like most, they do not think of community in terms of official boundaries, seeing instead colloquial neighborhoods (Hunter 1974) defined by networks of people bound by shared traits. Neighborhoods and towns thus defined, they do not see themselves neglecting those outside old-timers' community. This is redolent of claims others have used to restrict the presence of nonwhite others in ethnic enclaves, and this dimension of social preservation should not be overlooked (see, e.g., Rieder 1985; Hirsch 1998; Suttles 1968; and Kefalas 2003). In this sense, preservationists are not as generous with their privilege as we might wish them to be, and few would term this *virtuous* or celebrate this outcome. Indeed, if preservationists were more aware of their selection process, they might share these criticisms, for they regard their *marginality*, not their selection process (of which they are largely unaware), as virtuous.

Setting aside questions of preservation's efficacy for the moment, this chapter demonstrates the consequences of definitions of authenticity and community, not just for sociologists, but for those we study as well, for they shape behavior, specifically, in preservationists' case, lived experience of community. Social preservationists' association of community with others encourages their dismissal of their own communities: their networks, participation, ideological and elective communities, and relationship to the old-timers they admire. This suggests that authenticity is of consequence for authenticators and that to capture the state of community we must look to both ideology and practices; neither attitude surveys nor quantitative measures of participation would fully capture social preservationists' complex and somewhat contradictory relationship to community.

As a result, I offer two suggestions for further research. Preservationists' sentiments about community authenticity may be widespread, for they borrow from popular ideas about authentic people and place and are not alone in their residential mobility, which is essential to their sense of their networks' inauthenticity (di Leonardo 1998; Lamont 2000, 124). If their beliefs are part of a trend, the association of real community with others may contribute to beliefs about its decline. It is also plausible, given preservationists' familiarity with scholarly terms and debates, that theories and predictions of community decline contribute to the notion that some are ill-equipped to be a part of warm, supportive networks.

Finally, this chapter invites further inquiries into the expanse and implications of virtuous marginality. We can imagine others—likely also highly educated, residentially mobile individuals—who have a similar relationship to the authentic, who wish to shut the door behind them and even on themselves, such as tourists who visit an African village but hope to leave it untouched and environmentalists who seek the wilderness but wish to leave only footprints. Like social preservationists, such individuals borrow from an enduring tradition in which the privileged articulate a sense of personal inauthenticity and an admiration for those without their resources (see, e.g., Cantwell 1996; Davila 2001, 216; di Leonardo 1998, 94; and Roy 2002).[38] Furthermore, virtuous marginality suggests that some contemporary actors are more aware of their social position, specifically of their privilege, than we suppose. This calls us to consider the origins and breadth of such self-consciousness and how it shapes beliefs and behavior in a variety of contexts.

Locating Social Preservation

This chapter explores how those who articulated the social preservation ideology and engaged in related practices while I was in the field are similar to and different from other gentrifiers. Specifically, it compares social preservationists' demographic and cultural traits with those of the other gentrifiers in my sample. In so doing, it provides a portrait of the social location of the preservationists I interviewed as well as of the historical and cultural location of social preservation itself, of its relation to other sets of ideas and practices.

The prevailing explanation for why some gentrifiers do not neatly adhere to the frontier and salvation ideology is that criticisms of gentrification emerge among those who are "marginal" to the process (Rose 1984; Caulfield 1994). Specifically, scholars argue that gentrifiers who are socially marginal, have limited resources, or are themselves at risk of displacement are most inclined to criticize gentrification (see Rose 1984; Berry 1985; Caulfield 1994; Rothenberg 1995; Smith 1996/2000; Mele 2000; and Lees 2003).

Relying on a related logic, most gentrification scholars agree that there is a specified set of attributes that encourages participation in gentrification, such as residential mobility, childlessness, and a high level of educational attainment. For this reason, when studying gentrifiers, scholars examine traits such as occupation, education, age, marital status, race, and sexual identity. As I identified ideological and practical variation among gentrifiers, I began to collect data from my informants on these traits. However, this chapter explores several additional traits that I identified as the research progressed. For instance, during interviews, gentrifiers often mentioned films they had watched or books they had read. I began to wonder whether orientations to gentrification corresponded with cultural tastes or habits and came to ask about newspaper subscriptions, television and film viewing, and favorite books.

While my sample includes eighty gentrifiers from four distinct communities, because the questions I posed developed over time and, more important, because my sample is not random and includes a disproportionately high proportion of civically engaged gentrifiers, I regard my findings as tentative.[1] In places, I present quantitative data on the similarities and differences between social preservationists and other gentrifiers.[2] These data are meant to reflect similarities and differences only within my sample, not within the larger universe of gentrifiers.[3] However, drawing from a reasonably broad and diverse interview sample, they outline the social location of the preservationists I interviewed as well as of social preservation itself, which—like all ideologies—exists in relation to other beliefs and practices.

The chapter focuses on traits around which social preservationists and other gentrifiers diverge (albeit often subtly), on those that the gentrification literature suggests are most deterministic, and on those that social preservationists suggest inform or are informed by their ideology. However, I do not intend for this emphasis on difference to leave the reader with the notion that gentrifiers are situated in starkly contrasting social locations. Indeed, those in my sample share many traits: almost all the social preservationists and other gentrifiers I interviewed are white and highly educated.[4] In the aggregate, they are of similar ages and have lived in their neighborhood or town for comparable periods of time.[5] On the street, it is difficult to distinguish among them; their ideological differences are not apparent in their clothes or cars, and preservationists, pioneers, and homesteaders live alongside one another. Furthermore, social preservationists are not a uniform group with a solidified identity; they are individuals who, during the period of study, shared an ideological and practical orientation to gentrification. Thus, this chapter asks how demographically similar gentrifiers come to have divergent orientations to gentrification.

Finally, in contrast with prevailing scholarly explanations for gentrifiers' beliefs and practices, this chapter cautions against assuming an overly deterministic relation between demographic traits and ideology. Indeed, it demonstrates that preservationists' social location shapes their ideology but also that their ideology shapes some of their traits, such as (in some cases) the schools they attend, their courses of study, and their occupations.

I begin the chapter by reviewing the class characteristics of the social preservationists I interviewed. Following this, I explore their cultural tastes, residential histories, and sexual identities, aspects of their lives that these preservationists suggest share a close relation with their ideology. These, in turn, relate to a broader cultural ethos that encourages social preservation, an ethos I consider in the conclusion.

Social Class

When designing this study, I chose not to ask my informants to specify their income or wealth. I was interested in documenting and understanding gentrifiers' attitudes toward gentrification, and I wished to avoid questions that risked reducing the ease with which they shared their beliefs and sentiments. Instead, I asked a set of questions intended to serve as a rough proxy for class. I asked about their educational histories and their occupations as well as about their parents' careers and took note of the characteristics of their houses or apartments, cars, clothing, and other markers of social class. While such data cannot pinpoint informants' precise social location, cumulatively they paint a picture of preservationists' lifestyles.

I found that few gentrifiers in my sample have experienced great wealth or great poverty. By asking about their parents' occupations, and, secondarily, about their place of origin and educational history, I learned that most originated from across the spectrum of the middle class. Nonetheless, in terms of social reproduction, preservationists demonstrate a noteworthy trend that separates them from other gentrifiers.

Specifically, the preservationists I interviewed have experienced little social mobility relative to other gentrifiers. Thirteen percent have experienced upward mobility (as measured by occupation) and 8.7% downward mobility vis-à-vis their parents. In contrast, 41% percent of other gentrifiers have experienced upward mobility and 11.76% downward mobility. Thus, nearly 80% of the social preservationists in my sample have never experienced significant variation in their socioeconomic status, compared to only 47% of other gentrifiers.[6]

This difference in social mobility may help explain differences in cultural dispositions to gentrification. As preservationists suggest, their status security, specifically, their sense that they inherited their status, may be accompanied by a concomitant guilt for not having struggled for their economic position as well as by an aesthetic appreciation for financial struggle. In contrast, gentrifiers who have experienced upward mobility (41% of the gentrifiers in my sample who do not adhere to the preservation ideology) may be more protective of their social position and investments and have greater appreciation for the "upward mobility"—by gentrification—of the space in which they live.[7] Thus, a sense of having earned or worked for one's economic or social position may encourage support for gentrification, while, in contrast, having inherited status may facilitate generosity at least in relation to (nonthreatening) old-timers.

The stories that some preservationists volunteered about how their childhood class position shaped their attitude toward gentrification support the notion of a relation between social mobility (or a lack thereof) and orientation to gentrification. Most commonly, social preservationists suggest that the contrast between their own upper-middle-class childhood and the lives of the less privileged encouraged the cultivation of preservation practices. For instance, a Provincetown woman explained that struggling artists captivate her largely because she comes from an affluent family that contained artists. Sitting in her sparsely furnished but carefully restored home, she suggested that the discontinuity between the world of privileged artists she knew as a child and Provincetown's "working-class artists" intrigued her and encouraged her advocacy for those artists whom gentrification threatens.

More explicitly, a preservationist whose father was an insurance executive said: "I never really had to want for much. It just created a sense of responsibility for me, being fortunate." Another informant said: "I grew up in the suburbs near the water on Long Island, and I never had housing problems or housing issues." According to him, his late-in-life discovery that others face such issues encouraged his preservation concerns. Similarly, a woman from an upper-middle-class family argued that recognizing the contrast between one's own "good fortune" and the plight of the displaced encourages criticism of gentrification. These gentrifiers suggest what the numbers imply: that good fortune encourages attention to others' *mis*fortune.

In contrast, Michael, an Argyle homesteader, credits his working-class upbringing for instilling in him the belief that, with effort, the poor can "improve" themselves and that cultural assimilation is vital to such success. Sitting in his expansive living room, wineglass in hand, he said he views his father as a model: "See, my father came here [from Greece], and he really wanted to Americanize. He became a citizen. He didn't speak [Greek] to us when we were kids, which was a big downer. But I understand because he really just wanted to assimilate. I have a lot of respect for that. . . . [My father] worked his butt off and continues to work his butt off." In contrast, he suggested that his uncle was less successful because he failed to assimilate: "My father's brother, he came here . . . he [was] here thirty years, and the guy can hardly speak because they just speak [Greek]. Personally, I think that's wrong. If you're somewhere, fine, keep your culture and everything, but you gotta be able to communicate. I mean the guy can't speak!" This homesteader's beliefs about the relation between assimilation and economic success parallel his hope that local Asian business owners will improve their businesses to attract new clientele (people like himself). His belief in the importance of acculturation is likely related to the contrast between his parents' social position (a construction worker and a food-service employee) and his own as a lawyer. From his perspective, his father's assimilation facilitated his own social mobility. Social preservationists—most of whom in my sample lack this legacy of personal mobility—are less likely to turn to gentrification to pull themselves up by the bootstraps or to otherwise celebrate the upward mobility of neighborhoods or towns.

Social preservationists' status security is also apparent in, and, perhaps, supported by, their educational backgrounds, a key determinant of social location. While the gentrifiers I interviewed are uniformly highly educated—more than 88% of preservationists and 95% of other gentrifiers in my sample have a college degree, and many are working toward or hold advanced degrees—there are key differences among them in terms of course of study, type of school attended (e.g., private vs. public), and status of educational institution.[8] Specifically, the preservationists in my sample differ from other gentrifiers in the content and context of their education.

The social preservationists I interviewed are less likely than other gentrifiers to have acquired the degrees required of elite professions: only 4% have law or medical degrees, compared to 13% of other gentrifiers. Preservationists' courses of undergraduate study are also distinctive. Thirty-two percent of those who attended college have degrees in English or writing, 16% have degrees in education, and 12% studied one of the social sciences.[9] No significant patterns emerge among other gentrifiers, whose majors included history, sociology, math, physics, and medical technology. Given the small, nonrandom sample, we should not overinterpret these findings. However, the popularity among preservationists of humanistic and other degrees that do not neatly translate into lucrative employment opportunities may reflect their status security. It is also plausible that such courses of study cultivate appreciation for local characters and other markers of social authenticity (see Holt 1998).[10]

Second, social preservationists are more likely to have attended a liberal arts college. Twenty-four percent of social preservationists in my sample did so, compared to 9% of other gentrifiers.[11] Finally, the preservationists I interviewed are more likely to have a degree from an elite school: 35% have a degree from a school that *U.S. News and World Report* lists as a "top twenty" institution, while this is true of only 9% of other gentrifiers. By the same measure, only 16% of preservationists attended a third- or fourth-tier university, compared to 35.2% of others.[12]

While most studies of the relation between education and ideology analyze years of education, these findings suggest that, at least when it comes to gentrification, one must also consider the content and context of education.[13] While an argument about causation is beyond my scope here, the educational trends apparent in my sample raise a few questions. First, might

TABLE 9. Gentrifiers in employment categories (%)

	Social preservationists $N = 35$	Other gentrifiers $N = 29$
Arts and writing	17.14	10.3
Small business owner	17.14	10.3
Private-sector finance and marketing	5.7	17.24
Private-sector technology	0	3.4
Public- and voluntary-sector	22.9	6.9
Law	2.9	3.4
Education and research	8.7	24
Skilled, technical	5.7	10.3
Administrative/secretarial	5.7	0
Routine occupations	14.3	3.4
Long-term unemployed[a]	0	10.3

[a] This reflects the fact that several informants were disabled because of AIDS.

social preservationists encounter seeds of their ideology on certain types of college campuses and, perhaps, even in certain classrooms? If they do not encounter social preservation per se, might they discover a set of orientations that share a relation with it, such as a cosmopolitan worldview, cultural omnivorism, or a model of self-reflexivity about one's social position apparent in the commitment to virtuous marginality? Regardless of the answers to these questions, it is plausible that attendance at an elite school reflects and contributes to the sense of status security that encourages adoption of the social preservation orientation.

This same status or sense of privilege may encourage preservationists' engagement in certain professions. As table 9 details, the most popular field among preservationists in my sample is the public and voluntary sector, although a high proportion work in the arts and writing and own small businesses, such as a Dresden printing press and a Provincetown affordable-housing firm.[14] Among gentrifiers who do not adopt the preservation orientation, the most popular fields are education and research (half in this category are retired professors), followed closely by such private-sector fields as marketing and finance.

While social preservationists and other gentrifiers are nearly equally likely to be members of the "creative class" (Florida 2002), they cluster in distinct creative occupations,[15] suggesting that this class of workers may possess divergent orientations to place and social difference. Most striking is the relatively high proportion of social preservationists who work in the public/voluntary sector as social workers, government employees, and health care providers (22.9% vs. 6.9% of other gentrifiers) and the relatively low proportion who work in the private sector as marketers and property managers

(5.7% vs. 17.24% of other gentrifiers). This suggests that broad categories such as the *creative class* color over more subtle occupational differences that may share a relation with ideology. Indeed, there are few formidable occupational differences among gentrifiers: most are members of professions that are associated with the professional-managerial class and that minimally require a bachelor's degree.

However, my interviews and observations suggest that these seemingly small occupational differences between preservationists and other gentrifiers may have important ideological effects and, in turn, that ideology may affect gentrifiers' occupational choices.[16] Attending to social preservationists' reflections on the relation between work and ideology provides a window into a dimension of their lives not yet explored and, secondly, warns us against presuming—as many gentrification scholars have—that there is a unidirectional relation between demographic characteristics and ideology, that certain demographic traits produce the frontier and salvation ideology.[17]

Those social preservationists whose occupations lead them to work with individuals or groups at risk of displacement often draw from their work experiences to articulate trepidation about the process. For instance, when a Provincetown French teacher expressed concern about old-timers' displacement, she provided examples from her work: "That's the other thing that's changed tremendously is the number of students: the enrollment. The enrollment has really dwindled." Her awareness of Portuguese students' displacement, cultivated through her work, consequently drew her attention to the plight of the town's elderly: "Of course, me being a teacher, I'm more protective of the kids. But . . . I really want our old people to stay here. It's such a special thing to have access to the old people too. The young and the old in so many communities are so disconnected. This is an ideal village in that way, that we still have an intact family. I definitely want the old people to stay here. It would be very sad [if they weren't able to]." Similarly, an Argyle man who works as a child advocate on Chicago's South Side transferred concern for his clients to concern for Argyle residents. And, over breakfast in a Swedish restaurant, Linda, a young, energetic social worker who serves mentally ill clients, worried about their displacement from Argyle as well as about Andersonville merchants' displacement.

However, my interviews also reveal that occupations more popular among pioneers and homesteaders can encourage the development of a preservation ethic. In the mid-1970s, Joe—an Irish American graduate of a New England state university—began work as a Boston realtor. He worked in the South End, which, at the time, was primarily home to poor and working-class African Americans and Puerto Ricans. He soon found himself simultaneously "intrigued" by neighborhood diversity and appalled by other realtors' blatantly racist practices. He also began to notice the gentrification of the neighborhood, primarily by gay men like himself: "It was so diverse. You had people with money moving in, but you had the people who were living there prior to the '70s for a long time. Granted the neighborhood was run-down.... But it was that element that was there, it was really working-class, and some very low-income black and Puerto Rican households, and a lot of single, gay people, and what's happened over the past thirty years was starting back then." Recognizing that gay gentrifiers and racist realtors threatened much of what he appreciated about the South End—the neighborhood's working-class and low-income African Americans and Puerto Ricans—he decided "that being in the real estate business was not the right fit for me because I was seeing what was going on." Shortly thereafter, he left real estate and, for the past thirty years, has worked in the affordable-housing field. Today, Joe owns a Provincetown firm that constructs affordable housing. While his brief sojourn as a realtor does not alone explain what made him take notice of at-risk Puerto Ricans and African Americans (he also cites as influences his mother's volunteerism and the Catholic Church), it illustrates that exposure to a profession more common among pioneers can contribute to the adoption of the social preservation ideology.

Similarly, Ann, another Provincetowner, earned a master's degree in conservation, helped establish a private Provincetown museum, and worked as a historic-preservation consultant. While helping commemorate the past, she discovered a desire to preserve the town's contemporary social attributes: "[I was] realizing that this community was undervaluing itself in terms of what it had to offer." She explained that she has committed herself to preserving "community character," a "sense of place . . . intangible value," and the "indigenous" population. Likewise, standing at an easel in his studio, Sean—a gay man in his forties who devotes himself to both historic and social preservation—explained how he adopted concern for old-timers through his work. He recalled: "I've seen [historic preservation encourage gentrification]. I worked for Boston, and it happened in the South End. . . . That was a diverse, funky neighborhood when I worked there, and it's anything but that now. I encouraged it there."

As Joe's, Ann's, and Sean's stories demonstrate, *social preservationist, homesteader,* and *pioneer* are not monolithic categories. Some gentrifiers adopt a social preservation orientation after promoting gentrification. In this sense, it is not surprising that, while we are beginning to see some patterns of difference among the gentrifiers in my sample, those differences are no more obdurate than the categories themselves.

Finally, some preservationists use their work to advance their concerns. First, some *implicitly* select work that coheres with their ideology. For instance, a Dresden preservationist, a middle-aged consultant, used a political position she once held to try to protect farmers:

> We tried a few years ago to require realtors to identify active farmland, and when someone was purchasing a house, they would have to sign something saying, "I understand that I'm purchasing land next to active farmland. There could be dust, there could be chemicals, there could be noise early in the morning, insects, smells, all the stuff that goes along with agriculture." . . . Ultimately, it would discourage [newcomers] from suing, which is what's going on. Most people will come up and look at the land in the summertime when the cows are eating the clover, and they'll say, "That's wonderful." Then, in the fall, when the manure gets spread, it's not so wonderful anymore, or at four in the morning when someone is getting ready to cut hay, it's not so wonderful anymore.

Likewise, an Andersonville artist uses his work to celebrate local Swedish residents, and several Provincetown artists use their work to speak against gentrification.

Second, for some, the preservation ideology *explicitly* informs occupational choice. For instance, a Provincetown preservationist said of her work for the Provincetown affordable-housing company: "This was my dream job really, to do what I did as a volunteer for years and get paid for it." Indeed, she has long been an advocate for affordable housing and a voice against gentrification. Perhaps the most blatant case of the preservation ideology shaping professional choice is that of Joe, the informant who illustrates the possibility of discovering social preservation through work. Joe's realization that he did not want his work to contribute to gentrification led him, eventually, to establish an affordable-housing company, thus taking him full circle: from a person whose work contributed to his preservation ideology to a person whose professional choices were informed by that ideology.

However, professional exposure to displacement or at-risk populations does not ensure that an individual will become a social preservationist. For instance, at a crowded Argyle community meeting about a proposed supportive-living facility for the elderly, a social worker joined the chorus of newcomers speaking against the proposal, arguing: "We're looking at concentrated poverty in this area. . . . We have the highest concentration of mentally ill in the city." In great contrast to the social preservationists (including a religious leader who evoked his own profession), who argued that local Asian residents should "be able to stay where they're comfortable," the social worker kindled fears that "people without a work history" would reside in the building. We do not know why this social worker uses his professional authority to advocate for development while others do the opposite. Nor can we be sure why Joe decided to leave real estate while many others—who are equally aware of displacement—remain.

Thus, while some distinctions among gentrifiers' occupations are apparent in my sample—most notably the fact that preservationists are more likely to engage in the arts and the public/voluntary sector and that pioneers and homesteaders are more likely to work as educators and in the private sector—they do not alone explain why one adopts the preservation ideology and another does not. Work matters, both as a producer and as a product of ideology, but, in theoretical terms, a more noteworthy conclusion is that preservationists' occupational opportunities and choices and their ideology are related to a broader set of orientations tied to their social location. As we have seen, for many preservationists, this location rests on a sense of status security.

Together, data on preservationists' social mobility, education, and occupations contradict arguments that those who are themselves threatened by gentrification, such as "marginal gentrifiers" (Rose 1984), members of the "marginal middle class" (Caulfield 1994), and punks and artists (Smith 1996/2000; Mele 2000), are more likely to resist gentrification. Indeed, if a pattern emerges among my informants, it is the opposite: relative to other gentrifiers, preservationists are more likely to have experienced status security, to have attended elite schools, and to regard their occupations and courses of study as opportunities to encourage social change rather than advance their own social location. This suggests that social preservation is not predicated on concern for one's own ability to remain in gentrifying space. Rather, it relies on the sense that one has resources—monetary or otherwise—to offer those in danger of displacement and to use to invest in local authenticity. Finally, it hints at a broader cultural orientation related to social preservationists' social location and, relatedly, to their discomfort with their privilege.

Cultural Tastes and Consumption

In my last year of interviews, I began to recognize patterns of difference in the clutter evident in social preservationists' and other gentrifiers' homes. On a social preservationist's kitchen table, amid a few unopened pieces of mail, I might find the *New York Times,* several art books, a local magazine, and a well-worn novel. Pushing the books aside for coffee mugs, the preservationist might offer pithy commentary on the novel and, in the next breath, unabashedly refer to the latest episode of *American Idol.* In subtle contrast, on a homesteader's table I might find the *New York Times,* the *Wall Street Journal,* or *Crain's Business* and a single book, perhaps a novel or a travel guide. Most homesteaders and pioneers were hesitant to mention television habits and, when asked initially, emphasized devotion to CNN or other news programs.

My snooping uncovered a variety of books and magazines (to my disappointment, none of my informants had copies of texts on gentrification lying about). Preservationists do not uniformly read the same books or watch the same public-television programs. Furthermore, while I noted patterns, they were not neatly predictive. I did not always accurately identify preservationists by their reading material or television-viewing habits. Yet interviews affirm the distinct consumption patterns that I observed and point to a cultural orientation that corresponds with the preservation ideology.

First, the social preservationists I interviewed are avid readers. They read, on average, 3.6 newspapers or magazines, twice as many as other gentrifiers.[18] *All* social preservationists asked reported that they were currently reading a book, while this was true of only three-quarters of other gentrifiers, and preservationists more frequently reported reading multiple titles.[19] When asked for recent titles, preservationists often retreated to a bedside table, returning with multiple books in hand. It is plausible that this propensity for pleasure reading is a product of the liberal arts background that many social preservationists have or the high proportion of English majors among them.[20] However, it is also indicative of a broader disposition, specifically, of membership in the "reading class" (Griswold 2000, 2001).

As avid readers, social preservationists are well positioned for exposure to criticisms of gentrification. For instance, a content analysis of articles on gentrification published by nine papers in seven U.S. cities with a population of over 1 million found that those articles are more likely to offer criticisms of gentrification—through discussion of its costs for original residents or the loss of community and authenticity—than support for it (Brown-Saracino and Rumpf 2008). Furthermore, as several articles literally suggest, and as the geographer Neil Smith (1996/2000) notes, *gentrification* has become a dirty word. For instance, a 2002 *Chicago Sun-Times* editorial argues that gentrification is predictable and dangerous: "Soon, the urban pioneers venture in. Then, just as the neighborhood becomes a place fit to live, the housing prices skyrocket and the yuppies stampede. Before you know it, they are kicking out anyone who can't keep up with the Joneses. . . . It's no wonder gentrification has become a dirty word" (Richards 2002, 43). Papers also report newcomers' criticisms of gentrification: "'Gentrification, for us, has almost become a dirty word,' [one said]. . . . 'I moved from the Upper East Side to Park Slope because I wanted to be in a more diverse neighborhood'" (Yardley 1998, B1).

Similar criticisms are apparent in popular fiction. For instance, Dennis Lehane's novel *Mystic River,* which was adapted for an Oscar-winning film, depicts old-timers' perspective on gentrification. One character bemoans his imminent displacement: "And when I get that white-collar job, I'll move us out of here, out of this whole neighborhood with its steadily rising rents and stadium deals and gentrification. Why fight it? They'll push us out sooner or later. Push us out and make a Crate & Barrel world for themselves, discuss their summer homes at the cafés and in the aisles of the whole-food markets" (2003, 373). On the other hand, David Payne's novel *Gravesend Light* explores the social preservationist impulses of an anthropologist studying a Southern fishing community. Joe, the anthropologist, worries that newcomers' environmentalism will disrupt old-timers' traditions. Payne writes: "The stacks of rusty crab and eel pots everywhere, the hedges draped with nets, the old skiffs and wheelless, vintage pickup trucks on blocks in backyards accompanied, now and then, by an apiary . . . Joe loved it all. To him, Little Roanoke was like a rare and precious human ecosystem in a state of fragile balance that had taken generations to create. When he looked, that was what he saw: simplicity and character, old folkways, seafaring traditions dating back to Devonshire" (2000, 41). Likewise, in *Fortress of Solitude,* Jonathan Lethem provides an account of a newcomer's effort to resist Brooklyn's gentrification. Her son reflects on his boyhood: "I saw the changes here in terms of [my mother's] war on the notion of gentrification, which had been conducted mostly in the battlefield of my skull" (2003, 427). Finally, Susan Sontag's novel *In America* provides an account of social preservation. Referring to an affluent couple's relationship to residents of their summer village, she writes: "[They] . . . made a tacit contract of benevolence . . . that went far beyond the infusion of cash their annual presence brought. . . . The villagers' part of the contract (to that they'd not consciously assented) was: not to change" (2001, 66).

This is but a small collection of widespread criticisms of gentrification and models of the social preservation perspective that preservationists might encounter through their reading. We cannot be sure that preservationists have read these particular texts (in fact, it is likely that many have not). Instead, I wish to argue that, given the ubiquity of such criticisms of gentrification and the fact that the preservationists I interviewed are avid readers, they have likely come across depictions of gentrification and of gentrifiers similar to those offered above.

However, preservationists do not spend all their leisure time reading. Their film and television preferences are also distinctive. They are more likely than other gentrifiers to favor independent, foreign, or cult films, such as *Donnie Darko, Blow Up, Blue Velvet,* and *A Room with a View.* They also watch a greater number and broader range of television shows and networks. For instance, pioneers and homesteaders watch CSPAN, *David Letterman, E.R., The Apprentice,* and *Will and Grace,* while social preservationists view *Inspector Gadget, The Dave Chappelle Show, Meet the Press, The Gilmore Girls,* MTV, *News Night, Bill Maher,* and CNN.

In this sense, preservationists are cultural "omnivores," a trend that Richard Peterson and Roger Kern (1996) noted among those who previously preferred high culture.[21] That is, social preservationists combine *Meet the Press* and *Antiques Road Show* with the family series *The Gilmore Girls,* in contrast to other gentrifiers' steady diet of middlebrow programs such as *Will and Grace, ER,* and *David Letterman.* Peterson and Kern suggest that this broadening of cultural consumption among elites is related to "value changes concerning gender, ethnic, religious, and racial differences. . . . The change from exclusionist snob to inclusionist omnivore can thus be seen as part of the historical trend toward greater tolerance" (1996, 3).

Given preservationists' appreciation for the underdog, their omnivorous habits are not surprising. Many studies find the highest levels of tolerance among the most educated segments of the population as well as among those who belong to occupational clusters of which many preservationists are a part (see Ladd and Lipset 1975; Knoke and Isaac 1976; Ladd 1978a, 1978b; Weiner and Eckland 1979; Davis 1982; Brint 1984, 1985; Zipp 1986; DiMaggio 1987; Bobo and Licari 1989; and Pascarella and Terinzini 1991). Preservationists' omnivorous orientation is another reflection of a set of ideological and practical habits associated with their educational backgrounds and occupations.[22] Furthermore, their interest in old-timers and cultural difference may encourage their "wider repertoire of culture" (Erickson 1996, 221–22).[23]

Social preservationists' cultural tastes and consumption patterns are of importance, for cultural objects (Griswold 1986) shape or reinforce their ideology and practices and serve as a resource for the articulation of their ideology, particularly for their discussion of their place of residence, old-timers, and community change.[24] In a few instances, social preservationists even volunteered that specific cultural objects influenced their approach to old-timers and their communities. A Provincetown preservationist, a recent graduate of an elite, Midwestern liberal arts college, explained how a short story by Alice Walker shaped his actions when old-timers' interests conflicted with his employer's. After graduation, the social preservationist—Ryan—moved to Provincetown, where he worked both as a housekeeper at an inn owned by an affluent gay man and at a shop alongside Portuguese old-timers, including a woman whose husband grew up in the house that, after several changes of owners, had now become the inn. One day, the woman asked Ryan: "'Could we come by and take a look at [the inn] 'cause he doesn't remember? He was six years old when they moved to another house [in town].'" Ryan enthusiastically agreed to give them a tour.

During the tour, Ryan realized with trepidation that he was revealing the transformation of the old-timer's childhood home: "They came, and he was like, 'I don't remember any of this. This wasn't here, and this wasn't a room.'" Ryan felt responsible: "I felt like, Oh my God, we've taken your home and changed it around so it is different rooms and basically [only] the old wood . . . and the fireplace is the same." An ancestor of the old-timer, a sailing captain, had once resided in the house, and the innkeeper had restored it to the period of the captain's residence. He had decorated the walls with evidence of the house's history, such as banknotes and maritime paraphernalia. For this reason, Ryan grew alarmed when the old-timer's wife mentioned: "We have all of [the captain's] original plates and cups." When I interviewed him, Ryan explained that his alarm was rooted in the fact that "the inn owner] would die to get his hands on that stuff, but then I [knew] I would never, ever tell him because [of the Alice Walker story]."

The story, "Everyday Use," describes the return of an educated African American woman—who has changed her name from Dee to Wangero Leewanika Kemanjo—to the South to visit her impoverished mother and sister. When she arrives, she carefully photographs her family standing with their small home behind them: "She never takes a shot without making sure the house is included." She admires the benches her father made when they could not afford chairs: "'I never knew how lovely these benches are. You can feel the rump prints.'" She tells her mother that she "needs" the top from their churn, saying: "'I can use the churn top as a centerpiece for the alcove table'" (Walker 1973, 53, 55, 56).

She also asks her mother for two quilts made from her grandmother's clothing. The mother is hesitant, having promised them as a wedding gift to the daughter with whom she lives: "[Wangero] gasped like a bee had stung her. 'Maggie can't appreciate these quilts!' she said. 'She'd probably be backward enough to put them to everyday use.'" When the mother learns that Wangero plans to display, not use, the quilts, she pulls them from her hands and secures them in Maggie's lap. Angry, Wangero tells her family: "'You just don't understand . . . [y]our heritage'" (Walker 1973, 57, 59).

Ryan worried that, like Wangero, the innkeeper would view the old-timers' heritage as a decorative accessory and, in so doing, divorce emblems of a family's heritage from that family; he would value the cups as a symbol of the town's *past*—or of its "ghosts" (Bell 1997)—while, for Ryan, their value rested in the unification of that past with its *living* representatives. For this reason,

Ryan did not tell his employer of the dishes, a decision largely informed, as we have seen, by "Everyday Use." This suggests that, for some, cultural objects—in this case, an object likely encountered through an elite liberal arts college—inform the preservation ideology and directly shape practices.

A few others report that the arts awakened concern for old-timers. For instance, a gay Latino merchant indicated that, while he did not move to live near old-timers, a play encouraged him to reconsider Argyle's value: "I went to see their play down the street, and it was all about the people that moved to Uptown and their history. And I was like, Oh my God, all those children, and then you see all those people, all those backgrounds. When I moved here— you never move because of that. I never knew it!" A middle-aged affordable-housing advocate described how a play about realtors' exploitation of old-timers affected her:

> This play was about how this sort of disingenuous—sort of this ingenue who's a real babe comes to Provincetown. . . . I think she gets into a lesbian relationship and a straight relationship. You know the whole thing. She gets a job with a real estate agency. . . . And in this play [a realtor] trained [the new-comer] to go around to all these little old Portuguese ladies and say to them, 'Let's see this apartment, let's see what we could do here. Well, we could carve this. . . . You could get $300 a month for this closet,' basically. 'Let's put a little bed in there.'

Watching the production, the social preservationist found that it captured the town's social and economic dynamics: "[That is] what they did. They went around to these little innocent Portuguese ladies and operated on their instincts for greed." It affirmed a pattern that she had begun to note and encouraged her affordable-housing activism.

While not directly related to the arts, an Argyle social preservationist reported that his involvement with the Chicago Field Museum, with its anthropological content, changed his beliefs about the American melting pot: "I've heard people say, 'There's too many Asians here.' That's ridiculous. One of the things about urban anthropology—I've been involved with the Field Museum for years; I know about urban anthropology—is that we are not a melting pot. That was a lie that was perpetrated years ago. We are a tossed salad, and as such there are certain congregations of very ethnic communities, and that's how they express their history and their culture, and it's perfectly legit to have this in the community." This anecdote, combined with those about books and theater, suggests that certain cultural objects contribute to the development or articulation of the preservation ideology.

While books, plays, and museums directly influenced the social preservationists quoted above, others' tastes for particular television shows or books parallel an appreciation for certain community forms and individuals, particularly for local characters and the spaces that contain them. For instance, an Andersonville social preservationist reported that Charles Dickens is his favorite author: "I like . . . things that are strong in character, and so I also like biographies." Similarly, two Provincetown preservationists independently reported that their favorite film is *Grey Gardens,* a documentary exploring the life of Jacqueline Kennedy Onassis's aunt and cousin, who reside, along with several raccoons and flea-ridden cats, in a crumbling East Hampton mansion. While images of the pristine estates surrounding the timeworn mansion suggest that the film may appeal to preservationists' taste for contemporary representations of the past as well as for stalwarts in increasingly upscale communities, the greater parallel is between social preservationists' taste for characters and the film's portrayal of two iconoclastic women. It follows that preservationists, who are particularly attuned to the social characteristics of those around them as well as to old-timers' character, would also be attracted to books and films containing strong characters. This interest in characters may also relate to the high proportion of preservationists with a degree in English or writing (32% of those who attended college). The preservationist with a taste for Dickens appreciates the PBS program *Antiques Road Show* for the same reasons he appreciates living alongside old-timers: "You get regular human beings talking about their family, their history—who they are—and taking pride in that." In this way, certain cultural objects reflect and reinforce the preservation ideology.

They also serve as a linguistic device for articulating an approach to community and gentrification (Radway 1984). For instance, an Andersonville preservationist wrote: "I always think of Andersonville as my own 'Gilligan's Island.' Remote, but, . . . sooner or later, everyone drops in." Similarly, a preservationist told me that she can best describe her place in Provincetown through the words of a local author: "'You'll never be a native if you're not.'"

In still other instances, cultural objects provide a model for preservationists' expectations for community, old-timers, and place. For instance, an Andersonville preservationist drew parallels between his fond memories of the neighborhood's recent past and the community life depicted on the television program *Northern Exposure,* which portrays interactions between affluent (e.g., a millionaire who is a former astronaut) and highly educated (e.g., a doctor) newcomers and old-timers in a tiny, Alaskan town: "I miss being able to go into a store where nobody is, and talking with the owner, and hanging out for a while. You know, what used to be more like *Northern Exposure,* the TV show. *Northern Exposure,* that's what living here was like. There was the conversation mentality, there was an openness mentality, and, you know, it wasn't about fancy cars, it wasn't about fancy clothes, it wasn't about fancy hair. And now it's about visible, visible money, and it's too bad."

In this way, cultural objects, such as television shows, inform preservationists' expectations for their place of residence and sometimes explicitly shape their ideology and practices. However, such objects—ranging from books to films—also reinforce a preexisting ideology and provide a set of images that some employ to articulate their beliefs.

In summary, many of the social preservationists I encountered are cultural omnivores who avidly read a broad range of books, magazines, and newspapers and view an array of television shows and films. As they report, they find criticisms of gentrification and models of appreciation for old-timers in the cultural objects they consume. Social preservation reminds us that such encounters are of consequence for those in the practice of imagining others' lives through cultural representations. Most important, social preservationists' accounts of the cultural objects that influence them provide a rough outline of a broader cultural orientation that shares with the social preservation ideology an ethos that values a sort of cultural democracy that embraces a familiarity with low-, middle-, and highbrow cultural objects alike—that celebrates the idiosyncratic character of people and place and emphasizes reading and cultural difference. Below, I explore a few other orientations that have a similar relation to social preservation.

Place Matters and Cosmopolitanism

Given social preservationists' passionate concern for preserving place character, it is reasonable to suspect that there might be a relationship between their place of origin and their beliefs about community change. One can imagine a scenario in which social preservationists disproportionately hail from the suburbs, encouraging an adult quest for authenticity (Allen 1984, 33). Alternately, they might be lifelong urbanites, dismayed by the gentrification of their birthplace.

The residential histories of the social preservationists in my sample are not very different from those of other gentrifiers. On the whole, all the gentrifiers I interviewed possess the capital that enables residential mobility,[25] and all have lived in their place of residence for comparable periods.[26] Notably, preservationists are only marginally more likely than other gentrifiers to have participated in gentrification's first wave, suggesting that their ideology does not emerge from exposure to early-stage gentrification. And, as table 10 illustrates, besides a rough 10% difference between preservationists and other

TABLE 10. Place of origin: Percentage of gentrifiers by community form

	Social preservationists N = 38	Other gentrifiers N = 26
Rural or small town	22.6	17.64
Suburbs	19.4	29.4
Small city	11.1	11.8
Major city[a]	16.2	5.6

[a] A major city is defined here as one with a population greater than 1 million.

gentrifiers' childhood residence in suburbs and major cities, place-of-origin differences are small.

However, this is not to suggest that place does not matter. Indeed, it does, but not in the ways one might anticipate. While few patterns emerge in social preservationists' childhood residential patterns, many nonetheless report that their place of origin influenced their orientation to their current place of residence and to gentrification. The explanations for how this happens are disparate. Some suggest that their childhood neighborhood or town demonstrated the value of particular place characteristics, which they now seek to protect. Others suggest that a privileged suburban childhood made them mindful of the disparity between their own and old-timers' residential experiences. The pattern that emerges from this is simply preservationists' insistence that place matters, that exposure to particular places shaped their expectations for their current place of residence and even their self-identity.

For instance, some appreciate old-timers because they represent the distinction between their current place of residence and their less authentic childhood home. A Chicago social preservationist, originally from a small Connecticut town, explained that his taste for cultural difference is rooted in his desire to live in a place that is the opposite of his homogeneous, rural hometown. Similarly, an Argyle social preservationist who is currently working on her Ph.D. said: "I grew up in a [New Jersey] suburb, and I've been slowly moving my way towards a city, and that's why I was excited about moving to Chicago and why I wanted to live in the city." She reported that she consciously sought a neighborhood distinct from her suburban hometown: "I wanted there to be life on the streets, [which] was important to me. . . . I wanted to live in an area that was going to be ethnically diverse. . . . I knew there was a large Middle Eastern as well as a large Asian population." Similarly, a Provincetown preservationist suggested that the stark contrast between his affluent suburban hometown and the cities in which he lived as a young adult drew his attention to old-timers: "[Housing issues] were something that I came to care about the more I was exposed to what the problems were outside my suburban upbringing." Thus, some social preservationists seek a diversity of place experiences and celebrate and seek to preserve the distinction between their current neighborhood or town and the one in which they were raised.

In contrast, social preservationists who believe that they grew up in a place that was somehow authentic sometimes express a desire to re-create this experience by living alongside old-timers. For instance, nostalgia for her place of origin and for residence alongside her large Catholic family infuses an Argyle social worker's appreciation for local old-timers: "There's a sense of community, especially in the Asian community, . . . which is the way I grew up. I grew up in this kind of Catholic ghetto in L.A., and that's just the way it was." Similarly, a Provincetown preservationist—a gay artist who serves on the town's historic district commission—suggested that he could never move back to the small, industrial New England city in which he was raised: "I've thought about it, and, honestly, it will never happen because it's not the town I grew up in, to be honest with you. [Provincetown] is more like the town I grew up in even though it's a lot smaller. That sort of cohesiveness of community doesn't exist where I grew up anymore. They're all dead or moved away." In this sense, he measures Provincetown's authenticity against his 1960s Connecticut hometown.

Social preservationists sometimes explicitly seek places that embody characteristics of their place of origin. For instance, a preservationist who appreciates Andersonville's Swedish elements is nostalgic for his own urban Swedish relatives, and Sarah, a Dresden preservationist discussed in chapter 5, identifies a parallel between Dresden old-timers' "live-and-let-live" sensibility, which she very much admires, and that of farmers in her native Midwest. Finally, a gay man who grew up in a small New England industrial city with a large Portuguese population believes that his town offers lessons for Provincetown:

> I come from a town that, when I was a kid, it was great. You'd go to [the] shoe store. The owner ran it. When you went to buy your shoes for the new school year, he was like, "I know you're in fourth grade. You must be this shoe size." He just knew. I'd go to the store, and it was all people you knew. That is all gone. [I want to say to people in Provincetown]: look where I come from. We had what you have now. We had it then, and it's all gone, and it's not for the better.

For some people, Provincetown's version of a New England Portuguese town may be more hospitable than their own hometown, yet they hope that Provincetown will maintain lost and much-lamented elements of their birthplace.

Others hope that a certain type of social exchange that they recall from a childhood in a working-class neighborhood or town—particularly the close proximity of family members and informal street interactions—can be rediscovered through proximity to old-timers. For instance, one man longs for the working-class Polish neighborhood of his childhood: "My grandparents still live in the Polish neighborhood. So I really identify with that neighborhood. There used to be the Polish grocer, and all the Catholic churches [with Polish mass], because the immigrants didn't speak English. . . . All my great-grandparents worked in the mills. . . . [My town] used to be like that. You know [it] had a certain identity. There were events that happened, traditions that we did, that were only [our town's]. Gone." Because of his sense of loss, he values Provincetown's ethnic and working-class traditions. For instance, he appreciates the continuation of the Provincetown Portuguese Festival: "We have the Portuguese Festival, and we celebrate what the Portuguese have brought to the town."

Finally, preservationists sometimes demonstrate a sense of sophistication about place-specific processes, particularly gentrification, garnered through childhood exposure. For instance, one suggested that growing up in New York sensitized her to old-timers' plight by highlighting displacement: "You get that kind of sensitivity when [you] have either been displaced or been exposed to it in youth. . . . Living in New York you see displacement constantly. I mean I was in New York through the Dinkins administration. There were more homeless people in New York than residents practically." At community meetings, this particular preservationist warns others about how gentrification altered places like New York. Exposure to the process in other places lends her authority to coax others toward social preservation.

Given their devotion to the preservation of place characteristics, the centrality of place in social preservationists' imaginations is not surprising. Indeed, their claims about how their place of origin shaped their approach to gentrification and to old-timers speak more to the way they use place as a discursive and creative resource than to the concrete implications of childhood residence in particular community forms. This discursive celebration of place relates to a general cosmopolitan orientation—to which I turn below—that values distinctive places and the preservation of their authentic qualities.

In numerical and theoretical terms, the most striking residential difference between social preservationists and the other gentrifiers in my sample is preservationists' propensity for residence outside the United States. More than 31% of the preservationists I interviewed have lived outside the United States as adults, compared to 11.5% of others. Also, a slightly higher proportion of preservationists lived outside of the U.S. as a child (9.67 v. 5.6% of other gentrifiers).[27]

This propensity for living outside the United States relates to preservationists' broader adherence to a cosmopolitan orientation, with its attention to the value of place distinction and cultural difference, and devotion to preserving authenticity through social distance. We have seen in previous chapters that preservationists share with other cosmopolitans a taste for authenticity and a commitment to virtuous marginality with which they negotiate a tourist's distance from and a resident's familiarity with old-timers.[28] While, given their common mobility, all gentrifiers are cosmopolitans in the traditional sense (i.e., nonlocals; see Merton 1968; and Zimmerman 1938), those who adhere to the preservation ideology are more cosmopolitan according to contemporary definitions, which locate cosmopolitanism in worldliness or a global orientation (see Hannerz 1990). As Craig Thompson and Siok Kuan Tambyah write: "The aesthetic and cultural interests of contemporary cosmopolitans must now be transnational in scope" (1999, 216).

Borrowing from a colonial legacy of valuing "invigorating and rejuvenating effects of encounters with the native," present-day cosmopolitanism stresses the consumption of "differences in a reflective, intellectualizing manner" (Thompson and Tambyah 1999, 218, 216; see also Hannerz 1990).[29] It encourages adherents to feel "betwixt and between" or never quite at home (Hannerz 1990) and to organize their lives in correspondence with local practices (Thompson and Tambyah 1999, 227). As with social preservation, cosmopolitanism valorizes the local as the site of "authentic culture" (Thompson and Tambyah 1999, 237).

Most important, cosmopolitanism orients one to the consumption of difference: "Standing within the bourgeois circles of the global economy increasingly hinges on being a connoisseur of cultural differences" (Thompson and Tambyah 1999, 219).[30] Douglas Holt found that cultural elites, which he associates with cosmopolitanism, "favor cuisines from other countries, often the peasant variety, eclecticism (interesting foods), artisanry, and casual atmospheres." Cosmopolitanism draws one to the "exotic," "authentic," and "experiential": "[Cosmopolitans] seek out diverse, educational, informative experiences that allow them to achieve competence, acquire knowledge, and express themselves creatively" (1998, 11, 17).

While these taste preferences generally parallel preservationists' appreciation for the authenticity they associate with old-timers, they are particularly apparent in the manner in which they speak of time lived abroad and foreign travel. For instance, one highlighted the heterogeneity and individuality of two European cities: "Paris and Amsterdam. . . . What they have in common for me is there's a real acceptance of others. You don't have people who are all dressed alike trying to fit in. They're both places that celebrate the individual." Another expressed a desire to explore others' cultures: "I've never been [to Italy], but I'd love to be there. A lot of places in Europe. Thailand. If I could choose [where to be], it probably would not be in the States. I've done a lot of traveling in South America and Latin America, so I'd probably want to pitch over to the other hemisphere."

Preservationists' interest in others' cultures extends from their time abroad to their current place of residence. For instance, the travel preferences of a Provincetown journalist parallel his concern for preserving old-timers' culture and community:

> Of all [thirty] countries I've been to, India was totally my favorite. And why I liked it is because it was the first place I had been to that I really saw that Western culture and American culture was kind of an afterthought. Not that I don't like American culture as a whole or Western culture. It's just I hate how—and it's not just Western culture that can do it—I hate when things become so thoughtless. . . . When I was in Morocco, we were in Marrakech, and they have the Berber wine cellars and stuff, they're only wearing those costumes for the tourists. That's not what they really do. But, in India, when you saw the women wearing brightly colored saris . . . that's not for my benefit. They're going to do that whether I'm there or not. That's what I totally loved. I was like, "That is cool." Nothing was for my benefit. It was like, "This is India. This is how we do things. You don't like it? Oh well."

His words reveal an appreciation for authentic Indian culture and practices that mirrors his admiration for Portuguese fishermen. They also underline the parallels between social preservationists' ideology and their broader anti-globalization sentiment, a cosmopolitan concern that (elites') economic and cultural influences threaten authentic places and cultures.

Many value their place of residence for qualities it shares with a foreign place. For instance, an Argyle resident—a middle-aged administrative assistant—reports that she loves living near Thai, Chinese, Vietnamese, and Korean restaurants because they remind her of time spent in Japan. Another, a graduate student in performance studies, said: "I can't afford to travel right now, but living [in Argyle] is almost the same thing, aside from the geography. You meet people from all over, and you learn about things you just wouldn't otherwise." Sitting in a coffee shop, another Chicago preservationist—a young man who grew up in a rural East Coast town—said: "When you think of the old European cities where people are stacked on top of each other, that's what I wished for [in a neighborhood]. The energy level, the constant buzz going on."

For others, gentrification threatens their town or neighborhood's relation to a foreign place, as implied by the past tense that a social preservationist employed when stating what she appreciated about Provincetown: "[The town] *was* very quaint. It *reminded* me a little bit of Europe, because of the small streets, and it *had* that flavor" (emphasis added).

Social preservationists may encounter cosmopolitanism through (their disproportionate, vis-à-vis other gentrifiers) enrollment at liberal arts colleges as well as their study of the humanities.[31] It is possible that their markedly high attendance at elite, private, and liberal arts colleges accounts for some of their time abroad. Several lived outside the United States for college study-abroad programs, some later returning to live in their place of study. Certain schools or courses of study may introduce a seed of the preservation ideology in the form of cosmopolitanism.[32] However, (young) cosmopolitans or the children thereof may be attracted to certain schools and foreign travel, as well as to social preservation, and possess resources that enable them to act on their attraction.

As we have seen, preservationists are more likely than other gentrifiers to have lived abroad. In discussing their time outside the United States and even their current place of residence, many articulate a cosmopolitan orientation that shares many qualities with the social preservation ideology. That is, cosmopolitanism is apparent in many dimensions of social preservation—from concern for authenticity to virtuous marginality—but it is particularly evident in preservationists' reflections on foreign living and travel. Recognizing the link between the preservation orientation and a cosmopolitan worldview helps us understand preservationists' social location: as cultural elites who wish to concurrently preserve and consume distinctive places and the cultural differences they encompass. From this orientation arises a propensity for talking in terms of place, for relying on place to articulate beliefs about community, gentrification, their social position, and the privileges and responsibilities that they believe that position entails.

Sexual Identity and Self-Reflexivity

Social preservationists in my sample are more likely than other gentrifiers to identify as gay, lesbian, or bisexual.[33] Specifically, nearly 70% of homesteaders and pioneers I interviewed identify as *hetero*sexual, compared to only 50% of preservationists.[34] That is, half the preservationists I interviewed identify as gay, lesbian, or bisexual, compared to just over 30% of other gentrifiers. Of course, both groups contain a disproportionately large proportion of gays and lesbians. This reflects my selection of two sites (Andersonville and Provincetown) popularly celebrated as gay enclaves and destinations and gays' high rate of participation in gentrification.[35] The finding that gay, lesbian, and bisexual gentrifiers are more likely to adopt the social preservation orientation is surprising, given that scholars and the media often tout gays' role as gentrification's trailblazers and latter-day victims of late-stage gentrification, in short, as anything but critics of gentrification (see, e.g., Leff 2007).

However, as my informants' words and experiences suggest, there are several reasons why gays and lesbians are well situated to adopt a preservation ideology. First, as part of a desire to remain on a social or cultural edge, some relish their marginal status (believing in the virtue of marginality) and, therefore, prefer residence in places in which middle-class gays and lesbians like themselves are a minority. Second, in contradistinction to the former, many relate old-timers' marginalization to their own and, for this reason, empathetically seek to minimize their displacement. Finally, many are highly self-conscious about gays' role in gentrification. Some suggest that this arises from a personal history of participation in gentrification, while others fault popular and scholarly attention to gay men as the ultimate urban pioneers. This combination of factors encourages gays' and lesbians' keen awareness of their social location—of their marginalization *and* privilege—facilitating the adoption of a preservation orientation. While gays and lesbians are not the only social preservationists to articulate acute self-consciousness about their social position and privilege, they are especially well positioned to engage in a self-reflexive dialogue about marginalization and privilege, in large part because of the place of the gay man in popular and scholarly representations of gentrification.

As we have already seen, preservationists sometimes articulate appreciation for their own position as strangers in a place marked by old-timers (Simmel 1971; Caulfield 1994). More generally, they celebrate the virtue of their marginality relative to old-timers' culture and networks. Some report that this desire to remain on a cultural or social perimeter originates in their sexual identity. For instance, an Argyle informant said that, as a gay man, he is used to being "different" and enjoys living in communities where, as a white, college educated man, he stands apart from others: "My parents and brothers were hunters. . . . I wanted the attention of being different. I was really good in school, and, being gay, I was interested in things that they weren't, like playing the piano." This informant is among many gay preservationists who decry the homogeneity of gay ghettos and seek residence where they are the only English speaker in the building or the only gay man on the block. They resist what they regard as affluent gay newcomers' homogenizing force.

Social preservationists are not alone in this sentiment. For instance, a *New York Times* article on the movement of gays and lesbians into a predominately white, Christian Chicago suburb reported that a gay resident appreciates the town's "diversity": "'Of families,' he explained, after a brief pause. 'As opposed to city culture, where everything is around gay life, gay events. That's not really the way America lives'" (Zernike 2003, 12). Another article—in Atlanta's *Southern Voice*—suggests that increasing popular acceptance of homosexuality is responsible for such sentiments. Paraphrasing a real estate agent, the author writes: "With an increase in acceptance from society, gay men and lesbians in metro Atlanta don't always feel they need to live in Midtown to remain comfortable. . . . 'They used to feel that Midtown was the only place they would feel comfortable because it was more or less a gay ghetto at the time'" (Seely 2003, 13).[36] Ironically, while gay neighborhoods and towns are often celebrated (or shunned) for their distinction, social preservationists, and the articles cited above, suggest that some prefer to live in places they regard as more heterogeneous. For some gay, lesbian, and bisexual preservationists, old-timers' presence and traditions remind them of their own difference, providing a sense of otherness that they value. Of course, gentrifiers' social, cultural, and economic privilege enables appreciation for this sense of marginality.

Alternately, some older gays and lesbians prefer to live in a place they associate with old-timers because they are uncomfortable with the notion of sexual identity serving as a defining personal trait or place identifier. For instance, when I asked an eighty-year-old lesbian whether her sexual identity influenced her decision to move to Provincetown in the 1950s, she quickly responded: "No. It still doesn't. . . . Sometimes it annoys me because it's been taken over in such a way. I liked it in the old days. . . . People didn't care. All they cared about is they admired people that worked hard and paid their bills."[37]

While some gay preservationists celebrate their own otherness, they nonetheless worry about others' marginalization. In other words, they at once celebrate their marginality and worry about the costs of marginalization for others. Queer to the Left (Q2L), an organization based in Uptown Chicago (of which Argyle is a part) with which preservationists often ally, wrote in a treatise on gentrification: "'Queer' also embraces such struggles as that against *gentrification,* which is driven by an oppressive, racist logic, as part of the same larger fight for social and economic justice. Here, then, gentrification is a queer issue" (Queer to the Left 2004, 7). Q2L also casts gays and lesbians, as well as old-timers, as victims of gentrification: "[For example,] youth of color and homeless shelters (considered 'dangerous') or gay bath houses (viewed as 'dens of sin'), [are] no longer tolerated" (5).[38] Thus, for some, advocacy for old-timers is an extension of activism on behalf of their own identity group.

More frequently, gay and lesbian preservationists express self-consciousness about their participation in gentrification, a self-consciousness that personal experiences with gentrification and outside sources such as the media encourage. For instance, a young, lesbian Argyle informant asked bashfully: "You don't think we're urban pioneers, do you?" She acknowledged that a conversation with a reporter covering neighborhood change inspired her self-consciousness. Similarly, a young social preservationist articulated awareness of gays' role in Boston's gentrification. While he has never lived in Boston, he recognized parallels between its gentrification and that taking place in Provincetown:

> Listening to people talk, they're like, "The gays come in. They make it better. And then the straights come in because it's safe to live there." . . . And you saw it in the South End. When I was a kid, the South End in Boston was a place where you'd lock your doors and roll up your windows. It was a bad neighborhood. But then a lot of gay people moved in . . . a lot of artists, writers. . . . I don't believe in that bullshit that, because one group is discriminated against, they're going to be more sensitive to another group. . . . Because a lot of the gay people were white gay people, so they had more clout, the neighborhood got safer, cops paid more attention. Then suddenly the wealthier people move in, and they see sort of these poor people as less desirable. You see it [in Provincetown] all the time. All the time.

Chicago gay, lesbian, bisexual, and transgender residents formed Q2L out of similar awareness of gays' and lesbians' role in gentrification. Members marched in the 2002 Pride Parade under the theme, "There's no place like home . . . if you can afford one." At a meeting, they stated that they formed in reaction to "*hearing* about gay men and their role in the gentrification process" (emphasis added). They regularly advocate for affordable housing, protest against condominium conversions, and discuss gays' and lesbians' role in gentrification.

In print, Q2L vacillates between articulating a sense of responsibility for gentrification and arguing that claims that gays are the front-runners of gentrification have been overstated: "In no gentrifying neighborhood are white gay and lesbian professionals the majority of newcomers. More to the point, gentrification is being driven by the real estate industry, banks, and city government. . . . 'Yuppie faggots' could all die tomorrow, but that won't stop gentrification and its class and racial injustices" (Queer to the Left 2004, 10). As part of this dual response to discourse about gays and gentrification, in 2004 they sponsored a "Town Hall Meeting on Gentrification," at which they screened the film *Flag Wars,* which documents gays' and lesbians' gentrification of a neighborhood and the subsequent displacement of African American old-timers.

Discussing the film, many referred to their experiences as first-wave gentrifiers: "Look at the history of neighborhoods here in Chicago. . . . In the late 1960s and '70s, the gay community took credit for changing Boystown. We didn't really think about who we displaced. Now, in the late '90s, a lot of the gay community left [because] they felt [Boystown] was becoming another Lincoln Park [a highly gentrified Chicago neighborhood]. Now we're going to other neighborhoods. I think it is great that there are people thinking about what we're doing. . . . At least some groups are talking about this." Another audience member asked: "Should we all live in separate neighborhoods? Or how do we all live together? How do we get along?" At another meeting organized in response to a proposed high-end development, an Andersonville preservationist said: "This is what happened in Lincoln Park and Old Town. Gay men rehabbed, then moved out. How long will it take [here]?"

In my interviews, gay men like Joe, the former Boston realtor, and Sean, the former Boston historic preservationist, draw from personal experiences to articulate a similar sense of responsibility for gentrification. For instance, a gay artist who has lived in Andersonville for fifteen years talked about his decision to move away. He at once expressed distaste for Andersonville's affluent, gay newcomers and fear that he will contribute to the gentrification of the neighborhood to which he is moving: "I think [Andersonville] belongs to them now. I think the day of the neighborhood free spirits are kind of done. And we go to another neighborhood where can we fix it up, and it becomes the next thing. . . . It's happened all the way through my adult life. Every apartment that I've ever had, the rent just kept going up and up and up, and the neighborhood gets better and better and better." Similarly, another gay man—an Argyle professional—explained that his resistance to Argyle's gentrification is rooted in discomfort with the gentrification of Lakeview or Boystown, another Chicago neighborhood popular with gay men in which he once lived:

> Fifteen years ago, Lakeview was a lot like what [Argyle] is like now. It was a very diverse, very mixed community. . . . It used to be the place where a lot of students lived . . . a lot of Hispanic families that lived there forever. . . . At one time there was a very large Puerto Rican community in Lakeview. Mexicans, of course, they got forced out as things moved along. Then the gay community moved in, but it always had stayed a very mixed community, and it just wasn't anymore. It's just a real rich community now. . . . *The success of the community could almost kill it* . . . too many big cars, too many people, too many developments, just too much of everything. (emphasis added)

This reveals that, for some, the preservation ethic is developed through exposure to gentrification, and some gay preservationists (the lesbian preser-

vationists in my sample articulated these concerns less frequently) are particularly self-conscious about the fact that they helped gentrify multiple locales.
That said, it is important to reemphasize that, on average, the preservationists
in my sample have lived in their neighborhood or town for nearly the same
length of time as other gentrifiers and are similarly mobile. Furthermore,
my sample contained gay men who have participated in the gentrification of
multiple places but who do *not* adhere to the social preservation ideology.
Thus, we cannot simply attribute the popularity of the preservation ideology among gays and lesbians to participation in gentrification's first wave or
to their gentrification of a disproportionately high number of places. Many
gentrifiers—gay and straight, preservationist, pioneer, and homesteader—
have participated in early gentrification as well as in the gentrification of multiple places.

   Thus, preservationists are responding, at least in part, to the popular association of gay men with gentrification's first wave. Indeed, gay men are often depicted as either cut-throat or naive early gentrifiers. A *New York Times*
article about Harlem's gentrification reads: "Gays have often been at the forefront of gentrification in New York City and elsewhere in the nation, said
Charles Kaiser, author of *The Gay Metropolis, a History of Gay Life in New
York.* . . . 'Gays are pioneers,' he said. 'They are outsiders already'" (Williams
2000, 49). Jonathan Lethem's aforementioned fictional account of gentrification bemoans early gay gentrifiers' naïveté: "It was the same space that
communists and gays and painters of celluloid imagined they'd found in
Gowanus, only to be unwitting wedges for realtors, a racial wrecking ball"
(2003, 508). When PBS aired *Flag Wars,* the gentrification documentary discussed above, I received clippings from an aunt and e-mails from colleagues
and read a *New York Times* article about it. Many also encouraged me to view
an episode of the NBC show *Will and Grace* centered on gay men, historic
preservation, and gentrification as well as the 2006 film *Quinceañara,* which,
with the exception of a few minor characters who articulated social preservation concerns, portrayed gay gentrifiers as coldhearted speculators (see, e.g.,
Hall 2006; Antani 2006; Klawans 2006; and Taylor 2006).[39] Thus, popular
imagination holds the gay man as either a ruthless or a dangerously green
pioneer, two distinct roles that preservationists carefully avoid.

   Social preservationists—gay, lesbian, straight, and everything in
between—often paint an image of the gay gentrifier that, like many media
representations, reflects only a narrow slice of the group's economic, racial,
and cultural diversity. Their descriptions focus on toned men who labor
in Andersonville's Cheetah Gym or Provincetown's Muscle Beach (another
gym) and well-dressed men who flirt over coffee in Andersonville's Starbucks

or host catered Provincetown parties. Even when my informant was himself a gay man who defies this stereotype—for instance, a heavily built Argyle secretary who asked me to meet him in a coffee shop unpopular with gay men—he conflated (and condemned) himself with this caricature of the consumerist and increasingly powerful white gay man for being at the center of gentrification. Ironically, even in Andersonville—a neighborhood that, by most accounts, lesbians first gentrified—preservationists are quick to offer the affluent, white gay man as the nail in the coffin of local authenticity (and not, e.g., an image of a callous, bourgeois lesbian).

In this sense, many social preservationists seem to regard contemporary gay male identity as a force that threatens the authenticity they value. Like the popular culture sources offered above, they have come to associate a certain "destructive" lifestyle and attendant mentality with affluent gay men and gentrification. For instance, a preservationist said:

> Since most of the people moving to town are gay, it's like gay paradise. This exclusive ghetto mentality: this is a gay town. They're missing the subtleties of it doesn't matter and that Provincetown is not about identifying your sexual preference; it's about who you are as a person. That's being lost. So I think it's about upscale, nouveau riche, showing your money, and wanting it to be gay. You know, mostly gay . . . it's a very specific economic identification . . . upwardly mobile, white, fast cars, nice homes, travel. The upwardly mobile, upper-middle-class lifestyle. That's one little element of being gay. Most people don't—most Americans, let alone most gay people—don't live that way.

It is plausible that such complaints about gays' role in gentrification—made by gays and lesbians—relate to a broader discomfort with upward mobility, with gays' increasing wealth and power, particularly as it takes shape in places like Provincetown and Andersonville. For many gay preservationists, the sense of entitlement that they associate with this upward mobility is particularly problematic, for they believe that it does not adequately acknowledge their history of subjugation: "I think a lot of gay people, particularly the ones who are coming here more and more, feel this sense of entitlement because they feel so marginalized in their communities where they come from that they come here and it overreaches. They think: I can do anything that I want here. . . . Everybody has the capacity to discriminate regardless of your background." Despite recent popular attention to gays' displacement from their enclaves by late-stage heterosexual gentrifiers, social preservationists rarely report this concern, worrying instead about their own impact on old-timers, and, more generally, wondering whether they are equipped to wield newfound wealth and power in what they regard as a socially responsible manner.

The Sexual Organization of the City, a multiauthored study of sexuality in Chicago, argues that cultural economies enable and constrain certain sexual opportunities and organize "sexual relationships and identities" (Laumann et al. 2004, 349), suggesting that sophisticated gay consumers like preservationists would celebrate places like Provincetown and Andersonville as critical venues for sexual encounters and the construction of relationships, romantic and otherwise. Alternately, work on gays' movement out of enclaves (lesbians' residential choices are grossly underexplored) suggests a new apathy toward or an outright rejection of such venues (Brekhus 2003). Still others suggest that gays are displaced from enclaves in late gentrification, misidentifying them as only early-stage gentrifiers and early-stage gentrifiers as only gay men.[40]

Social preservationists present a more complicated relation to the gay bars and guesthouses that populate gay-friendly space than any of the perspectives outlined above allow. While many gay preservationists express discomfort with the ghetto, they nonetheless select residence in a place popularly linked to gays or lesbians (e.g., Provincetown and Andersonville) or in close proximity to one (e.g., Argyle). Only in Dresden do they live without immediately accessible gay bars and institutions. This is not to point to the hypocrisy of preservationists who decry the increasing presence of those like themselves only after establishing residence in or near such places. Rather, I seek to isolate their competing impulses: to live in or near gay space but to prevent those like themselves from destroying place characteristics they value and deem authentic.

It is tempting to surmise that increasing popular acceptance of homosexuality accounts for these dual impulses. In the context of increasing residential freedom, many Andersonville lesbians relocated, purchasing homes in largely heterosexual and less upscale suburbs and neighborhoods, and several gay Provincetown preservationists, weary of gentrification, left for rural Vermont and Maine. In other words, the world outside the ghetto is increasingly open, and this may reduce the ghetto's appeal and encourage criticisms of gay gentrification.[41]

Yet I was surprised to find that none of the social preservationists I interviewed articulated this perspective. Instead, while they shape their choices in the context of gays' and lesbians' new horizons, when it comes to residential choices, their concern is less about new opportunities and more about the cost of their increasing power (or, at least, the increasing power of the archetypal gay man) for others, particularly old-timers. In the minds of many gay preservationists, the enclave is no longer an escape from homogenization and oppression but *a force thereof,* a force that their own experiences and the popular press explicitly link to gentrification.

In general terms, gay and lesbian preservationists' relation to their sexual identity is entangled with their approach to their place of residence. While queer identification shapes attitudes toward gentrification, amid a public discourse that disproportionately holds gays accountable for gentrification attitudes toward gentrification also shape perspectives about one's sexual identity, or at least reactions to the presence of other gays in gentrifying space. Without making an argument about causation, I will close this section by noting the parallels between preservationists' perspectives on gays' decreasing marginalization and their views of gentrification.

While we often presume that social actors seek opportunities for power and influence, in the gentrification context some social preservationists are hesitant to do so, even (or especially) those like gays and lesbians who are cognizant of their social group's marginality. They work to carve out space for old-timers, rather than for themselves, despite parallels between gays' increasing local status and broader social advances, parallels that we might expect them to celebrate. Relatedly, while we often presume that social outsiders seek integration, some gay and lesbian preservationists relish their outsider status, and further research might explore whether this extends beyond their neighborhoods and towns to business or politics. Despite such seeming self-sacrifice, as a whole this chapter implicitly reveals the self-interest that accompanies social preservation. After all, preservationists' beliefs and practices are not simply intended to help those whom gentrification threatens. They affirm how preservationists wish to see themselves and be seen by others (i.e., as someone other than the power-wielding gay man unaware of his role in gentrification). Finally, even gay and lesbian preservationists are well positioned to be generous with their privilege. They do not depend on the accrual of local status to the same extent as those with fewer economic, social, and cultural resources. This serves as further evidence of how social preservation rests on a sense of privilege and status security, even among those we do not often think of as particularly powerful and who are cognizant of their own marginality.

This invites the question of whether other social groups that are at once aware of their traditional marginalization and of their increasing power—such as upper-middle-class or wealthy African American, Latino, or female gentrifiers—are also prone to adopt the preservation ideology and related practices. There is one reason to suspect that this is not the case. Put simply, such groups do not inherit the legacy of serving as the poster children of gentrification that so heightens the self-consciousness of many gay and lesbian gentrifiers. It is plausible that there are spaces in which other groups inherit this self-awareness vis-à-vis gentrification, such as upper-middle-class

African Americans in gentrified Harlem, and others might pursue this question. In the conclusion, I offer further speculations on the relations between marginalization, privilege, and social preservation.

Conclusion

This chapter provided a rough map of the social location of the social preservationists I interviewed by contrasting their characteristics with those of the other gentrifiers—pioneers and homesteaders alike—in my sample. We have seen that the preservationists I studied are more likely than others to have attended an elite college or university, have a degree in English, identify as gay, lesbian, or bisexual, work in the public sector or as an artist or a writer, and have experienced status security. These demographic differences relate to a set of cultural orientations to which many preservationists adhere. Many are omnivores (Peterson and Kern 1996) with a taste for reading and a cosmopolitan sensibility. Gay and lesbian preservationists articulate a self-reflexivity about their marginalization and about their privilege relative to old-timers that reflects the commitment to virtuous marginality common among all preservationists.

Contra standing arguments that resistance to gentrification concentrates among those at risk of displacement, the preservationists in my sample are not marginal to the middle class, nor are they marginal gentrifiers (Rose 1984; see also Caulfield 1994). Instead, while other forms of resistance to gentrification, such as longtime residents' self-advocacy, originate in fear of personal displacement, social preservation predicates itself on a sense of privilege. Specifically, it roots itself in dual forms of self-consciousness about one's agency: awareness that, by moving to a place, one contributes to transformation and disruption and that one possesses resources to counter this.

Picking up on a question posed at the close of the discussion of sexual identity, this relation between privilege and social preservation may tell us something about which gentrifiers may be less likely to adopt the ideology and engage in social preservation practices. Among those least likely to engage in social preservation may be those uncertain of their privilege or social location. As my data on social class suggest, those who regard themselves as occupying betwixt-and-between social positions may guard their position vis-à-vis old-timers and seek status and financial gain through gentrification (see Pattillo 2007). Deviations from this pattern may occur when, as with gay and lesbian gentrifiers, marginalized groups are made acutely conscious of their power and privilege within the gentrification context.

That said, the differences among gentrifiers in my sample are small. It is not enough to suggest that those who are highly educated are most likely to become preservationists. Nor is it sufficient to say that all members of the "creative class" (Florida 2002) appreciate old-timers, for smaller occupational differences are of greater import. Furthermore, there is much diversity among social preservationists, and individuals tell competing stories about the causal relation between traits like work and ideology. Thus, it would be a mistake to overstress the power of demographic characteristics to shape ideology and practices. Insofar as preservationists are members of a loosely defined group, it is shared ideology and practices—not a narrow set of demographic traits—that bind them together.

We also do not know whether the demographic differences more common among the social preservationists in my sample point to traits that directly produce the preservation ideology or whether they relate to a broader set of cultural orientations that, in turn, encourage the adoption of the social preservation ideology and shape related life choices. For instance, in their educational institutions and workplaces, preservationists may not encounter direct instantiations of the social preservation ideology but may instead uncover a related concern about authenticity, the underdog, and change processes. Furthermore, there is some reason to believe that preservationists' cultural orientation—their cosmopolitanism, membership in the reading class, and taste for marginality—helps determine their precise social location, for they may contribute to occupational, educational, financial (e.g., homeownership) and even relationship choices.

Ironically, having explored social preservationists' social location, this chapter reminds us that individuals can cohere primarily along the lines of ideology, rather than in relation to shared demographic characteristics, and that shared ideology does not depend on uniform demographic traits. While social scientists often examine ideological trends among demographic groups, my findings encourage us to think of ideology and demography as less neatly coupled. Those who share an ideology can, in demographic terms, be heterogeneous, and a large group like gentrifiers—bound by a relatively narrow set of demographic traits and residential and financial choices—can possess wildly divergent and even conflicting cultural orientations. Indeed, there is great ideological heterogeneity among gentrifiers, a group we mistakenly took to possess a common ideology because of shared material and demographic traits. Thus, while this chapter has identified trends among social preservationists, we must resist the temptation to turn to demography, rather than to ideology and cultural practices, as a primary marker of ideological alignment. After all, this very error helped disguise important differences among gentrifiers, leaving social preservation long unexplored.

Self-Representation:
Old-Timers' Perspectives

This chapter explores the response of those whom social preservationists regard as old-timers to gentrification. Drawing from interviews and observations, it documents their perspectives on community change.[1] The following pages reveal that, in spite of their well-founded attention to old-timers' struggle, preservationists' assessment of gentrification does not always correspond with old-timers'. Social preservationists are highly attentive to gentrification's costs and, therefore, notice certain facets of the process: disenfranchised old-timers rather than those who serve on boards; the growing number of Andersonville shops that serve gay men rather than thriving Swedish ventures. Most old-timers are aware that, in the aggregate, they are in a losing battle with gentrification. However, their collective response is more wide-ranging than preservationists'. While much of this book documents preservationists' view of gentrification (which they presume to share with old-timers), this chapter highlights the differences between their account and old-timers'.[2]

The chapter also emphasizes (as did my interviews) old-timers' perspectives over those of other longtime residents whom preservationists do not seek to preserve. This is not because other longtime residents' perspectives are less worthy of our attention. Among the eighty longtime residents I interviewed were approximately a dozen whom preservationists do not include in the *old-timer* category, and their sentiments and experiences inform this chapter.[3] However, I highlight the discord between preservationists' view of gentrification and that of those they seek to shelter from the process. Because most preservationists seek to speak and advocate for old-timers (rather than for the full body of longtime residents), the discrepancies between preservationists' and old-timers' perspectives on gentrification are most evocative for a full understanding of social preservation and are, therefore, the focus of this chapter.

I do not present these differences to debunk social preservationists. After all, their criticisms of gentrification resonate with much of three decades of scholarship (see, e.g., Atkinson 2000; Berry 1985; Chernoff 1980; Gale 1980; Kasinitz 1983; Marcuse 1986; Mele 2000; Quercia and Galster 1997; Smith 1996/2000; and Zukin 1987). Their sentiments also correspond with my findings about the cumulative costs of gentrification for longtime residents. Furthermore, preservationists and old-timers' concerns are often parallel. Both are attentive to old-timers' physical displacement as well as to their cultural, social, and political displacement (Chernoff 1980). Sometimes they work together to preserve old-timers' influence and to symbolically mark old-timers' presence. In short, social preservationists get it right much of the time. That said, having heard their perspective, this chapter emphasizes the distinctions between preservationists' and old-timers' accounts as a reminder that social preservationists are not old-timers' direct representatives. They do speak, not *for* them, but rather *of* them. Necessarily, their representations better mirror the experiences of some than of others.

This chapter reveals that response to gentrification varies substantially within and across the sites, from Argyle old-timers' relative enthusiasm for gentrification-induced improvements to Provincetown old-timers' vehement resistance. It also demonstrates that an individual old-timer's social position—as a homeowner versus a renter or as a business owner versus a laborer—influences his or her take on gentrification. The following pages detail this variation as well as old-timers' patterned departures from preservationists' perspectives on issues such as public space, community vitality, middle-class and "voluntary" displacement, gentrification's potential economic benefits, self-preservation, and newcomers' character.

Who Owns Space?

Scholars have long argued that a primary consequence of gentrification is the disruption of longtime residents' communities, of networks that provide, not only psychological support, but also social and economic support (see Bailey and Robertson 1997, 563; Henig 1984, 171; Logan and Molotch 1987; Zukin 1987, 133; Smith 1996/2000, 32–33; Atkinson 2002, 10). A parallel concern for the preservation of old-timers' community largely motivates social preservationists' resistance to gentrification.

In interviews, I asked old-timers about their networks, participation, and sense of community. With the exception of Argyle old-timers, most report community decline. However, my findings depart from gentrification literature's focus on displacement as chiefly responsible for community disruption and from community theorists' attention to institutions, participation, and networks as key to community vitality.[4]

Instead, many old-timers suggest that newcomers' usurpation of public space and the subsequent changing character of such space—specifically, its increasing cultural heterogeneity and the *vitality* of others' communities—lead to community decline.[5] Surprisingly, while they report the displacement of friends and family, they rarely regard this as the primary threat to community. Rather, they believe that newcomers' public presence endangers community.

While social preservationists seek to minimize their intrusion on public space by remaining marginal in places they believe belong to old-timers, they nonetheless think that community is best preserved by preventing physical displacement and maintaining old-timers' institutions. In this sense, some preservationists underestimate the basic effect of their presence on old-timers' sense of community and belonging.

Old-timers offer several explanations for how loss of space contributes to community decline. First, some speak of community decline as intricately connected to privatization and to changing norms about the use of space. For instance, to illustrate Dresden's weakened community, an old-timer contrasted the fact that a neighbor made a quilt for his wife when she was pregnant twenty years ago to a *new* neighbor's posting of private property signs across acres he had long used for hunting. For him, community depends on a communal approach to space. Other Dresden old-timers concur, simultaneously decrying community decline and the freedom to hunt on others' land. While Dresden preservationists often side with old-timers in conflicts over privatization, they do so to prevent old-timers' political displacement, not because they recognize the centrality of a communal approach to space for old-timers' sense of community.

Provincetown women frequently conflate community decline with a reduction in places "where kids can be around." Joanne, who is in her forties, recalled: "[When I was young], if the boogey-man started chasing you down the road, you could knock on anybody's door, and they would let you in, and they knew who you were, how old you were. I mean, you'd just stand there, and they'd say, 'You're so and so.' But that started changing [in the 1980s]." Another, Barbara, recalled a similar sense of spatial belonging: "What I miss most was growing up in [the traditional Portuguese neighborhood], which was a little community. There were kids across the street, kids around the corner. . . . There's not a lot of that left . . . places where you can run across the street, where kids can be around and run across the street and go to your neighbor's house." For Barbara and Joanne, community depends on the proximity of acquaintances and family and, as with the men discussed above, a sense that one belongs in communal and even private space.

Similarly, Provincetown Portuguese men frequently contrasted the (physical) freedom of their childhood with contemporary restrictions, a discourse suggesting that community declines through loss of power over space.[6] For instance, Albert, the assistant harbormaster, reported that the town's most substantial change was the transformation of state-owned (and communally used) wooded dunes, which border the ocean and constitute much of the town, into the Cape Cod National Park in the 1960s, a change that shortly preceded early gentrification (Shorr 2002, 99–101). He said: "All that was available to us in a way that it is not today. [You'd] think being a national seashore, all that land, all that open space—the dunes, the woods—would be open to us. In many ways, we feel like it's not open to us anymore." As a child, his experience of the dunes was essentially communal, and concern about the loss of childhood playmates infused his language about increasing park restrictions. He recounted learning how to drive a four-wheel vehicle in the dunes and "a whole network of trails" that he and his friends would explore: "That was our opening to that world." He shared a favorite childhood memory: "Can you imagine how I felt going out to the woods when I saw this wall of sand? I came down into the woods, and I said, 'What is that?' And I crawled up it and saw the dunes for the first time. [Recently,] my cousin . . . I hadn't talked to him in a long time . . . said, 'All I remember is that you and I saw the water for the first time together and how beautiful it was.'"

Provincetown women also contrast the sense of ownership over public, commercial space that they experienced in their youth with the current state of affairs. One said: "I go downtown to buy my shoes, and when I don't want to be around seeing all of that foolishness: I know other spots to go to 'cause sometimes it gets kind of sad because those spots are being taken over by them."[7] *Them*, of course, refers to gay and lesbian tourists and residents. Portuguese women report that they shop at malls in Hyannis—more than an hour's drive—to avoid Commercial Street, and, at a 2004 Community Visioning meeting, residents sought to ensure a year-round bus traveling to and from Hyannis.[8]

Thus, for many, community loss occurs in concert with or because of loss of power and space, specifically, because they feel uncomfortable in space they once felt they owned. They mourn the spaces—hunting blinds, dunes, and shops alike—in which they built and renewed ties. Preservationists sometimes defend such spaces, but they do so primarily because they regard them as markers of authenticity, not because they recognize their centrality to old-timers' community.

What is somewhat surprising about old-timers' claims about loss of space is that, in each site, old-timers maintain institutions and spaces primarily devoted to their group. For instance, many Provincetown old-timers are members of St. Peter's Catholic Church, drink coffee each morning at a drugstore counter, gather at the Knights of Columbus Hall, converse on the pier, sunbathe and fish at a particular beach, and sip beer in one of two bars.[9] While many community theorists emphasize the import of such spaces and institutions for the maintenance of community ties, they underestimate the significance of the content or characteristics of public space, specifically, according to old-timers, of its homogeneity.[10]

The fact that many Portuguese women feel that they must leave town to purchase shoes while Andy, the affluent gay Portuguese man, reports the opposite illustrates this point: "When I was a kid, it was like, 'We're going to Hyannis.' . . . Now a lot of the shops we have are so much more superior than when I was a kid. There was only a handful of clothing stores, one shoe store. When you had to do any kind of shopping, you had to go up Cape. Now it's not that way. It has become more of a small city."[11] Thus, it is not the literal disappearance of shoe stores that leads women to Hyannis. Embedded in talk about changing physical space is a response to the changing cultural character of space, and not all old-timers perceive such change as a threat to community.

However, many old-timers—chiefly in Provincetown and Andersonville, but also in the other sites—experience the increasing *heterogeneity* of public space (i.e., the growing presence and power of affluent newcomers, particularly gays and lesbians) as a threat to their ownership of space and, therefore, to community vitality. In Provincetown and Andersonville, the two sites with a highly visible population of gay and lesbian newcomers, many argue that gay newcomers are responsible for declining community because they have taken over public space.[12]

Specifically, old-timers suggest that they no longer feel comfortable in public space. For instance, after speaking wistfully of her Provincetown childhood, Joanne expressed sorrow for her daughter: "My daughter didn't have that when she was growing up. All of that stuff was gone. There was no place to hang out." Echoing the men quoted above, she said that she feels that sites of community that once were old-timers', such as shops and beaches, now "belong" to gay men. She said that she has "seen so much ruined," particularly "the national park": "[Gay men] go to the beach . . . , and they go trucking across the marsh, and they've ruined the vegetation . . . , and here are these people—the foot traffic is unbelievable—and they stopped letting us [old-timers] go onto the beach on the buggies [the four-wheel-drive vehicles many old-timers used to drive on the dunes]. . . . Gay people, they walk right across the strip, leave their condoms, everything, anything. Nobody went out and wrote 'em a ticket. . . . It just isn't fair; it isn't right. It's not. It's not." Similarly, a fisherman avoids taking his children downtown after, during a family ice-cream stop, witnessing men engaged in a sex act. He expressed vehement discomfort with the new face of Commercial Street and anger that it no longer feels accessible to him. Ingrid, an elderly Swedish woman, also hesitates to bring her family to Andersonville's Clark Street because of the annual Dyke March—with its bare-breasted participants—that takes place there.[13]

However, old-timers respond, not only to what they perceive to be gays' outlandish public behavior (e.g., topless lesbians, public sex), but also to their very presence, to subtle indicators of changing public norms and culture. For instance, an elderly Swedish man reported that pride stickers on Andersonville stores make him feel unwelcome in the commercial district. And, when I asked a Dresden old-timer why he believes community is weakened, despite the fact that the town is not popularly recognized as home to a large population of gays and lesbians, he said that gay newcomers have "replaced" heterosexual families: "It used to be families in town; now there's a lot of the gay community that lives in town. A lot of them. . . . A lot of them."

Using delicate language, Joe, an Andersonville religious leader who championed the inclusion of gays and lesbians in his congregation (once almost entirely composed of Swedes), nonetheless expressed concern about the changing character of public space. In reference to questions about community, he said that he worries that growing attention to gay and lesbian consumers makes "families uncomfortable." He spoke against a window display in a lesbian-owned erotica store that opened in 2005, not asking that the store close, but requesting a discreet exhibit. He recounted his concerns: "Just before [it opened], things they were selling were visible from the street. And at that section of Clark I know there's a lot of people with children. . . . But also there's some nightclubs . . . that are moving in and [a gay-owned hamburger franchise], and the concern there is, it is not very family oriented." He concluded: "I have a little bit of a concern that the increasing influence of the gay and lesbian community may [tip] . . . Clark Street into some places that are not what I consider very desirable for families." Thus, for him, community viability depends on the presence of heterosexual couples with children.

Such examples reveal the centrality of public space to many informants' sense of community and, implicitly, their belief that increasing heterogeneity threatens their sense of belonging. They also illustrate the extent to which communities overlap and interact within space. The informants quoted above indicate that what threatens their community is evidence of the *vitality* of gay and lesbian culture and public life. This suggests that some talk about community decline may not be generalizable to the multiple communities within a place—that within shared space some communities may flourish while others struggle—and that the belief that another community is flourishing may produce perceptions that one's own is in decline. This helps us understand the absence of talk about community decline within Argyle, for, despite Argyle old-timers' concerns about gentrification, they are fairly confident about the sustainability of their communities.[14] This is likely because their networks remain relatively isolated from newcomers—because linguistic, cultural, and physical barriers help maintain social boundaries.

Aside from those in Argyle, old-timers argue that gentrifiers' appropriation of public space, the increasing heterogeneity of such space (i.e., changing place character), and, implicitly, the vitality of others' communities lead to community decline. While they report that rising rents, property values, and local taxes have displaced friends, family, and neighbors, they rarely suggest that this influenced their perceptions of community vitality. This underlines the extent to which gentrification produces, not only physical, but also cultural and social displacement (see Chernoff 1980, 204; Levy and Cybriwsky 1980, 139; Spain 1993, 162; and Zukin 1987, 133).

It also suggests that, while institutions, participation, and networks are important measures of community vitality, so too is a less tangible sense of ownership of space and comfort with place character. Second, much like Tönnies (1887), Durkheim (1897), and Wirth (1938) (and preservationists), old-timers associate community with mutually dependent individuals who share common characteristics and physical space. Cultural heterogeneity, particularly that made public, threatens their sense of community.[15] That said, it is apparent that old-timers find certain forms of difference (e.g., gay men) more threatening than others.

Thus, old-timers are chiefly nostalgic for space in which, "when one leaves one's immediate personal or private space . . . , one moves into a world of acquaintances, kin, friends, enemies, and so forth, *with whom one shares a culture and a history*" (Lofland 1998, 9 [emphasis added]; see also Gusfield 1975). In a sense, they fear the public realm: "a world of many unknown or only categorically known others . . . many of whom may not share one's values, history, or perspective (cultural strangers)" (Lofland 1998, 9).[16] That old-timers rarely cite the displacement of members of their personal networks as responsible for community decline, emphasizing instead the disruption of public space, suggests that it is the parochial order or realm composed of acquaintances that they most miss (Hunter 1985).[17] However, it is also plausible that it is within public space and through cultural change that physical displacement becomes apparent.

In either case, old-timers tell a different story about community disruption than social preservationists and most scholars of gentrification and community would predict. While social preservationists are certainly aware of their intrusion on old-timers' space, they primarily seek to preserve community by preventing displacement and maintaining key social institutions. Such efforts likely benefit old-timers, but, by peopling space to accomplish such goals, they nonetheless contribute to spatial, and, therefore, community, displacement.

Assessing Community Vitality

There is also some discord between social preservationists' and old-timers' assessments of community vitality. In preservationists' minds, before gentrification old-timers' community thrived: vibrant institutions, traditions, shared identity, and mutual dependence nourished ties. Preservationists firmly believe that gentrification disrupts such communities by displacing people, institutions, and traditions. While most of the old-timers I interviewed concur, in fact emphasizing the costs of community disruption over and above those of physical displacement, not all did. Some report that community has not changed, and others suggest that gentrification strengthens institutions and increases interaction. Notably, most who report community fortification are middle-class and support other gentrification-related improvements.

For instance, Christina, an elderly Swedish woman who lives with her husband on Andersonville's northern border, said: "I think there's more community now than there was then. I really think so. I think people are much more friendly with one another now. I mean you knew your neighbors, but now we have phone numbers, and we visit at social gatherings, and we never did that when I was younger." Her sense of community vitality largely depends on a block club in her section of the neighborhood as well as on a community-policing group whose monthly meetings she began attending after a burglary. Both forums are popular with gentrifiers, and several elderly residents attended the meetings I observed. Together, middle-aged pioneers, homesteaders, and elderly old-timers complain about panhandling, homeless men and listen carefully to officers' reports of recent crimes. Christina's words indicate that her sense of community largely depends on such formal neighborhood activities.[18] Despite social preservationists' abiding appreciation for old-timers' community, formal organizations—like block clubs and community-policing groups—are much more popular with pioneers and homesteaders, and none of the preservationists interviewed recognized them as a site of community for old-timers.[19]

It is not surprising that some old-timers regard such organizations as community building blocks, for homesteaders and pioneers consciously join such groups and devote resources to them to nurture community (although many such gentrifiers regard old-timers as marginal to the community they hope to build). For instance, at a 2001 meeting of the East Andersonville Residents' Council, white middle-aged and elderly residents discussed a plan to place signs on local streets reading "Welcome to East Andersonville." Over tea in an antique-filled living room, one member asked: "Why do we need signs?" To this another replied: "For a sense of community." At the same meeting, the group planned a wine and cheese party to be held at the Swedish American Museum. One woman suggested that the party helps recruit members, while another suggested that the goal is to build "a connection to place." Similarly, members of a West Andersonville block club argued that their garden tour is not merely a fund-raiser: "My reason has nothing to do with money and everything to do with people. It's for the neighborhood to come out and look at itself . . . to get to know each other." In this sense, it is not surprising that some old-timers regard such organizations as sites of community. In contrast, preservationists regard them as mechanisms for gentrification and, therefore, as threats to "organic" community.[20]

Andy, the gay Provincetown Portuguese business owner mentioned above, also credited affluent newcomers with strengthening community by investing time and money in local institutions. This departs from preservationists' association of true community with those *without* the resources this old-timer believes strengthen ties: "The community just gets better and better and better. . . . There's a lot of people who come . . . they've had their share of big cities . . . , and they want to come someplace where they can have a sense of community. . . . Financially people have put a lot of money into the community that have moved here in the last twenty-plus years." Michael, a Provincetown mechanic, also reports that newcomers have strengthened community, in large part because they possess the human capital to do so: "There still is [a feeling of community]. I think the people that come here from away love the way they can feel welcome, they can get right into their community. . . . They need so many volunteers. They are always advertising for the planning board or the zoning board or the selectmen. There's so many of 'em. They need help. They need people. . . . A lot of people come here from away, a lot of educated people come here, and they contribute to the town." Indeed, during the interview, conducted at the height of the summer season at a busy intersection while the informant waited to lead a dune tour, passersby—newcomers and longtimers alike—frequently greeted him. Unsurprisingly, given his emphasis on politics as a measure of community vitality, he identified many by political position.

Similarly, another elderly, lifelong Andersonville resident reported that newcomers strengthened her community. Like the aforementioned Andersonville resident, she recognized the institutionalization of what were, in her childhood, informal neighboring practices: "Years ago we never used to have block parties. . . . Forty years ago. I don't think the block party was invented then. But everyone knew each other. People used to stroll around the block." (Indeed, several Andersonville block clubs hold summer events, renting inflatable play equipment for children or offering garden tours.) However, this resident, while an active member of an Andersonville Lutheran congregation, also emphasized the import of *informal* interaction for her sense of community. As the neighborhood "improved," as she put it, she was pleased to witness the return of families and informal street life she recalled from childhood, before many Swedish families left. She saw the departure of those she grew up with and the recent influx of affluent professionals, not as indicative of broad local transformations, but rather as part of a continuous cycle: "More families are moving back. A nice young man who's a doctor and his wife moved two blocks down with two little kids, and they joined our church. That's lovely to see. To me, *I don't see a lot of change because I've been here so long.* . . . A lot of people in this neighborhood have dogs, so they do walk around, so you see people" (emphasis added).

Such informants constitute a minority of Provincetown and Andersonville old-timers. That is, most report community decline. In contrast, Argyle old-timers—primarily first- and second-generation Vietnamese and Chinese American business owners—rarely express concern about the sustainability of their communities. While they articulate both trepidation and excitement about gentrification's economic and aesthetic implications, they do not attend to its consequences for their networks. When they did (infrequently) mention their community membership, they referred to ethnic business associations. For instance, a restaurant owner alternately used *community* and *Vietnamese Chamber* to refer to his association. Similarly, the leader of a business association conflated the establishment of businesses and of the community, only once referring to a larger *Vietnamese/Chinese community*.

In fact, some use *community* to refer to *newcomers'* networks. For instance, a Chinese merchant whose family has owned an Argyle business for much of her life made little mention of her *own* community (once referring to herself as part of a *merchant community* and another time as part of a *business community*). However, after speaking angrily about gentrifiers, she referred to them as *the community*. Referring to a block club's efforts to support businesses with improved facades, she said: "After we redid our facade we had more community business." She also referred to a pair of Vietnamese brothers who are members of the block club as "active in the community." Thus, in Argyle old-timers' lexicon, *community* is used to refer to the other (gentrifiers) or to formal institutions (e.g., the Vietnamese Chamber).[21]

Perhaps, much like social preservationists, who, despite their strong networks and civic participation, associate real community with old-timers, Argyle old-timers also associate community with others, specifically newcomers. As Schmalenbach suggests, community so imbues itself in daily life that we are seldom conscious of its existence (1922/1961, 334; see also Hunter 1975, 82; and Erikson 1976, 187). We are more likely to see *others'* communities (Schmalenbach 1922/1961, 334). Substantial cultural, language, and physical barriers insulate Argyle old-timers' networks from gentrifiers. Specifically, because they are relatively recent immigrants and small business owners, and because their neighborhood is at an early stage of gentrification and has a high proportion of old-timers, gentrifiers have not fully disrupted old-timers' networks. Furthermore, despite gentrifiers' residential in-movement, Argyle's commercial district remains segregated by race. Banners advertising Argyle as an "Asian Marketplace" and a pagoda atop the neighborhood el station differentiate "Asian Argyle" from the greater area. This physical distinction, combined with insulating cultural differences, may help prevent or delay the disruption of Argyle old-timers' community.

Thus, some old-timers depart from social preservationists' assessment of the effects of gentrification on community by arguing that it strengthens community by creating or revitalizing institutions. Others report that their new middle-class neighbors encourage informal interaction. Notably, those who most welcome gentrifiers' impact on community are the middle class, business owners, or the elderly and are more likely than others to welcome other improvements such as crime reduction and beautification.

This reveals a persistent theme: namely, how preservationists' parameters for distinguishing between old-timers and other longtime residents restrict their ability to perceive and predict the breadth of old-timers' perspectives. While, for instance, nearly all Andersonville preservationists conceive of old-timers as Swedes, they likely underestimate some old-timers' financial security. Their image of a real Swede does not encompass elderly women who move comfortably among middle-class newcomers or who celebrate the status of their new neighbors. For, as part of their construction of old-timers as bound to community, tradition, and place, preservationists emphasize economic insecurity and mutual dependence. Thus, while they do not explicitly exclude the middle class from the category, their image of the real old-timer underscores financial hardship. By associating economic struggle with community (and vice versa), preservationists are oriented toward a particular brand of individual that fails to encompass all old-timers and many other longtime residents, and this accounts for some of the discord between their own and old-timers' assessment of community.

Middle-Class Displacement

For similar reasons, social preservationists underestimate the concerns of property owners, many of whom worry that rising property taxes will lead to their displacement. While preservationists do not turn a blind eye to rising taxes, they primarily work to create housing for renters and homeownership opportunities for first-time buyers and to resist the conversion of affordable units into condos. Mirroring the gentrification literature, social preservationists' primary concern is for those whose displacement is imminent.

However, a variety of old-timers frequently articulate displacement concerns. Such concerns are particularly widespread among Provincetown's Portuguese homeowners. For instance, Suzanne, a middle-aged, working-class Portuguese woman said: "I love it here. I would never move. If I don't have to, I wouldn't. But, you know, it's getting too expensive to live here. Taxes, you know. The water bill. It's because they're having so much building going on." In her kitchen, a Portuguese woman in her eighties explained that her property tax bill nearly surpasses her social security income, forcing her to sell whale-watch tickets in the summer. Another said: "The only thing that concerns me is whether or not we're starting to outprice ourselves. . . . I hope that's not going to happen. . . . I am concerned about that." She explained: "We bought our [land] and built on it twenty-five years ago . . . before [prices] went way up. It was still expensive, but we really put a lot into it and bought the land and built the house. And it's become a struggle. My husband's job has changed [he is no longer a fisherman], and over the years it's become more of a struggle with the increase in property taxes." A middle-class Dresden widow expressed similar concerns:

> I just don't like what [gentrification] is doing to property values. . . . I mean the prices people are willing to pay! And we're one of the families affected by these horrible increases. . . . My husband bought a blueberry farm. . . . It was in his family since 1927. . . . We raise blueberries and sell them commercially to a freezer company. . . . It's not huge, but it's substantial. And that farm [has] a great view. You can see Mt. Washington . . . , and they're taxing us on that view now. It's a small Cape [Cod] that's not two rooms deep. It has little, tiny bedrooms, two big front rooms, and three little bedrooms across the back. But it has a summer kitchen and a huge two-story shed attached which is unfinished. It looks like a monster, but it's a small house. . . . The taxes are now

more than this house [the ranch she lives in].... When we first bought, taxes
were $700 or $800, and now they're $1,800! And that's in ten years.

She expressed particular concern about passing this tax burden to her
children.

Many blame newcomers for rising property taxes that put them at risk of
displacement. For instance, a Dresden old-timer reported:

> What messes us up is people come along and buy property from out of state.
> You know, they've always had better jobs than we do here and more money
> and everything, and buying property along rivers and paying big bucks, and
> that really ups our [tax evaluation].... I mean, they just move in here from
> out of state—doctors or lawyers or somebody who's always had a good job
> and made big money—that don't mean nothing to them. One hundred dol-
> lars to them wouldn't mean more than a dollar for us up here in the woods.
> That's the trouble.

Similarly, another said that rising property values and accompanying taxes
are "mostly from idiots from away that are willing and have the means to
pay extraordinary amounts of money for land. I'm of the opinion that, if
you come in here and you pay $100,000 for a few acres on the Eastern or the
Kennebec River, you ought to pay the town an impact fee of about $100,000
because what you're doing is driving everybody else's taxes up."

To illustrate the threat of displacement and the significant changes they
have witnessed, old-timers often contrast the purchase price of their home
with its current value. For instance, a Portuguese Provincetown woman near-
ing her ninetieth birthday recalled purchasing her home as a young woman:
"So we came and looked at this, and the woman said, 'Two thousand five
hundred dollars for this house.' Two thousand five hundred! And I thought,
'Oh my god, I can never do this!' ... At that time, it was sixty-eight dollars
for the taxes for a whole year and twelve dollars a year for water. The taxes,
now, on this house is more than what I paid for it, all these years later.... It's
not easy." Similarly, an eighty-two-year-old, Francis, recounted purchasing
a Provincetown house in the 1950s for $10,000. She said: "[It] recently came
on the market for $998,000. ... Imagine $998,000 for my $10,000 house!"
Recent owners transformed the small home into a luxury property, complete
with landscaping and guest quarters.

Some property owners offer policy suggestions to prevent displacement.
Bill, an elderly, Swedish Andersonville resident, said: "I bought my building
in 1964. We paid $27,500. But condos are going for $379,000, so therefore
my taxes go up. I think we should adopt [what] they have in California. If
you've lived in a place for twenty-five years or more, if you've owned it for

that many, then your taxes should be based on those rates. Because I think that's another thing that's driving older people out, because they can't afford the taxes. . . . I've seen a lot of them give up their homes." A Dresden couple made a similar suggestion: "It's not fair. If I paid $35,000 for my house and someone will pay a whole bunch more now, I should still be paying taxes on $35,000, especially if you're old. . . . Let the taxes go up when old people die or . . . sell. But people who have lived here for a long time and are old shouldn't be paying the same taxes as new young people who paid a lot more for their homes."[22]

These sentiments underline the fact that, while social preservationists are generally highly attentive to gentrification's risks, they underestimate their own effect on middle-class residents. This again arises from their association of real community with economic struggle and, therefore, from their devotion to working-class old-timers. As a result, preservationists underestimate the displacement anxiety of middle-class old-timers and, as we will see below, what some old-timers regard as their "voluntary" displacement.

"Voluntary" Displacement

For preservationists, gentrification's ultimate cost is old-timers' displacement. From their perspective, gentrification victimizes old-timers by forcibly uprooting them and leads to the disruption of the networks and traditions that lend meaning to their place of residence.

While displacement is a basic fact of gentrification and many old-timers bemoan this, some describe their friends' and family members' relocation in less stark terms, depicting them as reflective agents who deftly negotiate the opportunities and constraints that gentrification and their social position offer. Particularly in Provincetown, many report that some voluntarily relocated as gentrification took hold. As preservationists accurately anticipate, some report that this voluntary displacement results from political and cultural displacement, from the sense that the neighborhood or town is no longer home. However, others describe relocation, particularly by homeowners, as economically rational, reporting that some take advantage of rising home values or otherwise regard gentrification as an opportunity for mobility. The disjuncture between preservationists' explanations for relocation and old-timers' accounts likely arises from preservationists' limited attention to middle-class old-timers as well as from their assumption that old-timers are their opposite and, therefore, prioritize family, community, and tradition above the pursuit of external opportunities and economic gain.

The discrepancy between Provincetown old-timers' renderings of the town's boardinghouse past and preservationists' accounts of the same illustrates this point. For much of the early and mid-twentieth century, Provincetown Portuguese women rented rooms or cottages to summer residents, typically male artists, writers, and actors, many of whom were gay. Even when recounting local history, preservationists rarely mention this intermingling of new and old Provincetown. On the other hand, many old-timers fondly recall their interactions with boarders, celebrating their exposure to the world off the Cape and the promise of upward mobility. A few recount particular songs learned or books borrowed from boarders, the discovery that one was a famous composer, and the hope that a boarder might help a family member attend college or a conservatory. It is unsurprising that preservationists do not dwell on this moment when exposure to middle-class boarders promised to expand old-timers' horizons, minimizing the distinction between old-timers' perceived isolation and the middle-class urbanity of their boarders.

However, old-timers do not universally paint an image of newcomers' cultural capital as a resource from which they benefit. Indeed, some report that cultural displacement—a sense of being out of place in one's hometown—contributes to the decision to relocate. This feeling is particularly acute in Provincetown, where feelings of cultural displacement strongly relate to many old-timers' discomfort with the presence and power of gays and lesbians. For instance, an otherwise soft-spoken middle-aged Portuguese woman forcefully said: "My parents were driven out. . . . They lived here of course all of their lives, and [then] the neighbors bought this little tiny historic home. They were not nice neighbors. I mean, it's just people feel like they're being pushed out, and I think some of them are." A lesbian, this informant nonetheless believed that the neighbors—a gay couple—consciously sought to make her parents uncomfortable: "The two men that lived next to them, they were terrible, they were horrid. They were just nasty. They didn't want to live next to a straight, old couple. And they drove them out."

Several Provincetown old-timers suggested that gays' and lesbians' presence and growing influence induce old-timers to relocate; for some, the town is no longer culturally familiar. A middle-aged, working-class Portuguese woman said: "There's a lot of natives that are moving out of Provincetown. Just because of all the hoop that's going on, you know?" A more affluent Provincetown Portuguese man elaborated further: "The school system has a lot of gay people in it . . . which is one of the reasons why there are so few kids left. The parents don't let their kids go to that school. They move out of town and go to Truro Elementary. They stay away. I think there are only two hundred and fifty kids left in that entire school system. What they do is they vote with their feet: they leave town." Old-timers voiced similar concerns in 2004 when the town manager proposed using a portion of a school field to enable the privatization and expansion of the town-owned nursing home. At public meetings, old-timers suggested that this was the physical equivalent to the cultural displacement that they believed was afoot in the schools (see Faiman-Silva 2004, 170). A woman said: "Pretty soon they'll have us out of town!" Another offered a litany of concerns about the effects of the land seizure on children before stating: "I was born here, and I will be buried here, and my parents will be too!" Despite officials' argument that the move would allow elderly old-timers to remain in town (by keeping a facility open), many old-timers worried that it was another intrusion into one of the last bastions of straight Provincetown residents: the schools.

However, some depart from this narrative about cultural displacement, arguing that friends and family members sold and relocated to benefit from rising property values (Krahulik 2005, 197). In this, they follow a trend that Sharon Zukin noted in the 1980s when she suggested that some old-timers, particularly homeowners, seek to benefit from gentrification: "In economic terms, they forsake sentiment, or attachment to community, for exchange values" (1987, 133).

Many Provincetown old-timers recount stories of those who relocate for a profit. For instance, a successful merchant noted: "People come here, and, when the values go up so much, people just could see their houses [go up] and move to Truro or Wellfleet or anywhere else and build a bigger house for free. It's just economics."[23] A working-class old-timer was more critical: "They wave a price of money in front of you and want to buy your house. And they do it. Anything for a buck. It is almost like selling your soul. . . . That's why the town is going to hell because everybody's moving and nobody goes to town meetings . . . including myself."

Most are less judgmental, suggesting that rising property values provide homeowners with opportunities they would not have otherwise had. For instance, an elderly Portuguese woman described how rising prices benefited some of her working-class peers who long dreamed of retiring to Florida: "As prices started to go up, some of the people wanted to sell. They were making money. Now that was mostly if they were going to [move] to Florida [for retirement]."

These narratives stand in sharp contrast to social preservationists' interpretation of old-timers' relocation. Essentially, they believe that, when it comes to relocation, old-timers have little agency, and they hold newcomers responsible for this. One preservationist said:

> The Portuguese community sold out. They really did. . . . They sold their children out. They sold their grandchildren out. . . . That's not uncommon when you have people with a lot of money walking in and saying, "I'll give you some outrageous figure" that these people have never considered in their lives "if you move to Truro and give me your house." They weren't going to say no to this. "Oh, OK. We'll go to Truro. I'll get a new house." . . . They sold out, but it's like really, really who do you blame: the people who sold out, the people who were the tempters, you know, who were like way more sophisticated and knew exactly what they were doing? I mean who was the worst to blame? . . . [It's like] I'm going to tempt you, I'm going to dangle something absolutely irresistible in front of people who have struggled all their lives. . . . So, anyway, the Portuguese sold out, [and] there were different reasons for that. . . . It was like, "Well, I could send my kid to college. I can have a house." . . . But they're not thinking, "Oops, we lost any ability for them to have a home here."

Another concurs, saying: "You don't have a choice to say no to someone who would offer you that much money, but at the same time you really value your ethnic roots here." She elaborated: "Developers and people who want to build condos to the moon are coming in and saying to somebody who's lived here for a hundred years, giving them an offer that they literally can't refuse, like, 'I'll give you $4 million for your dumpy house.' I mean, how do you say no?" Thus, from preservationists' perspective, old-timers' marginal economic status, combined with newcomers' growing hegemony, left many old-timers with little choice but to sell out.

In contrast, old-timers articulate a greater sense of personal agency than preservationists would predict and suggest that gentrification offers a ticket out of town for those with few opportunities. Anthony, a sixty-something Portuguese resident who was a successful journalist off Cape for many years, argues that old-timers, particularly young people, sell, not only to make a profit, but also to pursue prospects outside town:

> I think the worst thing that's happened is the flight of young people and their families. . . . I'm not faulting anybody. It is an economic thing more than anything. I used to think it was obvious: they're leaving Provincetown because it's getting too gay for them. That's not what it was. People always used to say that the Portuguese drove the Yankees out of Provincetown. How the hell did we drive the Yankees out? We didn't drive the Yankees out of town. Every hometown in America, the kids want to get away. They want to leave their hometown and go elsewhere, almost all of them. Other people in Provincetown couldn't find a job. We were driven out by our parents, who said, "You have to get out of town. There's no opportunity here for you. You must go elsewhere." . . . So we left. And the same thing happened to the Yankees. . . . Coming back to Provincetown was a humiliating thing for those that had to come

back. They couldn't make it in the outside world, so they came back with their tail between their legs, and they were whipped, they were humiliated.

He recalled that his parents encouraged him to leave: "In the '50s when I grew up, the fishing industry was already starting to have troubles. We were driven out of town, really. Our parents all told us [to leave]. Boys like myself born into working-class families were first getting an opportunity to go to college." When he returned to town and purchased a business, some asked him: "'How could you possibly come back to this place?'"[24]

Similarly, when I asked a ninety-year-old woman who waited tables for most of her life (until she was eighty, when her doctor ordered her to stop) why her children live outside town, she said: "Well, actually, there's not much work. Especially the stuff they do. They're all doing very well. . . . They earn a much better living elsewhere, and they all have good jobs." Another recalled why many of the people she grew up with voluntarily relocated: "That first, second generation of fishermen's houses [opened up to newcomers]. The families weren't staying here because they were growing up, getting married, having their own homes, or going to college and not coming back. Their families would sell the stores, the houses." Angela, who is middle-class, said that, while she misses her daughter, who attends college off the Cape, she hopes she will not return: "I'd like her to be where there's a lot more for a young person."[25]

Informants suggest that young homeowners also relocate in search of opportunity, monetary and otherwise.[26] For instance, a man explained that his adult daughter's Provincetown home was too small for her growing family. After realizing that she could not afford a larger home in town, she opted to relocate. He said:

> My daughter built a house for $125,000. . . . You can't do that now. There's no land. She had an offer for $600,000. If she sold that place now, what would she be able to buy for $600,000 in Provincetown? Not something suitable for her family, and her kids would still have to go to Provincetown schools. So they're moving up Cape to Eastham, where they could buy a full Cape [Cod], a three-bedroom Cape with two baths, that type of house. . . . She'll be able to sell the house, get rid of her line of credit, pay off her car, buy her husband a new truck, move to Eastham, get a bigger, nicer house, and have better schools to send the kids to.

While he supports her choice, he was sad to see her leave, in part because it disrupted a family tradition he valued. He said: "[My granddaughter] is the fifth generation of our family here."

In this way, informants who suggest that relocation can bring opportunity are also aware of its costs—that friends and family members forsook "sentiment" and "attachment to community" to pursue economic opportunity (Zukin 1987, 133). Social preservationists are relatively unaware of how such choices, made by old-timers, are strikingly similar to those they have made: to follow career, residential preferences, and economic opportunity away from their family and place of origin. For instance, despite deep-seated admiration for old-timers' permanence, a preservationist acknowledged that he enjoys the freedom associated with his *impermanence:* "It's nice to have an opportunity to explore the newness of the moment, which you don't have where you've [lived before] because where you've been you're still who you used to be. . . . There's an opportunity [in a new place] to redefine yourself." Indeed, it is in part this sense of impermanence that draws preservationists to old-timers, those they imagine to be rooted in place. Paradoxically, social preservationists celebrate old-timers' permanence just as some old-timers— buoyed by gentrification—are experiencing their first taste of the freedom that preservationists at once enjoy and regard as inauthentic. It is plausible that preservationists fail to recognize old-timers' agency (i.e., the fact that some relocation is voluntary) precisely because they conceive of old-timers as those unlike themselves, making it difficult to discern shared traits.

Before concluding this section, I wish to address the question of why discourse about voluntary displacement is more widespread in Provincetown than in the other sites as Provincetown is not the only place from which old-timers choose to relocate. In the other sites, old-timers refer to others who have left of their own will, such as Vietnamese residents who live in the suburbs or Andersonville Swedes who moved to live near their grandchildren. Nonetheless, Provincetown old-timers' narratives about relocation are prolific and well formed. The fact that the town's Portuguese population dropped from 1,323 in 1990 to 691 in 2000—a much greater loss than that experienced in the other sites during the period—partially explains why displacement is at the forefront of residents' concerns.

However, Provincetown residents are also, in the main, adept speakers about gentrification, and they are ready with opinions about the process. While residents in the other sites sometimes hesitated before answering my questions, this was rarely true in Provincetown. This is likely because Provincetowners, more than those in the other sites, are practiced speakers about their town. This is a product of the town's tourist economy, which, for more than a century, has brought residents into contact with outsiders, encouraging them to articulate the character and concerns of their town to outsiders (and, perhaps, before tourism, the fishing industry similarly encouraged sailors to do so). Furthermore, Provincetown, unlike the other sites, is frequently the subject of books and scholarly articles. Thus, Provincetown residents are seasoned informants who have much experience with ethnographers and other authors, and this likely generates a unique self-consciousness about their place of residence.

For instance, on a snowy morning, I arrived at an elderly woman's home for an interview. After chatting about her recent flu, she led me to her living room and asked where I would like her to sit. "Karen had me sit on the couch," she said, referring, to my surprise, to Karen Krahulik, a historian who had conducted oral history interviews in town (see Krahulik 2005). I told the informant that she could sit where she liked and began asking my usual questions. She settled into a chair and responded with insight and good humor. However, our conversation was peppered with comments like, "As I told Karen . . . ," or, simply, "Like I told her . . . "

After the interview, she served a lunch (to this vegetarian) of tuna and pickled beets and then announced that we would watch a videotape of her oral history interview. On viewing it, I was struck by the similarities between the answers she provided on the tape and the answers she provided to me. As we watched, she occasionally said: "See, like I told you . . . " Apparently, for her, the interview was very much a performance, a performance she had had the opportunity to rehearse, review through the videotape, and, apparently, memorize.

While this interaction offers the most pointed evidence of Provincetowners' sophisticated approach to interviews, it was not the only one of its kind. On reading another account of the town, I found that an old-timer who tearfully offered me an account of her father had offered a similar story to another ethnographer. I also found myself awkwardly balancing what I learned from my interviews with two informants with what I had already learned about them from detailed profiles in another book. Observing a campaign meeting for a candidate for state office, I found that I was not the only researcher mingling with residents on the hosts' harborside deck: a group of Harvard students were researching and filming a documentary about gay marriage. Finally, Provincetown residents sometimes seemed mildly disappointed or even offended if I did not ask to interview them, saying: "I was wondering if you were going to call!"

In this sense, ethnographers (myself included) and other authors encourage Provincetowners to cultivate their stories and provide opportunity for their rehearsal. In addition, many have read others' accounts of Provincetown's gentrification, and this may help them develop strong opinions about the subject. Indeed, Provincetown residents avidly discuss written material on the town. For instance, at the end of 2002, the *Provincetown Banner* recounted the public reception of Peter Manso's *P-town: Art, Sex, and Money on the Outer Cape,* published that year: "Manso's book caused a furor even before it was released and he rode the resulting publicity throughout the summer. Hardly anyone admits to purchasing the book but many residents seem to have read it to see what the fuss was all about" (Harrison 2002b). Manso's was not the only book about Provincetown to appear that year; Michael Cunningham published a walking tour of the town and Kathy Shorr a collection of short stories about local history. Finally, a 2005 *New York Times* article, "Rich Gay, Poor Gay," on the town's gentrification generated much public discussion (Colman 2005). This is all to say that Provincetowners are not alone in their attention to voluntary displacement. Instead, they are adept at articulating their experiences with and thoughts and feelings about gentrification, particularly those that call for a discerning analysis of gentrification's risks and opportunities.

In summary, preservationists' and old-timers' analyses of voluntary displacement are not entirely disparate. Many concur that cultural displacement and financial pressures conspire to encourage old-timers' relocation. In this sense, few regard old-timers' choices as truly voluntary. Yet many old-timers—particularly the middle class and homeowners—would object to preservationists' suggestion that they are purely at the mercy of gentrification and universally devoted to residence in their neighborhood or town. Social preservationists' perspective arises in part from an apt analysis of the socioeconomic conditions of gentrification as well as from their understanding of old-timers as working-class individuals rooted to place and community.

Economic Benefits

Social preservationists worry that gentrification-related economic revitalization jeopardizes authenticity and leads to displacement. Therefore, it would surprise many preservationists to learn that some old-timers welcome gentrification because they believe that it will generate economic opportunity. In every site, business owners are the most ardent supporters of gentrification; however, Argyle merchants express particular optimism. That said, even Argyle old-timers' optimism about economic change is not unadulterated. Concern about political, cultural, and physical displacement dampens some of their enthusiasm.

Many merchants speak of gentrification in hopeful terms. For instance, a gay Portuguese inn owner believes that it has introduced positive change: "The people that . . . decide to become residents of Provincetown take a huge interest in the community, gay or straight. People that have fallen in love with

Provincetown, they move here, they really embrace the community and have a lot of interest in what goes on."

The same informant responded to a question about the most positive change he had witnessed by saying: "Just the growth." Similarly, Kathy, a Dresden old-timer who owns a bottle-redemption business, welcomes development. In fact, her primary concern is ensuring that commercial growth keeps pace with population increases, a position that both social preservationists and homesteaders object to. Sitting in her trailer, she said: "Growth is good, but we need to expand the town with it. I would like to see the town grow. It is going to happen eventually. We need to do something to make it more attractive." She criticized newcomers and old-timers who wish to limit growth: "So many people—some older people—don't want to see business in town. People who have been here a long time are used to it being rural and try to keep it that way. Some newcomers too. The town needs commerce and industry." Her business ownership likely encourages her positive view of commercial development. She labeled those who try to prevent change *conservation people,* recognizing their efforts to preserve markers of their families' local legacies. She described conservation people as "older people who can remember when the town had the ice industry, when it was flourishing and friendly and everyone knew everyone. They are afraid to take away from the beauty and culture of the town. As far as younger people from out of state, they don't want it to be a commercial state."

Support for economic change is particularly high among Argyle old-timers, many of whom (according to preservationists' conceptions) are entrepreneurs. As shop and restaurant owners, they believe that they will benefit (financially) from gentrification.[27] However, Argyle old-timers' enthusiasm also rests on the fact that, of the sites, Argyle contains the highest percentage of households earning less than $15,000 (32.94% of Argyle's population in 2000). Furthermore, residents believe that the crime rate is high (in fact, in 2004, the police district of which Argyle is a part had the lowest crime index and rate in Chicago) and associate this with poverty.[28] Together, these characteristics of early urban gentrification encourage interest in gentrification-related revitalization.

In Argyle, every old-timer interviewed noted neighborhood improvements, such as crime reduction. For instance, a religious leader suggested that most of his congregants—primarily elderly immigrants from the Ukraine—welcome gentrification. "[Now] they can go to the grocery store without getting mugged," he said. "Now their main concern is to avoid getting run over on the sidewalk by a bicyclist."

In typical fashion, a young Vietnamese American business leader expressed satisfaction with Argyle's changes as well as hope that they will continue. She emphasized crime reduction and development, saying: "In the '80s, it was a rough neighborhood, and everyone had [burglar] bars. . . . So now we're in 2005, and I think this area has changed quite a lot, and we've got condominiums, a lot of new development going on, and it's a little safer than it was before. . . . I think it's time to make the change now and give our area a better appearance." A Vietnamese business owner concurred: "I know this neighborhood more than twenty years. It used to be about 5 p.m. the Argyle district scare everybody. Very empty. But lately, like [in the] last ten years now, a lot of change." Another longtime resident, seated at a mural-planning meeting, said: "I was waiting and waiting for the neighborhood to become the next Lincoln Park [a highly gentrified Chicago neighborhood that many social preservationists consider inauthentic], and now it is! There's a lot of reinvestment in the neighborhood." Similarly, a Chinese business owner suggested that Argyle was once "dead": "This neighborhood used to be shuttered. . . . There was more garbage. . . . [There's a] difference between a living place and a dead place. . . . In a living place, there's a lot happening, there's a lot of activity . . . , whereas, in a dead place, there's business, but after a certain time it kind of flatlines. . . . When I was younger, this was a pretty dead place, especially after 5 p.m. . . . People didn't come out. . . . When they heard the gates come up, the rodents came out."

A young Chinese American business owner recounted Argyle's changes: "This was a very bad neighborhood. It was major, major bad twenty years ago . . . a lot of arguing, fighting, a lot of drunk people, a lot of drugs." As a child in the 1980s, she was embarrassed to tell suburban classmates that her father owned an Argyle business: "I always knew this was a dangerous place, so I never tried to paint it as a bad neighborhood when I told my friends about it . . . [because] they'd go home and tell their parents. . . . Their parents wouldn't bring them. . . . When I was little, we never mentioned the fighting and the junking and the drunkenness and the drugs." Because the neighborhood has improved, she now proudly tells friends where her family's businesses are located. Thus, for personal and economic reasons, she celebrates change: "I've watched [it] go from this really dumpy, nobody wants to come to your rat infested hole to now everybody wants to come. I think at our highest point we had like two thousand people here on the weekends. So this is a really exciting neighborhood."

Despite her appreciation for change, her account does not entirely contrast with social preservationists' analysis. Like preservationists, she celebrates changes that she believes Asian old-timers induced and criticizes those that she believes gentrifiers have forced on them, reporting that the "most negative change" is "yuppies that are moving in, these condos that are coming up." She is concerned because, as she puts it, "[yuppies] are very, very demanding": "They want change after only having lived here less than five, ten years. . . . We [Asian business owners] are the golden goose. We did lay this golden egg, and that's why people have moved in, to collect on the golden egg. . . . They moved here because we have made this community thrive. Under the worst conditions, under the most terrible ordeals, we have survived, and we have made this neighborhood. . . . There are cars up and down here on the weekends. There are people here during the weekdays too." She may have borrowed language about the *golden goose* from a social preservationist who had recently used the term at a block club meeting to speak against a pioneer's suggestion that coffee shops and pancake houses replace Argyle's Asian businesses.

Several other Argyle informants stated that affluent, white newcomers do not deserve credit for improvements and express concern that improvements will lead to the displacement of old-timers and their businesses. For instance, the head of the Vietnamese Chamber of Commerce at once celebrated Argyle's changes and worried about potential displacement: "The Vietnamese people and the merchants, they're really happy that there's a lot of new development, and they're proud to be in this neighborhood where they started." However, like her Chinese American counterpart, she offered a note of caution: "[The changes are] pretty positive, but they hope with new residents coming in that they understand that Argyle Street is such a unique street. . . . They just hope residents are open-minded when they come into the community."

Thus, in all the sites, but particularly in Argyle with its high proportion of low-income residents and crime concerns, old-timers who own businesses express an appreciation for gentrification that would surprise many preservationists. Specifically, they believe that they will benefit from economic renewal. However, as Argyle old-timers' words reveal, the sense that gentrifiers control the direction and tenor of revitalization efforts mutes their enthusiasm. Even so, at least in Argyle, old-timers are more supportive of redevelopment than are preservationists, many of whom hope to preserve, not only the presence, but also the authenticity of old-timers' small shops and restaurants. It is plausible that, as gentrification advances, Argyle old-timers will become less enthusiastic about gentrification's economic potential, for in Provincetown and Andersonville—where gentrification is advanced—such enthusiasm is rare.

More commonly, old-timers express appreciation for particular facets of economic revitalization, such as aesthetic improvements, enhanced public and commercial services, and cultural amenities.[29] Of course, these are common products of gentrification. Gentrifiers often renovate dilapidated housing and typically create new foot traffic, encouraging "housing investment and . . . additional retail and cultural services" (Freeman and Braconi 2004, 39). Notably, these are changes that preservationists resist for fear that they will encourage gentrification and minimize the distinction between their neighborhood or town and other, less authentic places.

Paradoxically, some old-timers suggest that aesthetic improvements help preserve local history by returning a place to its heyday. For instance, a middle-aged Portuguese woman said: "A lot of places are really being taken care of. A lot of homes that twenty years ago might have been run down are also being brought back to life and rebuilt, and a lot of it is put back to the way it was historically . . . which is wonderful."

Others articulate general support for economic upgrading. For instance, an elderly middle-class Swedish woman said: "There's been a lot of upgrading in the neighborhood. . . . There's been a lot of upgrading. There's a lot of yuppies in the neighborhood." Her husband framed these changes in explicitly positive terms: "In the last five or ten years, particularly in the last five, there's been a lot of improvement: renovation and modernization and such." In turn, when I asked whether she would freeze the neighborhood in a particular moment in time, the wife replied: "Now."

Some middle-class old-timers benefit from amenities enabled by gentrification. For instance, a young Swedish professional recognized that she is more enthusiastic about commercial changes than are some *newcomers* (likely social preservationists): "[Andersonville] has changed a lot. When Starbucks opened here a few years ago, it was a big hoopla with some of the smaller shops and some of the residents who were against corporate. I can understand their point, but Starbucks started somewhere too. . . . It has attracted more people because they might come from one shop and they go for another next time. And it's the same thing with the restaurants. The more restaurants, the better." She welcomes upscale shops that she, unlike many other old-timers, can frequent: "Better quality. The best businesses, more money, more expensive, nicer, more high-end places. Some cheaper stores are no longer here. . . . On a Saturday all the restaurants are jam-packed."

Despite complaints about overregulation and rising taxes, a Dresden old-timer admits that gentrification has improved town services: "[It] has probably improved some. Roads are paved that didn't used to be." An Argyle Vietnamese business leader also welcomes aesthetic change, lending her support to gentrifiers' vision: "The treasurer or the secretary from the block club wanted to see a fountain. That's pretty too. Some trees, everyone wants trees. So we're working with the alderman to get some trees down here."

Some suggest that gentrification has cultural benefits. A third-generation Swedish Andersonville resident praised gentrification for exposing her to groups with whom she would not have otherwise interacted. She said Andersonville has changed "not so much for the worst. For the better. It has become more diverse. I can walk down my block and go around the world." Another finds the presence of a leather bar amusing and suggests that gay men, like Japanese restaurants, contribute to local diversity: "There's sexual diversity. . . . It's the leather museum that is into S&M. . . . I think it's kind of interesting. . . . There are also two very, very good Japanese restaurants on our street."

A college-educated daughter of Dresden farmers suggested that gentrification had a more personal cultural effect. She recalled that exposure to the cultural capital of first-wave newcomers (back-to-the-landers and professionals) created opportunities for her:

> The town has really changed. It's not as poor as it used to be. There's greater diversity. I think it started to change when I was in high school. I think about the folks who moved in, who I got to babysit for. They were so different from the kind of families that I had grown up with. These were people who were college educated, doctors. . . . They really brought for me, I swear, a breath of life. I felt so fortunate to babysit for these people because they gave me a window to the outside that I really hadn't seen much. . . . When these people started moving in, they were helping me. They could help me with my college applications. . . . They did. They were absolutely fabulous. . . . [They were] encouraging my father that the choices I was making were good choices, that even though it wasn't the University of Maine nursing program maybe there were other schools out there that would be good too.

Circuitously, newcomers' capital helped her accumulate her own, including a degree from an elite university. In turn, this facilitated her appreciation for newcomers' culture and her ability to successfully run a Dresden farm that caters to newcomers' tastes. Together, these factors helped her avert displacement as other farms closed.

Thus, some old-timers welcome economic renewal and related changes, such as beautification and crime reduction. Foremost among these are merchants and middle-class residents who believe that gentrification will improve their quality of life. Some, like Argyle merchants, actively seek local economic reinvestment, while others simply welcome amenities they once associated with other, more affluent places. Both these perspectives depart sharply from those of preservationists, who contend that revitalization is an unmitigated threat to old-timers' influence and presence and rarely acknowledge that old-timers may welcome such change.

Conceiving the Newcomer

Social preservationists rarely distinguish among gentrifiers. They are universally critical of newcomers' effect on gentrifying neighborhoods and towns, and, while they may wish for old-timers to recognize their differences from other gentrifiers, they rarely do so themselves. In contrast, old-timers offer a more nuanced assessment of gentrifiers, welcoming some, and vigorously complaining about others. In this sense, they mirror preservationists' differentiation of old-timers from the larger body of longtime residents.

For instance, a Portuguese bank teller drew a parallel between herself and working-class newcomers, saying: "If they came here to work, they had the same feelings we do. We're here, we live here, we hold jobs, and we need to sleep and get up early." Yet some Provincetown old-timers express hostility toward Jamaican newcomers, who, with H2-B visas, perform labor during the tourist season. One said: "Businesses are sponsoring people by the hordes—I'm not prejudiced in any way—but I'm becoming that way. I can't say that I'm not prejudiced. Businesses are sponsoring mass numbers of these people—Jamaicans—to come here and work at their establishment. . . . So they're taking away good jobs that, when I was a kid, you used to get paid nine, ten dollars an hour for. . . . They're getting paid half that, if that." Thus, old-timers do not automatically align with those who are not a part of the gentry. In this they are much like social preservationists, who focus their attention on certain residents and largely overlook the same populations—such as Jamaican laborers—that old-timers do.

In stark contrast, a pair of retired Andersonville old-timers did not recognize middle- and upper-middle-class newcomers as new at all. Instead, from their perspective, newcomers are working-class renters, while middle- or upper-class newcomers do not seem out of place. Rather, in their view, they help the neighborhood regain equilibrium lost in the 1960s and 1970s when many middle-class Swedes left for the suburbs. Thus, old-timers' beliefs about the character of their place of residence (e.g., middle-class Andersonville or Portuguese, working-class Provincetown) shape their identification of certain residents as new or out of place.

Old-timers also differentiate between newcomers on the basis of their length of residence or the stage of gentrification during which they arrived. For instance, one distinguished recent arrivals from those with a longer history in Provincetown: "It got out of control, it really did. It really got out of control. Let's see. . . . Probably the '70s were OK. . . . I would have to say the mid-'80s [is when it got out of control]." A Swedish Andersonville woman holds early arrivals in higher regard than more recent newcomers:

We've always been an evolving neighborhood. . . . We had a number of Japanese, and we've always been open to that. And nobody has before tried to take over. There's a couple of people now that are saying [that a Persian business owner is trying to] get the Swedes out. He has openly stated that he wants to get the Swedes out. I guess he wants to take it over for his people. He's got a lot of property. . . . The Japanese have been [wonderful] neighbors. We have never had any clashes with them. In fact, when my kids were young, they would be playing with kids from India, from Japan, from Yugoslavia, from Germany, and the only thing they would be arguing about is who made first base. . . . Swedes tend to be open to that. They're not ones to make waves. And they keep to themselves. They won't say much.

Despite such claims, she later spoke disdainfully of her new gay neighbors. Implicit in her words are nostalgia for the neighborhood pregentrification, or at least at an earlier stage of gentrification, as well as a preference for groups that do not (explicitly) seek symbolic or economic ownership, therefore allowing Andersonville to remain Swedish.

Thus, just as preservationists consider only some to be old-timers, some old-timers identify only certain recent arrivals as newcomers. In Andersonville, Japanese Americans who arrived at an earlier stage of gentrification are not new, but their gay counterparts are. Similarly, Dresden old-timers more frequently characterize as new those from Boston and New York than those from other Maine towns. Social preservationists rarely make such distinctions. In their view, newcomers—regardless of their precise income level, place of origin, or even approach to old-timers—are problematic.

Old-timers occasionally distinguish between newcomers in a manner that suggests that preservationists stand apart from other gentrifiers or that they recognize components of the social preservation and homesteader orientations. For instance, a woman otherwise critical of Provincetown newcomers respects those who "stick it, stick here in the winter. They are tough people. They're caring people. They love what the town is, or what it *was*." Another differentiated between those who look for "cute" restaurants and those who "will just delve into it": "They'll sit down and talk to people in a club or wherever, find out what's what in a place. So I suppose there are those that are going to be knowledgeable no matter what."

Similarly, an Argyle Vietnamese American business owner suggested that newcomers who advocate for the replacement of Asian businesses with American establishments constitute only a small portion of newcomers, likely recognizing preservationists' and some homesteaders' resistance to pioneers' transformation efforts. She said: "[Most] don't want businesses to change; they just want the appearance to change. Everyone wants it to look better. . . . So I think that resident that made that remark [about wanting a pancake house] was just one out of [many]." Likewise, a farmer differentiated Dresden newcomers from others: "In juxtaposition, Dresden is a great place because [farmers] are supported. [Newcomers] are sympathetic [about] property taxes because they've seen what has happened in southern Maine. . . . And [most] have made a conscientious decision that, 'No, we want this land to be farms. We want it to be open space. It's irreplaceable.'" More subtly, some casually differentiate between those who seek friendship with old-timers and those oriented to other newcomers.

Therefore, while old-timers do not refer to gentrifiers by distinct names such as *social preservationist* and *homesteader,* they recognize intragroup variation. Indeed, they are more adept at this than social preservationists, who nearly singularly focus on the fact of gentrifiers' intrusion on old-timers' communities. While social preservationists are cognizant of newcomers' distinct approaches to gentrification, their criticism of gentrifiers (including themselves) leaves little room for the sort of differentiation that old-timers perform.

Self-Preservation

I am often asked whether old-timers join social preservationists' efforts to resist gentrification and preserve their presence. Indeed, despite some old-timers' support for change, in each site I encountered old-timers who work to prevent the displacement of their group and to preserve markers of place character.[30] Some old-timers' preservation practices parallel social preservationists'; they seek to prevent their own physical, political, and cultural displacement. However, the resistance to gentrification that drives social preservation does not always motivate old-timers' efforts. That is, some engage in preservation efforts without articulating opposition to gentrification. More frequently, they work, not to prevent physical displacement, but to preserve their mark on the local environment, regarding themselves as custodians of the built or natural landscape, history, and character. In so doing, they sometimes ally with social homesteaders. Below, I document old-timers' self-preservation practices, particularly those that depart from preservationists', and the characteristics of place and person that shape their efforts.

Like preservationists, some old-timers are concerned with maintaining their cultural and social presence. However, some—particularly merchants— also seek to preserve their economic power. For this reason, they are more supportive of reinvestment than preservationists and, thus, partially ally with pioneers and homesteaders. However, even such old-timers hope to preserve

local character alongside their economic power. For instance, a Vietnamese American business leader wants Argyle to improve without sacrificing
markers that attach Argyle to her group: "I hope it will look better [but] will
still have that Vietnamese/Chinese look to it—the pagodas or the dragons.
I think it will still have that Asian look. I would like to see it really neat and
clean and the awnings uniform. . . . It won't look like Old Town [a gentrified
Chicago neighborhood]. It will still have that Asian look to it, but at least it
will look prettier." Like social preservationists, she must determine which of
old-timers' artifacts represent neighborhood character. In interviews, many
Argyle old-timers suggest that they are caught between cultural preservation
impulses and a desire for upgrading. Specifically, they struggle with the question of whether security gates on Asian-owned businesses are part of local
character, as social preservationists argue, or are a remnant of an economically depressed past that they should escape, as pioneers and homesteaders
contend.

Some Dresden farmers engage in similar efforts to simultaneously support their businesses and preserve their symbolic presence in the midst of
the town's gentrification. With state funding, the Dresden Growers Association (composed of four farms) posted road signs marketing the "Dresden
Farmlands." According to a founder, the goal is, not only to market produce,
but also to celebrate local heritage and identity, to explicitly link farmers to
Dresden and Dresden to those farms. Like Argyle old-timers, these farmers
wish to benefit from gentrification-related commerce while also preserving
their presence.

A gay Portuguese old-timer, who is part owner of a historic inn, also
articulated a dual commitment to preservation and progress. As his words
about reopening the inn indicate, he considers himself a custodian of Provincetown's past: "It is nice to save a part of the town's history, and people have
been extremely supportive in everything that we've done to bring the place
back to life." On a winter afternoon, we spoke on the inn's sun porch, facing
the blue and gold of the harbor at low tide and, beyond it, the spire of Wood's
End Light. Eyes on the water, he repeatedly referenced the Pilgrims' landing
in the harbor (one of only a few to mention such).

For him, to own and restore the inn is to possess local history, given its
proximity to the site where the Pilgrims came ashore. His attention to the
past is deeply connected to an interest in progress: "I would have loved to
have been around in the early, early days when the [Pilgrim] Monument was
built and before all the roads and the horse and buggy. I would have loved to
have seen P-town then, *just so I would have even more of an appreciation for
how it has grown and how it grew*" (emphasis added). His words reveal that

a combined emphasis on history and progress is the domain, not only of homesteaders, but of some old-timers as well (of course, he is more affluent than most, which may account for his taste for progress).

Indeed, not all share his taste for preservation and progress. Sounding more like a social preservationist, an employee of Andersonville's Swedish American Museum revealed how gentrification-related change—specifically, threats to old-timers—inspired preservation. She described the museum's origin: "I think the first instinct was we shall not forget our past and history, and our impact on Chicago, and then also *holding onto everything* because this neighborhood was not very Swedish at that time. . . . The change was obvious, so that was why we wanted to maintain [the Swedish presence]." She added: "Any ethnic neighborhood has to stay where its original roots are."

Similarly, in the face of competing claims to neighborhood ownership, specifically over which group a proposed Andersonville streetscape should celebrate, Swedish American old-timers emphasized their claims to local history and tradition. As one said: "Because of its diversity, you can't say that this is a gay neighborhood or Middle Eastern because it's not. It's a very diverse community, and *traditionally and historically* it is a Swedish neighborhood. If you talk about ethnicity, this is the Swedish neighborhood in Chicago. And we have the Swedish bakery, and we have two Swedish delis, we have two Swedish restaurants and the museum. . . . You can eat Middle Eastern food anywhere, but you can't go to a Swedish museum anywhere, and you can't go to a Swedish bakery." The possibility that the streetscape would celebrate lesbians or Middle Easterners inspired old-timers' concerted effort to preserve their claim to local identity and character.

Indeed, another Swedish old-timer criticized the creation of a gay-themed streetscape in Chicago's Lakeview, or Boystown, neighborhood: "According to them, it was their neighborhood, and they wanted to build the towers that had the rainbows and so forth [rainbow-colored pylons representing the gay population]. . . . It doesn't give any recognition to the people that founded that neighborhood." Like the one just quoted, this viewpoint emerged during the streetscape-planning process, which old-timers regarded both as a potential threat to their place in Andersonville (should the streetscape recognize another group) and as an opportunity for the further institutionalization of Swedish character. While members of the streetscape task force engaged in much debate about which group the streetscape should recognize, the outcome—the blue and yellow banners and Swedish bells—indicates the success of Swedish old-timers' self-preservation discourse, with its claims to neighborhood history and tradition. While rarely explicitly referring to gentrification, at community meetings regarding proposed neighborhood

changes, such as the construction of a big-box store or proposed development restrictions, Andersonville Swedes regularly rely on local legacy claims
to preserve their symbolic and physical presence in the neighborhood.

An interview with an elderly Provincetown Portuguese woman revealed
a similar emphasis on old-timers' culture precisely because she believes it
to be threatened. After explaining that she does not attend the Blessing of
the Fleet because the diminished fleet makes her "too sad," she noted that
a younger generation of old-timers created a Portuguese Festival, held the
same weekend as the blessing (Krahulik 2005). She suggested that the formal
celebration of her culture became ever more important as its implicit celebration, the blessing, became less so: "Now it's a Portuguese festival. . . . [I] had
to see that. The parade is becoming very traditionally Portuguese, which is
very nice. They have groups who do Portuguese dances." She was less aware
than preservationists that, in 2002, those cooking Portuguese dishes included
Jamaican laborers and that lesbian tourists danced beside old-timers at a Portuguese folk concert.

Old-timers sometimes respond to a generalized sense of cultural and political displacement by advocating for the preservation of institutions with
which they identify. While preservationists sometimes join them, old-timers
articulate little of their cynicism about institutional representation (recall the
social preservationist who suggested that institutions are "monuments to a
dying culture"). For instance, in early 2006, Provincetown residents gathered
to discuss the town manager's formal inquiry into closing the high school,
which, with its dwindling student body, some homesteaders and pioneers
regard as a superfluous tax burden. Old-timers, like Andersonville Swedes,
evoked claims to local history and tradition in their opposition to the proposal. For instance, one alluded to the town's fishing village past, saying: "'If
you close the school, you might as well lay us out . . . and gut us like a fish'"
(Sowers 2005). Another characterized those in favor of closing the school
"'washashores'" (Sowers 2005). Others simply emphasized their enduring
relationship to the school: "'I was born here. I went to school here. . . . My
husband was born here. He went to school here. I had a couple of daughters
who went to school here. It's very necessary to keep our schools open'" (Sowers 2005). In response to such sentiment, the school board and town officials
compromised: the school reduced its funding request, and the town manager
dropped his formal inquiry into school consolidation.

Old-timers' language echoed a 2004 debate, discussed earlier, about
whether the town could appropriate a portion of a school playing field to
lure an investor to purchase the town-owned nursing home. The selectmen
anticipated that an investor might develop the land to create a "care campus."

At public meetings, many old-timers responded to the proposal by decrying a general sense of displacement. For instance, a woman said: "We don't want to be out on the highway. . . . Pretty soon they'll have us out of town!" A police officer echoed: "Once again we're shortchanging the kids. . . . We should not kick them out of town." He referenced how another institution associated with old-timers, the pier, was damaged in a storm after privatization. Eventually, the nursing home board, reporting that public sentiment influenced its decision, revoked its support for the land appropriation (Bragg 2004). In this way, old-timers work to preserve certain institutions (often with preservationists' support) and, to do so, often rely on claims to history and tradition as well as complaints about marginalization by gentrification.

Some old-timers, like social preservationists, emphasize the preservation of institutions because they serve as an important community site. For instance, an Andersonville old-timer suggested that the location of the Swedish American Museum—in what was once a Swedish-owned hardware store—is meaningful because the store was both an important Swedish institution and a time-honored site of community interaction: "[The] hardware store and this building—this was a place where people bought hardware, but it was also a social place . . . a resource center where a lot of people would use their space to get together to do things. So, when this building became available and some people wanted us to move out of the city, [a founder] said, 'No, we should not move out of the city because our past, our roots, our history is in Andersonville and the hardware store [represents that].'" Thus, like preservationists, old-timers sometimes emphasize community preservation.

In stark contrast, some old-timers ally with homesteaders rather than preservationists by emphasizing the preservation of the physical environment rather than social institutions or community. Among these are old-timers who serve on conservation commissions (aimed at preserving nature) and join historical societies. Others seek to return their place of residence to its heyday through beautification efforts. While social preservationists share some of these interests, they are more concerned than most old-timers that such efforts contribute to gentrification.

In general, while social preservationists often support old-timers' work to institutionalize and even market their presence, preservationists are more wary of the risk that such efforts—such as a streetscape celebrating Argyle's Asian residents or a Provincetown historic district—will encourage further gentrification. That is, many old-timers, like homesteaders, share preservationists' interest in place character, but their definitions of such character do not always align with preservationists', nor do all old-timers dedicate themselves to averting gentrification.

Which old-timers are most likely to voice preservation concerns? While my sample of old-timers who have lived away before returning to their place of origin is small, those I encountered disproportionately engage in self-preservation. It seems that living away can encourage commitment to preservation. For instance, the Provincetown inn owner who considers himself a custodian of town history lived in Boston and Florida. A Dresden old-timer, a retired military officer who has lived around the world, mused: "If you get out and go away, the yardstick for measuring the value of Dresden grows tremendously, and you come to value those basic things so much more." He explained that he wishes to preserve Dresden's "Americana" culture: residents' independence, roots, and volunteerism, qualities he finds insufficient in other places. Of course, residential mobility is also related to class, and a middle-class sensibility may encourage those who have lived away to engage in preservation.[31] Indeed, those who are without the resources to meet historic district codes or who depend on land sales are least likely to advocate for preservation. In this sense, some are better positioned than others to support preservation.

Self-preservation efforts among those who have lived away may also be a method for reaffirming a claim to one's place of birth, a motivation quite distinct from those that drive social preservation. For instance, I had the occasion to drive to Provincetown's airport with a Portuguese native, David, who now teaches at a West Coast university. On our drive, a woman on the sidewalk called out to him to slow down, to which David responded: "I'm *from* here!" She replied: "I *live* here!" Driving off again, and still maintaining a high speed, he muttered in response: "Where are *you* from?" He recalled a friend "driving eighty on this road." As we drove through the national park to the airport, he gestured to the dunes on either side of the road and, like any number of working-class old-timers, complained about park regulations. He recalled that his relatives "used to make and sell beach plum jam and rose hip preserves before you could get arrested for doing that." He associated this with the town's communal nature, which he nostalgically recalled from his childhood and linked to its early, renegade history. More generally, throughout the ride, he emphasized his lineal claim to Provincetown.

However, community-level economic and cultural factors also shape the likelihood that old-timers engage in self-preservation. For instance, Dresden old-timers' resistance to regulations—and, therefore, to policy changes that might aid their ability to remain in town—complicate their preservation instincts. A few informants were somewhat aware of this paradox. A working-class man named Jeffrey said: "It's not the community it used to be, but they're trying to keep it the way it used to be, but I think they're going about it in the wrong way [i.e., with conservation]." He added: "Do I like to see 'em [subdivisions]? No. Do I want to stop them? No." He explained that his resistance to development regulations, which the conservation commission frequently advocates for, is partially rooted in his tenuous financial situation: "I've got ten acres. I might want to [develop] it someday. If the wife gets sick or I get sick and we need money. Do I want [more development]? No. Do I want neighbors closer than what I've got now? No. But, if I have to, I have to." Thus, while he does not wish to see Dresden's rural character destroyed, his economic position and commitment to libertarian principles popular among Dresden old-timers complicate his preservation concerns.

Given that Argyle old-timers are also less likely to engage in self-preservation, it is plausible that the combination of early/midstage gentrification and high rates of business ownership among old-timers discourages self-preservation. Old-timers in places at early stages of gentrification sometimes believe that they will benefit from gentrification and, therefore, embrace initial changes, such as crime reduction, beautification, and an expanding customer base.

However, some old-timers in all four sites engage in self-preservation. Sometimes their practices align with preservationists', other times with homesteaders'. As with social preservation, the character and frequency of old-timers' practices vary by site, and, within each site, some individuals are more likely to engage in self-preservation. Middle-class residents who have lived elsewhere readily articulate preservation concerns, and, at the community level, overt threats to old-timers' physical presence or social authority—such as advanced gentrification, the proposed closure of institutions, or the desecration of traditional festivals—inspire preservation efforts. Therefore, we cannot simply conclude that old-timers and social preservationists have parallel commitments to preservation. While they share many concerns and sometimes ally with one another on particular projects, old-timers do not universally engage in self-preservation, and some of their efforts more closely correspond with those of homesteaders than with those of social preservationists.

Conclusion

This chapter reveals that old-timers' response to gentrification does not always neatly align with social preservationists'. For instance, social preservationists might find it surprising that some old-timers express support for economic and aesthetic revitalization, articulate a relatively strong sense of agency regarding their residential choices, and provide a different set of explanations for community decline than those social preservationists turn to.

This is not to suggest that old-timers believe that gentrification is without cost. Indeed, even those who hope to benefit from gentrification recognize its cumulative risks, and few express unmitigated enthusiasm for the process. Almost universally, old-timers' perspectives on gentrification are interlaced with concern about displacement and loss of power and space. Nonetheless, their responses to gentrification do not always correspond with social preservationists' view. In general, their perspective is less uniform than preservationists presuppose and less consistent than social preservationists' outlook.

The chapter also demonstrates that old-timers' responses vary within and across the sites. Four factors largely explain why old-timers' collective reaction to gentrification varies by site: the percentage of old-timers remaining in the town or neighborhood, stage of gentrification, old-timers' cultural and economic characteristics, and a place's urban or rural character. Importantly, none of these factors *alone* explains why old-timers in one site are more likely to support gentrification than are those in another site. For instance, Argyle and Dresden are both at an early/midstage of gentrification, yet Argyle old-timers are much more supportive of gentrification than are those in Dresden. This can be explained by Argyle's urban character and old-timers' crime and safety concerns as well by Dresden old-timers' libertarian resistance to regulations that they associate with gentrification. Similarly, despite their shared advanced stage of gentrification, Provincetown old-timers are much more opposed to gentrification than are those in Andersonville. This is rooted in the fact that Provincetown old-timers compose more than 20% of the town population (and, thus, have a sense that they can resist newcomers), while Andersonville old-timers compose less than 4%, and, relatedly, in the fact that a greater proportion of Provincetown old-timers are working-class and, thus, at greater risk of displacement.

Variation within a given site is often explained by an individual old-timer's economic position, with merchants and middle-class old-timers— particularly property owners— offering greater support for gentrification than do their working- or lower-class neighbors. Those who share traits with some newcomers, like gay Portuguese men, are also less likely to regard the relationship between newcomers and longtime residents as strictly discordant. In many cases, these individuals reside on the boundaries of preservationists' conception of the old-timer. As a result, preservationists are least attentive to those whose perspectives most conflict with their own.

While some old-timers are more enthusiastic about gentrification than are social preservationists, many nonetheless worry about the threat of physical displacement, and even more decry their "social displacement" (Chernoff 1980). In this sense, the chapter underlines the consequences of nonphysical displacement for *all* longtime residents. Regardless of the site or the individual, chief among old-timers' complaints is the sense that one no longer feels at home. This serves as a reminder that it is not enough to measure gentrification's costs by the number of individuals displaced or businesses lost. Even those who remain in a place as it gentrifies, and even those who celebrate some of gentrification's effects, bear the costs of transformation.

It is plausible that the emphasis that old-timers place on nonphysical displacement mitigates—from old-timers' perspective—the apparent differences between social preservationists and other gentrifiers. While social preservationists are highly conscious of their influence on old-timers' community and institutions, they cannot avoid contributing to the social and physical disruption associated with gentrification. Even their adoration of old-timers and their businesses may contribute to cultural change, and, while they are mindful of maintaining old-timers' authority, their practices nonetheless disrupt local political traditions. In this sense, because old-timers (as conceived by social preservationists) are the beneficiaries of preservation practices, which often focus on minimizing physical displacement and keeping old-timers' businesses afloat, they may be more likely than other longtime residents to identify social and cultural displacement as gentrification's greatest cost. As a result, with their attention drawn to social and cultural displacement, they may be less likely than other longtime residents to distinguish social preservationists from pioneers and homesteaders.

Finally, this chapter warns us against regarding social preservationists as direct advocates for old-timers and, therefore, against regarding social preservation as a neat solution to the problems endemic to gentrification. Components of the preservation ideology—such as association of real community with economic struggle—cloud adherents' ability to identify the full range of old-timers' perspectives on gentrification and even to recognize the perspectives of other longtime residents. This diversity of perspectives reminds us, much as the previous chapters have with regard to gentrifiers, that longtime residents do not possess a uniform outlook. Indeed, we have seen that even the minority of longtime residents whom social preservationists regard as old-timers vary greatly in their attitudes. This serves as a general caution against the presumption—made by scholars of and participants in gentrification alike—of ideological uniformity among the actors who people gentrifying places.

Conclusion

Hiding in the Open

This book introduced a heretofore unidentified and little-examined social process—social preservation—a bundle of ethics and practices rooted in the desire of some, who tend to be highly educated, to preserve the social authenticity of the gentrifying places to which they move. In four disparate sites—Argyle, Andersonville, Provincetown, and Dresden—social preservationists work to preserve the character of their place of residence. They do this by seeking to prevent old-timers' physical, cultural, and political displacement and through other strategies aimed at thwarting gentrification. Social preservationists orient their beliefs and practices in opposition to those of the pioneer, who seeks to "retake" space (Smith 1996/2000) and whom scholars have long regarded as the prototypical gentrifier.

This book demonstrates that perspectival and practical variation is widespread among gentrifiers and even among longtime residents. It emphasizes the fact that many gentrifiers deviate from the frontier and salvation ideology long held to be the sine qua non of gentrification.[1] Those who depart from the pioneer prototype include social preservationists and those I term *social homesteaders.* Like social preservationists, homesteaders are cognizant of gentrification's risks for longtime residents as well as for other forms of local authenticity. As a result, they share some of preservationists' criticisms of gentrification. However, homesteaders' criticisms are rooted less in concern for longtime residents or in an association of authenticity with old-timers' culture or community and more with preserving place characteristics, such as the built or natural environment. For instance, some Andersonville homesteaders support social preservationists' efforts to block the in-movement of chain stores, albeit to preserve the local streetscape and independent shops rather than the Swedes whom social preservationists wish to preserve. In Argyle, some support neighborhood improvements but are uncomfortable with pioneers' antipoverty rhetoric. Therefore, social preservationists are not alone in their departure from the pioneer prototype. Indeed, many contemporary gentrifiers take steps to distance themselves from the image of the ruthless pioneer central to popular and scholarly representations of gentrification (Butler and Robson 2003).

Nonetheless, this book has taken social preservation as its focus. It has outlined the tenets of the social preservation ideology and the practices that accompany it as well as how those practices vary by site. It has also examined social preservationists' admiration for certain longtime residents, the demographic and cultural distinctions between social preservationists and other gentrifiers, and how and why old-timers' assessment of gentrification departs from social preservationists'.

The presence of social preservation in all four sites, that is, the breadth of the phenomenon, is surprising given that for four decades scholars failed to fully identify the process. Indeed, this book suggests that preservationists likely reside almost anywhere other gentrifiers do and that they possess readily identifiable beliefs and practices. In the book's opening pages, I touched on why gentrification scholars left social preservation, and, for that matter, social homesteading, relatively unexplored. As I have argued, part of the explanation may be that discomfort with the urban pioneer prototype—driven by growing scholarly and popular criticism of gentrification as well as by gentrification's expanding scope—has, in recent years, increased and spurred the growth of social preservation. However, even if this is the case (and, for reasons I delineate in the introduction, it is doubtful that social preservation is altogether new), it does not fully explain why other scholars have not identified social preservation.

I believe that the reasons for this oversight encompass and extend beyond the trends in gentrification scholarship noted in the introduction. Specifically, I believe that they are threefold and rooted in several prevailing theoretical assumptions and preferences apparent in the social sciences—particularly in urban and community scholarship and cultural sociology—that have, in turn, informed gentrification research. These broad traditions include (1) scholarly allegiance to and advocacy for the "downtrodden," (2) urban researchers' nearly singular attention to political economy, and (3) the pervasiveness of two models in cultural sociology, the treatment of culture as product and producer of capital (a trend I mentioned briefly in the introduction) and the notion of culture as fragmented. The following pages detail how these traditions encouraged scholars to leave social preservation hiding in plain sight.[2] They also point to the scholarly interventions that social preservation encourages us to make.

First, it is likely that the political perspectives of many scholars (many of which I share) kept people from identifying social preservation or, when it was observed, from fully examining it. For instance, commenting on my *City and Community* article introducing social preservation, Tom Slater worries that to call social preservationists by a name other than *gentrifier,* and to attend to gentrifiers' intentions and practices, obfuscates the import of outcome. He writes: "Despite the efforts of the highly educated to protect the 'old-timers' from displacement, one has to question whose interests it serves to avoid using the term gentrification when it so clearly captures what has been happening extensively in the neighbourhoods of 'capital's metropolis'" (2006, 752).

While I disagree with Slater's analysis of my work—after all, I do not dispute that gentrification is happening or its consequences but rather call for a more expansive view of the ideology and practices that drive it—Slater is not alone in his desire to ensure that scholarship attend to the interests of long-time residents and specifically of those at risk of displacement. Indeed, many frame their scholarship as a mode of advocacy for those with few resources to advocate for themselves. In political and humanitarian terms, this desire to use scholarship to heighten awareness of gentrification's costs is commendable. Yet, for studies of gentrification, it has several attendant risks. First, in place of empirical questions, it encourages cynicism about the sincerity of actors' intentions, their ability to translate belief into practice, and the success of preservation efforts. More pressingly, if left unchecked, it impedes our scholarly objective (or at least mine) of accurately describing the social world in all its complexity and of explaining how it came to be as it is.

This inclination to use scholarship to advocate for the marginal closely relates to another broad tradition that regards the culture of the powerful, particularly of white middle- and upper-class Americans, as essentially uniform. As a result, many scholars devote attention to the structural position of the gentry rather than to their culture(s). This leads us to overlook cultural variation among the middle and upper classes, even while remaining wary of ignoring heterogeneity among the poor and the working class (see Small 2004, 176). We want to know about the cultures of the less powerful but too often leave that of the gentry little explored, presuming that it is enough to know their structural location and their related effect on the disenfranchised.

Correspondingly, ethnographers frequently seek to capture the perspectives of the disenfranchised but rarely attend to those who share scholars' traits. This arises from concern that depictions of the poor and marginal are diminutive or inaccurate, and the desire to remedy this, as well as from the belief that the gentry's viewpoints are familiar or predictable.

Arguably, these trends—advocacy for the disenfranchised, disinterest in the "known" cultures of the middle and upper classes, and a dearth of ethnographic studies of those who share scholars' traits—are rooted in a common political commitment to advocate for the less powerful. Ironically, scholars whose politics discourage attention to the perspectives of and variation among the elite may be surprised by the extent to which social preservation affirms their cynicism about gentrification as well as by preservationists' ready acknowledgment of such. By seeing through social preservationists' eyes, we learn how people construct a sense of authenticity that can trap them despite their best intentions. While acknowledging this, instead of blaming the affluent and ignoring their intentions or practices, I sought to document their beliefs and practices and to explain their origins and how they relate to those around them.[3]

In short, it is plausible that some of the very concerns that inspire social preservationists' attention to old-timers encouraged scholars to overlook social preservation or at least to leave it unexplored. Specifically, by emphasizing outcome and actors' economic positions, scholars forsook notice of a set of beliefs and practices that challenge the notion of the iconic pioneer whose culture and practices serve his economic interests and ensure gentrification's success.[4] Paradoxically, those who wish to advocate for old-timers may have missed opportunities to join forces with preservationists or to take social preservation into account when formulating policy.

These political concerns connect to a second explanation for why social preservation was long unidentified: urban scholars' long-standing devotion to the study of political economy. While it is imperative that we have an understanding of how politics and economies—from the local to the global—shape urban processes like gentrification (and, certainly, they determine much of its form), a lack of scholarly attention to culture helped leave social preservation in the shadows. Specifically, the assumption that a political-economic analysis tells us all we need to know about gentrification produced the notion that gentrifiers' ideology was superfluous.[5] Relatedly, when scholars broke from this and considered ideology, they too often assumed that, because gentrifiers share a structural and a geographic location, they must possess a uniform ideology (one neatly parallel and in service to their economic position).

The above products of urban scholars' theoretical preference for political economy contributed to the notion that the only newcomers truly critical of gentrification were those at risk of displacement or otherwise at the economic margins of the process, such as first-wave gentrifiers or renters. This book demonstrates, however, that social preservation depends on the assumption that gentrification does not endanger one's self, that one can use one's own privilege to try to shelter others from gentrification's consequences. Social preservation reveals how culture, specifically ideology, shapes choices, practices, and interactions. It also suggests that we cannot fully understand individuals, their relationship to others, or their cultural orientations by measuring their demographic characteristics or economic position. Finally, it demonstrates that it is not enough to understand the political-economic terrain of gentrifying places. For a full picture of gentrification, we must also explore the beliefs and practices of a variety of actors.

However, urban and community scholars' tendency to analyze actors in terms of their political-economic position rather than their culture is not the only reason social preservation was overlooked. Even scholars of culture overlooked social preservation, perhaps because of the notion that culture is a product of political economy that unambiguously serves the "growth machine" (Logan and Molotch 1987). Under this framework, individuals are foremost economic or instrumental actors and only secondarily cultural actors; shared meanings are resources we draw on for personal economic or status advantage. Also, by suggesting that gentrifiers' culture *always* serves their financial and other status interests as well as the interests of political and economic elites, this perspective implies that actors—even those who resist gentrification—have little cultural agency.[6] It is likely that this perspective discouraged attention to social preservationists, whose culture seeks to disrupt capital accumulation and is not directly oriented to the accrual of symbolic capital.[7]

Indeed, social preservation asks us to attend to how an ideology and set of practices complicate economic revitalization processes as well as personal financial and status gains. It conflicts with the assumption that all beliefs rest on pure self-interest or that all actors are simply trying to compete and gain advantage. In this sense, it calls us to consider culture for culture's sake because social preservation is value rational (Weber 1968/1978), or at least driven by less instrumental goals than those we typically associate with gentrifiers.

There are three ways in which social preservation complicates processes of capital accrual and, thus, challenges the culture-in-service-of-capital tradition in cultural sociology. First, few gentrifiers engage in social preservation to achieve financial gain; that is, preservationists do not seek to use their culture or ideology for material benefit. In fact, preservationists' behavior often impedes (while not entirely blocking) the maintenance or achievement of personal economic capital. Some forgo (further) profit by providing affordable housing, deciding not to sell for a profit, and working to prevent or stall economic revitalization.[8] Preservationists celebrate those who have fewer economic and cultural resources than they do and advocate for the reversal of local social, political, and cultural hierarchies from which they might otherwise benefit.

Second, while preservationists wish to distinguish their neighborhood or town from other, less authentic places, they do not seek to translate this distinction into cultural or economic capital.[9] They resist regarding their place of residence as a financial investment or measuring its value in economic terms. While their claims about authenticity and local culture sometimes contribute to the marketing of place, it would be a mistake to conflate outcome with intention or to assume that all claims about place character are of equal economic benefit.[10]

Finally, social preservationists do not emphasize old-timers' difference as a method of status acquisition. After all, social preservation rests on a sense of one's privilege relative to others. Preservationists' distinction from old-timers is a starting point rather than an end goal. For this reason, we must distinguish their appreciation for old-timers from the conspicuous consumption that others note among gentrifiers.[11] In fact, a commitment to *in*conspicuousness rooted in a desire to avoid the disruption of old-timers' community life characterizes social preservation. Furthermore, while preservationists orient their practices in opposition to those of the prototypical gentrifier, there is little reason to believe that this is rooted in a quest for status. Preservationists rarely seek recognition for their differences from other gentrifiers. Indeed, most identify as gentrifiers, thus providing little basis for such distinction.

Therefore, it would be an error to reduce social preservation to a strategy for the accumulation of cultural capital or distinction. If self-aggrandizement occurs, it is primarily located in the private moral universe of the individual preservationist. This is not because social preservationists have no desire for status. Rather, social preservation rests on the hope of keeping a place undiscovered. This prohibits the conspicuous consumption or claims to status that they might otherwise engage in.[12] Thus, social preservation demonstrates that we ought not to presume that behavior—even that rewarded with capital, as social preservation often is by virtue of gentrification's economic outcomes—is oriented toward material or status gains. At least for those with a sense of material security, sometimes authenticity or a sense of one's own virtue is its own reward. Scholars' keen awareness of the ways in which culture serves the growth machine (Logan and Molotch 1987) and of culture as a product of political economy discouraged awareness of nonmaterial motivations for gentrification as well as of the ways in which preservationists' culture complicates individual- and community-level resource accrual.

Another recent trend in cultural sociology—which sees culture as fragmented, as composed of disparate and even contradictory elements—may have also made social preservation difficult to discern.[13] This perspective is most notably demonstrated in Ann Swidler's work (2001) on love and marriage, which reveals how individuals draw from multiple and conflicting scripts about romantic love.

The culture-as-fragmented perspective may have encouraged those who noted evidence of social preservation to assume that it was but one repertoire or script that all gentrifiers call on (i.e., that social preservation may be an ideology but that no individual consistently evokes it), that the ideology was divorced from practice (Rose 2004), or that certain contexts encourage the articulation of the social preservation script while others encourage a revanchist approach.[14]

Of course, my findings do not altogether overturn this view.[15] After all, we have seen how gentrifiers and old-timers' practices and, to a lesser degree, their discourses vary by context. Homesteaders' ideology contains discordant fragments, such as a belief in gentrification's restorative power and concern that it threatens authenticity. In addition, as a group, homesteaders rely on a variety of scripts. Some emphasize the preservation of buildings, others of local diversity. In this sense, their narratives are neither internally nor externally uniform. Furthermore, the cultural elements that social preservationists draw from are not wholly consistent. At its foundation, social preservation draws from the incongruous desire to move to gentrifying space and to prevent gentrification.

That said, the social preservation ideology is otherwise surprisingly coherent.[16] Not only do individual preservationists reliably articulate the same scripts over time in the private and the public spheres; those who reside in disparate places provide remarkably uniform accounts of their resistance to gentrification and relocation and their taste for old-timers and other markers of authenticity. There is much homogeneity even in their discussion of the fundamental contradictions of social preservation. In short, social preservation embodies a cultural coherence that recent intellectual traditions in cultural sociology render improbable. Therefore, it is not surprising that some regarded social preservation as just that.

It is plausible that social preservationists' reactionary stance produces internal coherence and that the ideology's external uniformity is a product of the public nature of the scripts social preservationists react against as well as those they borrow from. Those in Provincetown and Argyle alike have access to *New York Times* articles decrying the closure of a fish market and the dispersion of the working-class men who populate it. A couple in Dresden, Maine, and another in Andersonville may watch the same *Will and Grace* episode about efforts to preserve an independent bookstore and its charismatic owner and may have simultaneously cringed when the lead characters forgo preservation on learning that an upscale gym popular with gay men will replace the store. Indeed, individuals in my sites may have shared a classroom in which they read scholarship on the urban pioneer and formed notions of who and how they did not wish to be. As I lay out in the introduction, criticisms of gentrification and its practitioners are abundant in the media, as are older narratives about the costs of urban renewal and of imperialism. Social preservationists evoke these (recall the woman who asked, "You don't think we're urban pioneers, do you?" and the man who compared Andersonville's gentrification to the colonization of Hawaii) and shape themselves in relation to them. This may be largely responsible for the coherence of their words and actions.

Regardless of its origins, somewhat paradoxically social preservation's internal coherence and external uniformity made it difficult for scholars to discern. Those schooled in the culture-as-fragmented tradition may not have thought to further investigate the pieces of social preservation that they encountered. Why look for social preservation*ists* among gentrifiers if one does not believe that any individual gentrifier possesses a uniform view of gentrification or that a social preservation narrative could be attached to practices? Why investigate ideological uniformity across sites if one is certain that context renders culture heterogeneous?[17]

For this reason, as well as for those discussed above, social preservation and even homesteading have been left unidentified and little explored. Scholarship that points to departures from the pioneer has done little to alter our view of gentrification, rooted as it is in the traditions outlined above.[18] In the following pages, I discuss social preservation's efficacy and, in so doing, invite further inquiries into the substantive consequences of this gap in the literature.

Does Social Preservation Work?

The following pages address the question—often posed, whether by readers or by those in attendance at conference presentations—of whether social preservation works. This is a question for which there are few easy answers. Social preservation's success is not easily quantifiable, and a thorough assessment would require a longitudinal study. Furthermore, it introduces more questions than answers. For instance, from whose point of view should we assess its efficacy? Should preventing *all* displacement be our criterion or the creation of heterogeneous communities? Should we work to ensure that long-time residents retain political power, or is preservationists' political capital a resource to be harnessed for their benefit? Despite these questions and our limited knowledge of social preservation's long-term effects, in what follows I offer a few speculative conclusions.

If we take as our measuring stick social preservationists' stated goal of limiting old-timers' displacement, there are several reasons to believe that social preservation works. First, preservationists' private practices help sustain old-timers' businesses. Because they wish to support old-timers, and because they appreciate their culture, many preservationists shop at their vegetable stands and delis and dine in their pho restaurants and Swedish breakfast places. Some offer informal advice to struggling merchants, while others use their occupations to establish policies to protect old-timers' businesses. For instance, a Dresden preservationist who held a state political office advocated for laws that would protect farmers from lawsuits over manure and pesticides.

Such advocacy relates to a second benefit that social preservationists offer old-timers—their influence over local politics. As both old-timers and preservationists note, newcomers bring to local politics their cultural and economic capital and often their free time (because of retirement or self-employment), and social preservationists aim to wield these advantages to old-timers' benefit. They sometimes support old-timers' candidacy for political positions (e.g., Dresden's board of selectmen), and concern for old-timers influences their voting choices. Some vote for those whom they believe have old-timers' interests in mind (e.g., a Chicago aldermanic candidate), while others consciously abstain from certain political acts for fear that they will contribute to old-timers' displacement. Through such means, they help keep some old-timers in power, such as a Portuguese woman who was long the sole old-timer on Provincetown's board of selectmen, and sustain old-timer-friendly political policies.

Third, in three of the sites, social preservationists explicitly engage in efforts to maintain or create affordable housing. In Andersonville and Argyle, they protest on behalf of affordable housing, and, in Provincetown, they successfully fought for policies that set aside funds for housing assistance. Some Provincetown preservationists devote their professional lives to the construction of affordable housing. Furthermore, I met landlords who rent their property below market rate or forgo lucrative summer rentals to provide year-round housing. Even in Dresden, where the topic has not entered public debate, preservationists implicitly support affordable housing by resisting efforts to restrict trailers and infrastructural changes that would contribute to rising taxes. While such practices do not protect all those at risk of displacement, they are of consequence for those who reside in Provincetown's affordable units and Dresden's trailers.

Social preservationists also contribute to the sustenance of old-timers' institutions and cultural events, which are essential to preservationists' and old-timers' sense of community vitality. Through donations and fund-raising, preservationists support organizations such as Provincetown's art association and Andersonville's Swedish American Museum. Symbolically, attendance of Argyle's Chinese New Year Parade and Dresden's Harvest Supper affirms the relation between old-timers and place identity and may encourage communitywide investment in old-timers' symbolic presence. For instance, symbolic gestures have the potential to contribute to popular understanding of place character. On the eastern edge of Andersonville, on the day of the 2002 Swedish Midsommar Festival, a young boy asked: "Daddy, what is Clark Street?" Without hesitating, his father replied: "Clark Street is where the Swedes live." Later, in a popular Andersonville Swedish restaurant, an eight-year-old asked her aunt about the neighborhood. "This is where the [Swedes] live," the white twenty-something aunt told her niece. She elaborated: "This is a Swedish and a Persian neighborhood. Isn't it nice?" These comments demonstrate the effective wedding of place identity to a social group. While this may not prevent physical displacement, the sustenance of institutions and rituals may counter old-timers' sense of cultural displacement and community dissolution.

The uniformity of social preservationists' ideology suggests that, despite the fact that they do not recognize the extent to which others share their beliefs (or, for that matter, how they are ideologically and practically distinct from other gentrifiers), they likely engage in conversation about gentrification. In this sense, they contribute to growing self-consciousness about gentrification. Through conversation at dinner parties or in coffee shops, the display of artwork criticizing gentrification, engagement in public advocacy for affordable housing, or the defense of old-timers on the town hall floor, preservationists may encourage others to rethink their approach to old-timers and gentrification. Indeed, homesteaders' self-consciousness about the urban pioneer prototype may be, in part, a product of social preservationists' discursive work.

Finally, while preservationists seek to avoid disrupting old-timers' communities, they cannot altogether circumvent interactions with neighbors. Their daily contacts with old-timers are distinguishable from those of many other gentrifiers. Imagine the preservationist who, highly conscious of the presence of his elderly old-timer neighbor, takes her soup when she is sick (as a Provincetown man does) or the parent who encourages his children to play with old-timers' children—perhaps the very children whom an Andersonville homesteader is comforted to know fear his dog. Such small practices may cushion old-timers as gentrification disrupts long-standing networks.

Thus, while social preservation is a part of gentrification, it is not without benefit for old-timers. Social preservationists shape the space in which they live. While they may not always shape it in the manner in which they desire or that we may hope for, their ideology and practices are not without positive effect. In other words, social preservation is not merely a set of contradictions; complimentary practices of real consequence accompany the ideology.

However, as with all modes of gentrification, social preservation comes with attendant risks. First, despite preservationists' desire to shut the door behind them, the very symbolic preservation efforts that buoy old-timers' institutions and businesses may encourage tourism (e.g., attendance at Provincetown's Portuguese Festival) or the in-movement of homesteaders in search of diversity writ large (e.g., the Andersonville man who likes that the neighborhood has "a little bit of everything"). Indeed, many politicians and boosters support symbolic preservation efforts. At a 2002 Argyle Streetscape meeting, an alderwoman proclaimed that the "competitive advantage is urban authenticity. People are craving something real. Not Disneyland or a suburban shopping mall. We *are* urban authenticity. They want to see a *real* Foremost Liquor and a *real* Asian Market." Such moments are confusing for social preservationists, who desire a real Asian Market and other forms of authenticity yet also believe that, when space is marketed as bona fide, it loses its authenticity. In addition, the tourists and homesteaders attracted to the markers of authenticity that preservationists help produce contribute to escalating costs—from housing to restaurant meals—and, consequently, to old-timers' displacement. We can imagine the social preservationist who boasts about his neighborhood's authenticity at the water cooler or over cocktails with friends. In this sense, they are like environmentalists who, by conserving land, inadvertently encourage the ecotourism that populates and taints the authenticity they value.

In a more straightforward manner, preservationists' presence in a gentrifying area—whether because of their race, relative affluence, or cultural capital—signals to others that it is safe and desirable and, thus, invites further investment by homeowners, businesses, and local government. Just as old-timers' presence is of symbolic import for social preservationists, preservationists' presence is of similar significance and value for other gentrifiers and investors. Thus, like all gentrifiers, social preservationists encourage further gentrification by contributing to rising property values, rents, and the in-movement of businesses geared to their demographic.

Fourth, social preservationists work to preserve only some residents. In this sense, they encourage the preservation of some at the cost of others, and we should be wary of the fact that, in some sites, they overlook residents of color in favor of white ethnics. For instance, Provincetown preservationists' identification of Portuguese old-timers and, secondarily, struggling artists as the real townies discourages attention to Jamaican laborers. In a manner that also borrows from existing racial hierarchies, Argyle preservationists' interest in Asian, particularly Vietnamese, longtime residents encourages support for their shops, rather than those owned by African Americans. We should also note that the groups preservationists most admire are generally small; in none of the sites do old-timers constitute more than a third of the local population. This pattern, rooted in a taste for the endangered, means that preservationists focus their efforts on a small portion of the local population and on members of those groups who possess certain traits. Thus, while social preservation upends many assumptions about gentrifiers, it shares some traits with efforts to preserve community for the benefit of some at the exclusion of others (typically nonwhites) that have shaped cities and towns for the last hundred years or more, and we cannot allow this aspect of social preservation to go unnoticed.[19]

Fifth, despite their efforts to remain marginal, social preservationists contribute to old-timers' sense of cultural displacement and community dissolution. Their practices—from the cars they drive to the size of their homes and even their self-imposed marginality—contribute to old-timers' sense that their neighborhood or town is no longer home. In fact, preservationists' very admiration for old-timers may be particularly pernicious. Despite their effort to leave old-timers' lives undisturbed, it may encourage a sense that old-timers are on display or are objects of consumption.

Less tangibly, social preservation risks an attachment to old-timers' struggle. For preservationists, old-timers' authenticity rests on their marginality, specifically on their economic struggle. In this sense, old-timers' social problems are evidence of the authenticity of their place of residence. This underlines an essential paradox of social preservation: the desire to save old-timers and to preserve their *plight* along with them.

It is also true that, in the process of preserving old-timers and their traditions, social preservationists preserve some destructive practices or support political positions they might otherwise regard as untenable. For instance,

Dresden preservationists' political abstinence sustains some old-timers' environmentally harmful practices, and Provincetown preservationists rarely reproach old-timers for their homophobia.

Finally, and in some ways most surprisingly, social preservation comes at a cost for preservationists themselves. Specifically, their commitment to virtuous marginality and their related association of authentic community with old-timers discourages recognition of their own networks, organizations, and relationships. While the satisfaction of living in authentic space may counter the cost of believing oneself incapable of membership in real community, we can imagine that self-imposed marginality is, nonetheless, of consequence for social preservationists. In this sense, this book underlines the costs of constructions of authenticity and definitions of community for authenticators, of romanticizing others in a way that devalues and deauthenticates one's own actions, commitments, and networks.

What do social preservationists have to say about the extent to which their practices work? As we have seen, they rarely distinguish themselves from other gentrifiers, holding themselves liable for gentrification. As I have argued, this is a cornerstone of social preservation; the desire to preserve local authenticity arises from the awareness that one threatens it. In this sense, preservationists are dually aware of their agency: of their role in gentrification and of their ability to engage in practices that, to some extent, minimize gentrification's risks for old-timers. Thus, it is difficult to imagine the social preservationist who would deny her complicity in gentrification or who would suggest that his practices are without positive effect. Social preservationists are keen social observers and, in a sense, students of gentrification. Perhaps for this reason, they largely recognize the costs and benefits of their efforts.

While the previous pages discussed whether social preservation works, I am wary of closing this book with a discussion of outcome, for scholarly neglect of the *process* of gentrification has discouraged appreciation of the variations, contradictions, ideologies, and daily practices that are part and parcel of gentrification. Indeed, I have argued that gentrification scholars' emphasis on outcome—on gentrification's costs and benefits—is partially responsible for the absence of research on social preservation. Specifically, by measuring outcome (e.g., population displaced or rising property values) rather than attending to patterns of interpersonal interaction and beliefs that do not neatly align with outcome, scholars failed to acknowledge a set of norms that influence local politics, commerce, and daily interactions in gentrifying communities. In fact, they neglected, not only preservationists, but preservationists'

influence on other gentrifiers and old-timers as well. To lump preservationists with other gentrifiers neglects an acknowledgment of how nearly half
the newcomers to Andersonville, Argyle, Provincetown, and Dresden understand themselves, their residential choices, and their neighbors as well as of
how their understanding shapes their practices and interactions.

Furthermore, to highlight social preservation's costs and benefits is also
to draw attention to the contradictions that drive social preservation. While
these contradictions should be acknowledged and examined (as social preservationists themselves do), an overemphasis on such inconsistencies may
undermine the serious inquiry the process deserves. It is easy to dismiss that
which appears contradictory or hypocritical, as we might dismiss an environmentalist who lives in a solar-paneled home but drives a car that contributes
to global warming. While we should not ignore the contradictions such a
person embodies, there is more to gain intellectually from an exploration of
why she prioritizes solar power over choices that might have reduced reliance on a car and, more generally, from an understanding of the beliefs that
motivate her actions and how they shape her daily life and the lives of those
around her.

We rarely condemn those who created our national parks for failing to
preserve vast stretches of the Western United States, nor do we criticize them
for peopling them (although some environmentalists, like preservationists,
are critical of their own presence in wild spaces). Nor do we fault historic
preservationists for preserving one block and not all or one home rather than
many (although the question of why they preserve one house or block and
not others raises questions akin to those explored in the discussion of social
preservationists' selection of old-timers for preservation).

So too we might regard social preservationists' practices without the expectation that to deem them successful we must find that they have wholly
preserved neighborhoods and towns or otherwise been without cost for old-
timers. In fact, we might even ask whether wholesale preservation is desirable. Those who criticize preservationists for their shortcomings might also
censure them for successfully (and artificially) freezing communities in time.
We should acknowledge that, despite social preservationists' intentions and
practices, gentrification displaces old-timers and other longtime residents.
Yet we must also recognize that preservationists' admiration for old-timers
may sustain individual old-timers as well as their businesses and, in so doing, delay, if not prevent, their displacement. In the face of nearly certain
displacement by gentrification, how do we measure the value of the extra
week, month, or year an old-timer might spend in her place of origin as a result of preservationists' efforts against the displacement of those they did not

preserve? I propose that this is a calculation that we should hesitate to make, for it risks placing us in a patronizing or judgmental posture rather than an inquisitive or evaluative one.

Finally, analysis of the risks and benefits of gentrification all too often relies on quantitative measures, such as old-timers displaced or rising housing costs. While these are undoubtedly pertinent, they do not fully capture the variety of forms of displacement that residents experience, or the symbolic significance of the Swedish restaurant that remains in business even as spas and antique shops replace convenience stores, or the presence of a protest against high-end development. In short, I am not confident that a study of social preservation's costs and benefits would capture the breadth of political, cultural, and social displacement or the full set of intentions, practices, and benefits that preservationists bring to their place of residence.

This is not to suggest that social preservation's costs and benefits should be left unexplored. Rather, it is a caution against repeating the mistake that left social preservation unidentified and underexplored. We should not be so attentive to outcome that we forsake the process of gentrification, specifically the practices and perspectival positions that shape everyday life in the central city and small town and lend meaning to the lives of their inhabitants. In this book, I sought to follow Jon Caulfield's directive to fill much of the "silence of 'gentrifiers' themselves in the scholarly literature" (1994, xi) and, in so doing, identified a range of voices, including preservationists' particular perspective. As Caulfield predicted, by failing to ask gentrifiers for their perspective on gentrification, we have neglected much variation from the ideal-typical pioneer and, in so doing, sacrificed a full understanding of our informants, the contexts in which they live, and gentrification itself.

This book asks us to consider that there may be other ideological orientations and related sets of practices that we have neglected to explore in the country and the city because of our devotion to outcome and quantifiable measures thereof. For instance, we have seen that admiration for others' communities drives social preservation, a concern that most studies of community vitality, with their reliance on quantitative measures and survey methods, would not capture. We have also learned that, within the field of gentrification, a taste for authenticity—of one kind or another—drives more actors than it does not.

Indeed, social preservation and social homesteading seem to be indicative of a growing concern for the preservation of qualities that actors regard as threatened and of a retreating taste for progress. In this, my informants borrow from the age-old concern for authenticity, but, within the field of gentrification, there is evidence of what may be a broad trend: growing concern that the place characteristics one cherishes are fragile. Social preservation in particular reminds us that a sense of authenticity and the desire to preserve it are not always explicitly associated with the past; preservationists wish to freeze old-timers and their communities in the present, before the rapid transformation that characterizes contemporary life further alters it. This, combined with homesteaders' trepidation about unmitigated gentrification and efforts to preserve the physical environment, suggests that early gentrifiers' enthrallment with progress and transformation may be yielding to a new set of concerns tailored for an age of globalization and rapid change. Foremost among these is a desire to preserve the authentic and fragile, whether a dilapidated Victorian home, a two-hundred-year-old landscape, or the faces, voices, and everyday presence of people seemingly detached from the mechanisms of change that many gentrifiers have come to associate with themselves.

Appendix 1: Research and Sampling Methods

The first chapter provides a descriptive account of my sites and an overview of my methods. Without repeating that account, this appendix details key aspects of my ethnographic research and the construction of my interview sample. I begin by detailing how I entered each field site. I then offer a brief discussion of how I decided which events, organizations, and public spaces to observe and which individuals to interview.

In each site, my approach to fieldwork was threefold. I sought to observe residents' informal interactions in public spaces as well as formal interactions in the context of key institutions, such as churches, town halls, and block clubs. In addition to these everyday formal and informal interactions, I sought forums in which residents explicitly discussed gentrification. These included instances and contexts in which gentrifiers strategized methods of gentrification (e.g., through economic development or marketing campaigns) or in which residents sought to prevent or stall the process (e.g., by rallying against development plans or for affordable housing). In this sense, following the model that Mario Small lays out in *Villa Victoria* (2004) of forgoing a traditional, holistic community ethnography in favor of a more theoretically driven and focused one, my goal was to conduct a sustained investigation of residents' reactions to gentrification and related struggles between new and old residents for control of politics, economy, social life, and culture. Below, I discuss the particular approach I relied on in each site, beginning with Andersonville, and closing with Dresden.

I had spent very little time in Andersonville before I began studying it in 2001. While I lived only a half mile from its main commercial strip, I was new to Chicago and had only just begun to visit the neighborhood—to have lunch with friends at a Middle Eastern restaurant or to shop for books at a feminist bookstore. I did, however, know that Andersonville was gentrifying and that it possessed a unique amalgamation of new and longtime residents grappling with its recent changes: lesbian newcomers who clustered in rentals or condominiums on the neighborhood's northern, southern, and western borders; affluent white heterosexual couples who owned Victorians in its most affluent section; and a commercial center marked by the presence of Swedish and Middle Eastern longtime residents.

When I began my study, the Chamber of Commerce and the Forty-eighth Ward alderwoman had recently completed a streetscape-planning process for the neighborhood's commercial district. A streetscape committee, which the city regarded as a mechanism for economic development, seemed an excellent starting point for a study of gentrification. It promised to provide a venue for examining relationships between power holders (e.g., representatives of the Swedish American Museum and business owners) and for capturing actors' perceptions of the neighborhood's identity and vision for its future. I was, therefore, disappointed to learn that the committee had completed its process just before I entered the field. However, my disappointment was assuaged when an aldermanic aide, with whom I met to discuss my research, shared his planning meeting minutes and a list of participants.

I contacted planning committee members, most of whom served on the committee as representatives of local organizations. I requested interviews with them and, subsequently, permission to attend their organizations' meetings. Thus, I began observing several block clubs, events at the Swedish American Museum, community-policing meetings, forums sponsored by the Chamber of Commerce, and public discussions about proposed developments at Andersonville's feminist bookstore, Women and Children First. These venues were the bedrock of my early field notes. Through the informants I met there, I learned of additional sites for observation: Swedish festivals, church services, garden tours, an open house at the Edgewater Historical Society, community-safety meetings, economic development forums, the Chamber's Easter procession, and protests against chain stores and high-end housing.

Gentrification was sometimes, although not always, a topic of discussion at such meetings and events. For this reason, I also sought forums that promised to discuss gentrification or, more generally, neighborhood change. For instance, a Lutheran congregation sponsored a home fair at which business owners and residents presented strategies for home improvement and renovation. Similarly, the Chamber of Commerce and the alderman's office sponsored several meetings on the expansion of economic development north and south of Andersonville and others regarding the preservation of independent businesses. At forums at Women and Children First, I learned of protests against a proposed Borders Bookstore and on behalf of affordable housing that I later observed.

During my first year of study, I continued to live just north of Andersonville, in Edgewater. A year later, I relocated to a street on the border of my two Chicago sites. This provided convenient access to the events I observed and facilitated observation during my daily rounds. Indeed, I observed some revealing interactions when I did not have a notepad in hand, such as when, walking home from the gym in Andersonville's center, I came across a gay couple posing for a photograph atop a blue and yellow horse that a Swedish city had gifted to Andersonville or when I overheard gay men mistake an iconic bar owned by a Swedish family for a male strip club. When a federal judge's family members were brutally murdered in her Andersonville home, grief was equally apparent at a block club meeting I formally observed (at which a news crew and cameras were in attendance) and on streets I walked to the grocery store, where grim-faced neighbors greeted each other and storeowners placed messages of solidarity in their windows. Similarly, Argyle's crime and safety concerns were made apparent as a graduate school friend and I exited a bar onto a quiet, snowy street only to be trailed for several blocks by a patrol car driven by officers presumably concerned for our safety as we walked toward Andersonville. After making such observations, I would record field notes, either temporarily on paper or directly into my computer when I returned home.

Of course, these informal observations reflect my personal tastes and habits: the particular route I take to the grocery store (no doubt influenced by my residence on Andersonville's less-expensive southern edge and even by my gender as I sought well-lit streets), the restaurants and bars I enjoy, and my taste for long walks and runs. That said, many of my characteristics—as a (then) twenty-something, white, childless, highly educated person—mirrored those of many newcomers, and I likely encountered scenes similar to those they witnessed. Furthermore, conscious that my rounds did not approximate those of all residents, I observed social scenes I would not otherwise have participated in, such as block club meetings at which homeowners predominate (I was a renter), Lutheran church services, and a Swedish Christmas fair geared to families with young children.

In 2001, when I met with the Forty-eighth Ward aldermanic aide about Andersonville, I learned that Argyle, which I had also selected as a site, had—inspired by Andersonville—recently started a streetscape-planning process, and I immediately began observing meetings of the planning committee. Interest among residents—primarily newcomers—was strong, and the committee divided into one focused on the streetscape and another focused on safety. I observed meetings of both groups and, as with Andersonville, of the organizations that committee members represented. This brought me into contact with the Chinese Mutual Aid Association (where both groups held their meetings) and a few other agencies that serve local Asian populations, an elementary school, several ethnic business associations, and four block clubs.

In addition to regularly observing meetings and events of many of these groups, I attended community forums that they sponsored or that their members attended. For instance, I observed several protests against gentrification organized by Queer to the Left, a meeting with single-residency-occupancy hotel managers at which block club members launched complaints against mentally ill residents, a meeting of a coalition of Uptown block clubs advocating for economic development, a city hall hearing about the proposed closure of a popular gay bar, a forum sponsored by a nonprofit organization about a proposed development, a meeting regarding the neighborhood's homeless population, and heated gatherings in an Argyle synagogue about proposed beautification strategies and plans to construct a retirement home for low-income seniors.

Additionally, I observed ethnic celebrations, a social gathering at the synagogue, and informal interactions on the streets and in parks, stores, and restaurants. I also regularly monitored the online forum of Argyle's most active block club, through which residents reported recent crimes, rallied residents to support Vietnamese merchants who improved their store facades, and debated whether to support additional affordable housing.

As with Andersonville, my residence near Argyle provided opportunity for informal observation. Visits to the Uptown post office over five years revealed population turnover as well-dressed white gay men took the place of some of the working-class people of color who once populated its infamously long lines. In my last year of data collection (2005), instead of being taken aback by the indifference of bystanders as a mother repeatedly and forcefully struck her child while waiting to buy stamps, I was surprised to encounter a newcomer documenting the length of the slow-moving line with his camera, perhaps to accompany a letter of complaint to the post office. On my return to Chicago after a year at Cornell (after I had completed formal observation), I noted organic products nestled alongside tortillas and pita bread in a convenience store at Argyle's and Andersonville's nexus. Change was less evident in other neighborhood spaces. For instance, throughout the period of study, when I entered a Vietnamese grocery, I was often the store's only white customer. Walking through the park to the lake, I would encounter elderly Ukrainian residents listening to portable radios or conversing on benches and Vietnamese men watching their grandchildren play. Panhandlers frequently asked me for money as I left an Argyle Mexican restaurant popular with newcomers, perhaps the same individuals about which many gentrifiers complained. Walking in the dark to a block club meeting, I almost always witnessed police cruisers circling blocks containing Section 8 housing. In this sense, as with Andersonville, my work in Argyle relied on the formal observation of organizations and events,

many directly pertaining to gentrification, but I also gathered data—somewhat inadvertently—
from observations collected during my daily rounds.

My time was shorter in Provincetown and Dresden than in Chicago, and, for this reason,
the study of my New England sites required a more concentrated approach. For instance, in
Chicago, I did not observe every single meeting of a given block club over the course of five years
of fieldwork, instead alternating between meetings of several different block associations across
the period of study or attending several meetings of one club in a given period and then turning
to another. In contrast, in the New England sites, I sought to observe every meeting of certain
groups and organizations that occurred during my period of observation.

I began my fieldwork in Provincetown in the summer of 2001. Because Provincetown is a
tourist town, its municipal committees and boards meet irregularly during the busy summer
season as many residents negotiate multiple jobs or devote many hours to their small businesses.
For this reason, I focused my early ethnographic data collection elsewhere: observing residents'
interactions and behavior in the town library, at church services, in museums, restaurants, galleries, and shops, and on streets and beaches.

Some of this observation was informal, conducted while walking from the grocery store
to my small rental near the center of town, ambling along Commercial Street, dining out, or
visiting the workplaces of a handful of residents I knew before beginning formal research. However, I also sought to systematically enter the varied social worlds of which Provincetown is
composed. I visited and took notes on museum collections and the interactions that took place
before them, spent time in the library and at bookstores (typically following the publication
of a book or the release of an article in the national press about the town), attended services at
religious institutions, and observed festivals and parades. In the summer of 2001, these included
the Portuguese Festival and Blessing of the Fleet, the Fourth of July Parade, and the Carnival Parade.[1] In subsequent years, I returned to observe other events, such as the town's 275th birthday
celebration, the Year-Rounder's Festival (which I observed twice), and the town meeting.

More frequently, I sought opportunities to observe and meet longtime residents, for I infrequently encountered them in the public spaces in which newcomers and tourists are increasingly present and influential. I made small talk with the operators of dune tours who hawk
tickets on a busy corner, and a driver offered me a place in the front of his buggy during a tour.
I ventured into bars popular with longtimers and conversed with fishermen and their friends.
Later, in 2004, I sometimes walked the pier, watching fishermen bring in their loads, and speaking with them about the town's changes.

Over the course of my research in Provincetown—primarily conducted in the summer of
2001 and the winter and early spring of 2004—I observed numerous social and political events:
a cocktail party at an inn, a book-signing party for two Provincetown authors, a campaign kick-
off party, a political fund-raiser, readings at the Fine Arts Work Center, drag queen competi-
tions, cabarets, pool parties, and fund-raisers for nonprofit organizations.

I generally sought spaces and events popular with newcomers or old-timers as well as communitywide occasions. Often a single event served as an opportunity to observe newcomers
and old-timers side by side. For instance, while newcomers dominate the Carnival Parade—
strolling Commercial Street in drag or riding on floats draped in rainbow flags—longtime residents who live or work along the parade route join crowds of tourists to gaze at the spectacle.
During such events, I sought to capture both the content of the parade and bystanders' reactions, from day-trippers who posed their children with drag queens to Portuguese old-timers who shook their heads at men dressed as nuns. Similarly, when I observed the Portuguese
Festival Parade and the Blessing of the Fleet Procession, I attended, not only to dancers from

Gloucester who took part in the parade and the fishermen who processed carrying a statue of Saint Peter, but also to the reactions of gentrifiers who mingled with tourists on the sidelines.

When I returned in early 2004 to conduct observations during the town's off season, I concentrated observation efforts on formal, political organizations. I regularly attended meetings of the Board of Selectmen, the Conservation Committee, the Historic District Committee, the Long Range Planning Committee, and the Planning Board as well as of the Community Visioning and Affordable Housing task forces. I also observed several meetings of the School Committee, public hearings on matters such as affordable housing and the municipal nursing home, a political forum for the town's senior citizens, and the annual town meeting. Sometimes I was among a handful of individuals in attendance at selectmen's meetings, but they often provided an opportunity to observe the behavior and decisionmaking processes of the board as well as interactions among a diverse array of residents. For instance, when the town pier was on the agenda, fishermen crowded the room, and, when affordable housing was discussed, a variety of individuals populated the space—from realtors to residents in need of housing—and some in the hot and crowded room called for the selectmen to move the meeting to a larger space.

In Dresden, more than in any other site, I had to rely on the observation of formal social settings because the town—which contains no sit-down restaurants, coffee shops, or bars—offers few spaces for informal public interaction. For this reason, I targeted institutions and organizations as initial sites of observation.

After using census data and a preliminary visit to select Dresden as a site, I met with the town librarian about my interest in Dresden. She spoke candidly of her concern about gentrification and introduced me to volunteers (primarily retirees, many of whom are newcomers), who, as I came to observe, are nearly always present in the library. In this sense, the library serves as an unofficial social center as well as a site for town committee meetings, and, over the course of my fieldwork, I spent much time in it.

Having settled into a borrowed house twenty miles from Dresden, I met with the town's administrative assistant, who suggested that I speak with the selectmen about my research during their office hours, held one evening per week. I returned on a Tuesday evening to find the three selectmen in the back of the one-room town office. The head or "first" selectman was conversing with the planning board chair—a thin, bearded man with a hint of a Maine accent—while another toyed with software designed to assist with property value assessment. A third selectman, wearing work boots and coveralls, sat behind a counter watching the door for residents to appear.

Introduced by the administrative assistant, I explained my research interests, and the selectmen heartily encouraged me to study Dresden, agreeing that an influx of newcomers was afoot, and expressing hope that I might help them address challenges such as how (and whether) to control development. In turn, they introduced me to the planning board chair, who told me that permit requests—particularly for construction along the town's rivers—were swamping his desk.

At the selectmen's suggestion, I began observing the meetings of all town committees that met weekly or monthly: those of the Select Board, the Planning Board, and the Conservation Committee. I also observed meetings of the Historical Society, the Snowmobile Club, and of a private charity. While the Planning Board and the Conservation Commission contained more newcomers than old-timers, some individuals from both groups served on both committees, and their meetings facilitated observation of interactions between new and old residents.

In addition to such weekly and biweekly meetings, I attended meetings that drew large crowds of residents. These included public hearings on matters like an effort to make the library handicapped accessible, a special town meeting on a proposal to purchase an archive for town records, and the annual town meeting, at which residents voted on, among other matters, the town and school budgets. I also attended a meeting at which residents debated whether to continue sending junior high and high school students to a regional school in a neighboring town as well the Historical Society's annual potluck and local history lecture.

In search of opportunities to capture nonmunicipal events, I observed a harvest supper held in the elementary school cafeteria and attended an open house at the historic Pownalborough Courthouse. I dined with Dresden residents and sometimes alone in neighboring Richmond at a restaurant popular with residents of both towns. I also spent time in Dresden's library, at the recycling center and swap shop, and at farm stands—all sites where residents gather. During interviews, or when visiting with informants, I observed interactions between neighbors in their private homes and yards. While my presence undoubtedly colored their interactions, I was able to learn something about the purpose of the visit and to identify friendships patterns. Such drop-ins were more common in Dresden than in my other sites, likely because Dresden contains few public spaces for casual social interaction. Similarly, a few residents gave me driving tours or led me on walks on back roads, and I came to recognize that the streets were themselves a site of interaction, as residents (selectively) waved to passing cars or even stopped their vehicles in the street to exchange greetings through open windows.

Along the same lines, early in my fieldwork I realized that Dresdenites rely on formal occasions—such as committee meetings or public celebrations—for informal social interaction and, therefore, made certain to conduct observation both before and after the meetings and events I observed. In these settings, one selectman chided another for failing to invite him to a social gathering, revealing a tension that had not been readily apparent (to me) during meetings, and Conservation Committee members spoke candidly of their distaste for newcomers whom they believed exploited the existence of long-abandoned roads to garner permits for development.

Thus, across my sites, I sought to observe the meetings, celebrations, and public forums of all key community institutions as well as public contexts in which residents promoted or resisted gentrification—from home fairs and neighborhood tours to protests against development. I devoted special attention to communitywide festivals and celebrations, observing seventeen in total. I also balanced observation of formal and informal interaction and observed spaces I encountered in my daily rounds as well as those I would not otherwise have entered had I not been conducting research.

In the early stages of fieldwork, I sought interviews with those I encountered through observation. I did so because interviews with those who populated the meetings and events I observed contextualized and added empirical depth to my field notes. In an interview, I could ask a Dresden political official to demystify the town's wetland ordinances or to explain why a selectman became disgruntled when a particular resident spoke. In this sense, interviews with those I observed provided a back story for the public occasions I witnessed. Thus, my early sample was largely composed of those active in local political and community life, such as block club members, religious leaders, politicians, museum board members, and merchants.

Of course, I did not interview everyone I encountered through observation. Instead, I initially prioritized those in leadership positions as well as a sample that represented the diversity of attitudes toward gentrification apparent in each observation site. My pilot study of Leyden revealed that gentrifiers do not possess a uniform ideology. Specifically, I found that their orientations to their place of residence—particularly the qualities of place that they value—varied, as did their thoughts about gentrification. Because I was interested in capturing a diverse array of perspectives on gentrification, I sought interviews with publicly engaged individuals whose positions seemed to span a broad gamut. For instance, while observing months of Argyle streetscape-planning meetings, I interviewed several members, such as an accountant who fervently complained about African American residents, an art professor who expressed appreciation for Vietnamese and Chinese culture, and a Vietnamese small business owner who advocated for economic development. I just as frequently sought interviews with those who opposed gentrification as with those who sang its praises, and I made certain to include in my sample those who did not seem to fall on either side of the ideological spectrum.

All my interviews relied on a single interview schedule (see appendix 2), with the one exception that I posed questions about cultural tastes (e.g., favorite movie or book) to newcomers but not longtime residents. While I treated the schedule as a guide and sometimes let my informants direct the order of questions and determine the tenor of our conversation, my interview transcripts reveal that I consistently posed uniform questions to a variety of informants and otherwise adhered to a relatively standardized interview structure.

I did not directly ask informants about their reactions to gentrification. In fact, I never introduced the terms *gentrification* or *gentrifier*. Instead, I asked informants to describe their place of residence and residential choices, to pinpoint the place characteristics they most and least value, and to comment on any changes afoot in their neighborhood or town. I then asked them to describe other residents' reactions to change. Given gentrifiers' nearly universal self-consciousness about gentrification, to directly ask them to comment on gentrification or to refer to them as *gentrifiers* might have encouraged more universal criticisms or defenses of the process. By asking open-ended and nonleading questions, I was able to learn which informants are in the habit of thinking and speaking about gentrification and how central the process is to their understanding of their place of residence. As Mario Small writes: "When residents are asked simply to describe their neighborhood, their answers will betray their framing of the neighborhood because they will either discuss its history or they will not, allude to its political significance or not, depict it as beautiful or not. In this sense, it is possible to obtain relatively bias-free descriptions of residents' conceptions of the neighborhood" (2004, 83). In my case, in addition to the above, I was able to document the varying centrality of gentrification and notions of authenticity to population groups as well as how gentrifiers wish to position themselves—to a researcher—in relation to gentrification.

Some months into my interviews, I realized that I could often anticipate which individuals I observed would be most likely to express concern about gentrification or to celebrate upscaling in interviews. Those with an interest in the preservation of houses or landscape clustered in historic district commissions and conservation committees, while those devoted to revitalization often populated safety committees. On the other hand, streetscape task forces included a diverse array of gentrifiers, and affordable-housing protests were a locus for those resistant to gentrification. For this reason, I sought to observe and draw my initial sample from a diverse array of groups. Indeed, if I oversampled any organizations, they were Chicago block clubs and New England municipal committees (because of their seeming omnipresence). Therefore, my sampling strategy does not account for the high number of social preservationists I identified.

After making initial contacts through observation, I relied on snowball sampling. I asked all my informants for suggestions of others I might interview, both new and longtime residents alike. Typically, they suggested neighbors, friends, family members, or those affiliated with local institutions, such as a museum or political office. Unless I had already spoken with the individuals suggested, I typically sought interviews with them.

To ensure that my sample included those who were not at all or only minimally engaged in local politics and other public aspects of community life, I asked informants for suggestions of those less involved with local institutions. My usual line was: "I am having trouble finding people who don't attend the meetings and events I observe. Can you suggest anyone I might talk with?" Once asked, informants almost always provided suggestions, and those they proposed I speak with rarely refused an interview.

To further diversify the sample, I also sought interviews through nonsnowball means. I sometimes sought informants by emailing a list-serv or making an announcement at a meeting. I requested interviews with those I encountered informally, such as an Andersonville artist whose Swedish-themed work I noted in a storefront or a fisherman seated beside me at a Provincetown bar. I also consciously sought undersampled groups. For instance, in Dresden, I rarely encountered through observation those who had moved to town within the previous two years. For this reason, after I collected the bulk of my data, I made a return trip to interview recent arrivals (largely relying on the town librarian for a list of new residents). Similarly, during my first extended period of fieldwork in Provincetown, residents consistently referred me to elderly Portuguese women. During my second stay, over the course of a winter, I focused on collecting interviews with male and younger longtime residents, an effort aided by the fact that fishermen were more accessible during the off season than during the summer, when I first conducted fieldwork. In each site, to capture the perspective of those who had been displaced or had otherwise relocated, I conducted a handful of interviews with former residents, new and old alike. Occasionally, those who did not live in my sites suggested contacts, and I used this to start a new snowball. Several of the newcomers I interviewed were writers or artists, and, whenever possible, I sought to learn more about them and their perspective on gentrification by reading their books or articles or viewing their artwork. To keep the ball rolling, I consistently followed an interview with a thank-you note, and in the small town sites—particularly Dresden—I often arrived at an interview with homemade cookies, pie, or bread.

My interviews with those I had not encountered in the public sphere, not only helped ensure a more representative sample, but also provided important insights. Sometimes I had not encountered an individual because of an illness or young children whose care conflicted with evening meetings. However, political conflicts and concerns kept others from appearing at meetings I observed, such as tensions with a particular political figure or a commitment to political abstinence. For instance, one preservationist ceased attending block club meetings because he had heard other members express racist sentiments.

While, as the account given above details, I developed several strategies for including less enfranchised individuals in my sample, my sample nonetheless contains a disproportionately high number of civically engaged individuals. One may, at first, surmise that this partially accounts for the finding that social preservationists universally wed their ideology to practice as well as for the coherence and uniformity of their ideology. After all, it is reasonable to conclude that those who engage in public debate about gentrification may be more likely than others to develop a coherent discourse about their ideology and that, by seeking informants at rallies, festivals, and meetings, I ensured a sample that contained many in the habit of translating belief into practice.

While there may be some veracity in this proposition, it has limited explanatory power. For, because of the manner in which my full sample was drawn, it cannot explain why homesteaders and old-timers have a less uniform ideology or why even those preservationists I did *not* find in the public sphere commit themselves to preservation practices. In short, my data best reflect the perspectives of civically engaged gentrifiers of every ilk. While it is plausible that participation relates to certain demographic and attitudinal traits that may separate engaged gentrifiers from those less active—such as access to resources that enable participation—this characterizes many in my sample, not just preservationists.

In short, it is unlikely that my sampling technique is responsible for my identification of social preservationists or for their high proportion in my sample. That said, future studies might explore whether the fifty/fifty split I identified characterizes the broader field. It is possible that my selection of older communities (vs. new suburban developments or exurban sprawl) and of at least one town renowned for its intellectual and activist tradition (Provincetown) partially accounts for their high number. However, the latter does not explain the presence of preservationists in less likely sites such as Dresden, which shares neither Provincetown's leftist political tradition nor late-stage gentrification. The diversity of sites studied should leave us confident that social preservation is present in many other older, gentrifying communities. However, the proportion of preservationists, homesteaders, and pioneers may be particular to my sites, and I therefore regard this finding as primarily descriptive and only cautiously predictive.

In the first chapter, I outlined my sampling strategy for recruiting longtime residents for interviews, acknowledging that I consciously oversampled those whom social preservationists regard as old-timers. I will not repeat that discussion here, but I do wish to address in further detail a set of particular challenges encountered in Argyle, where language differences thwarted attempts to ensure a representative sample of old-timers. While I easily arranged interviews with old-timers active in business associations, block clubs, and political initiatives, few Vietnamese, Chinese, Laotian, or Thai residents who were not already engaged in organizations that built coalitions with newcomers agreed to speak with me. This was not for lack of effort on my part. For instance, after several informants suggested I speak with a Vietnamese health practitioner, I made several telephone calls to his place of business. The receptionist always informed me that he was out, and I eventually stopped by his storefront to inquire about his availability. Despite my persistence, he never agreed to an interview, and, likewise, there were some ethnic associations that never granted me access. As a result of such disappointments, I can best address the perspectives of Argyle old-timers who speak English, own businesses, and collaborate with gentrifiers. As I note in chapter 7, in all likelihood this partially accounts for the enthusiasm for gentrification that my Argyle informants expressed.

A few residents in each site served as key informants, introducing me to friends and family members, answering practical queries such as how to find a back road, and offering candid thoughts about my research. For instance, two weeks into my Dresden fieldwork, a town official told me I had to meet Robin, a community activist. At a meeting in the school cafeteria, the official led me to Robin, a middle-aged man wearing a '70s-style beard and an old wool sweater who stood discussing the topic at hand: an effort to make the library handicapped accessible. Robin, who has a background in social science and has conducted historical research, immediately agreed to an interview.

Subsequently, on an unseasonably warm September afternoon, Robin introduced me to his goats and chickens and gave me a tour of his vegetable garden, fields, and woodlots, the stroll colored by tales of local history and the season's first changing leaves. As we stood in the center of his barn, he spoke of his decision to move to town two decades earlier, of his high regard for old-timers, and of his love for Maine's landscape.

For the duration of my time in Dresden, Robin served as a host or ambassador, frequently inviting me to meetings of the organizations he participates in, recommending informants, and providing an elaborate map of one social preservationist's perspective. Well respected by new and old residents alike, he served as an important reference when I requested interviews with others. I sometimes joined Robin and his wife for a meal or a stroll, and, in this, they were among those who alleviated the isolation of ethnography.

To balance multiple sites, I relied on newspaper reports and minutes of events and meetings I could not observe as well as on anecdotal reports via e-mails, letters, and telephone calls from those like Robin. E-mail updates about meetings and events—combined with letters from a Dresden old-timer and the town newsletter—were crucial when something occurred that I could not observe. I similarly came to depend on a Provincetown journalist who periodically called or emailed to notify me of affordable-housing meetings.

Worlds away from rural New England, an Argyle Chinese American business leader served in Robin's role, pulling me aside at parades on busy Argyle Street to meet politicians, merchants, and police officers, and encouraging reticent nonnative English speakers to talk with me. Similarly, Mary—the Provincetowner I met during a college summer—provided contacts as well as insider knowledge. Her insights were particularly valuable as she represented a population group—relatively disenfranchised, middle-aged working-class Portuguese women—I rarely encountered at the town hall.

Over four years of fieldwork, key informants came and went. Rising rents displaced Mary as well as an Iranian storeowner I relied on for occasional Andersonville gossip. A Chicago aldermanic aide was a crucial source as I sought interviews and invitations to meetings, but, after a year in the field, I found it necessary to deviate from his networks to capture an alternate perspective. In this sense, with only limited intention, even my sample of key informants diversified over the course of my research.

Any community ethnographer must seek access to the disparate worlds of a given neighborhood or town. Key informants play a role in this, but so too does a researcher's identity and self-presentation. Of my four sites, Andersonville and Provincetown particularly required that I put informants of very distinct backgrounds at ease. I generally knew that I had succeeded when an old-timer volunteered homophobic remarks or a gentrifier openly criticized old-timers. I was aided by the fact that my appearance (fairly feminine and petite, with chin-length hair) and demeanor (often reserved) lend themselves to multiple interpretations. With a quick change of clothes, I could go from observing Catholic mass to an interview with a gay activist. As Michael Bell notes, this trick of the trade relies on "some slight deceit," and I concur with Bell when he writes: "I was well aware that these assumptions on the part of [informants] contributed to my acceptability, something which I stood to benefit from" (1994, 244).

Such attempts to move between worlds sometimes made me uneasy, and they were not universally successful. In Dresden, I attended a Snowmobile Club meeting as part of an effort to observe old-timers' community events. The meeting was held in the underheated old town hall, and club members stretched across the back of the room on a long bench. Most were men, wearing barn jackets or plaid shirts. A few women sat at tables in the center of the room, where they had laid out cookies, brownies, and soda. Before the meeting began, I made small talk with a trim, twenty-something man seated to my left. He asked why I had come to the meeting, and, after I explained, he mused that I must like writing. In turn, I asked what he did for a living. Crossing one work boot over another, he said he fished from a coastal town. Without thinking, I asked: "How did you decide to do that?" He briefly fell silent, his expression one of hesitation and slight confusion. "Do what?" he finally asked. "Fish," I said. He momentarily fell back into silence and then, subtly tilting his head to the side, said: "Well, my father did it. It's what I do, I guess." Blushing, I recognized that I had posed a question better asked of middle-class informants, a line of inquiry out of place on that bench in Dresden's drafty hall. I was not surprised when he began a conversation with the man beside him.

This example notwithstanding, I was often impressed by my informants' need or desire to regard me as complicit in their perspective on their neighborhood or town and on gentrification. Many Andersonville Swedes assumed that I shared their complaints about the neighborhood's lesbian population despite the fact that I was a part of that population. Yet, with ease, I observed the neighborhood's annual Dyke March, against which Swedes launched many grievances. Conservative Provincetown residents almost universally presumed that I was heterosexual, and this may account for the fact that I found more homophobia in Provincetown than many other scholars. To an extent, I cultivated this impression by suppressing personal sentiments when homophobic remarks were uttered and emphasizing my femininity in certain interviews and settings.

However, my informants did much of this interpretive work without my aid. For instance, after I interviewed a pair of Dresden old-timers, the wife asked: "Which high school do you go to?" This surprised me, for at the outset I had explained that I was working on my Ph.D., and they had signed a Northwestern University interview consent form. Others, having directly asked about various facets of my identity, swiftly disregarded my honesty. For instance, shortly after having asked—in his small boat cabin—if I was straight (given our isolation, I was hesitant to answer but in the interest of honesty did so anyway), a fisherman nonetheless recounted times when he attacked gay men and bawdily flirted with me. Thus, I occasionally relied on "slight deceit" by neither affirming nor denying assumptions about my identity and politics. However, so strong was my informants' desire to see me as complicit in their worldview that, as with the fisherman, this was often superfluous. Through both methods, I continually sought to put my informants at ease and to enter—in the least disruptive manner possible—the varied social worlds that are my four sites.

Together, through these various methods, I cultivated a sample of individuals that approximates the demographic and perspectival characteristics of the gentrifiers and old-timers in my four sites, particularly of those who are politically and civically engaged. Similarly, my observations reflect a wide pool of political, civic, religious, and social events, including organizations with disparate missions as well as those that are home to informants whose perspectives on gentrification span the gamut, from the ardent pioneer to the committed preservationist, and from the displaced old-timer to those who engage in boosterism. Future research may explore the particular configuration of individuals in other sites, the tenor of their discourse, and the contour of their practices. However, I am confident that future studies that adopt a combination of interviews and observation and that rely on open-ended and nonleading questions about community change and residential choice will reveal a similar diversity of perspectives among gentrifiers and longtime residents alike.

Appendix 2: Interview Guide

Personal Background

- What do you do for a living?
- How long have you been in that occupation?
- Did you always plan to enter that occupation?
- Do you belong to or regularly attend a church?
- If so, which church? How frequently do you attend? How long have you belonged?
- If not, have you ever (including as a child)?
- What year were you born?
- Tell me about your family. What did your parents do for a living? Did you live near extended family? How many siblings do you have?
- What is your highest level of education?
- Are you married/do you have a partner?
- If so, how long have you been together?
- Do you have children?
- If not, do you plan to have children?
- As a child, where did you reside?
- Do you remember, during your childhood, where you imagined living as an adult?

Residential Patterns

- How long have you lived in the neighborhood/town?
- Where did you live before then [ask for full residential history]?
- Tell me about your decision to move to the neighborhood/town.
- What was your initial reaction to the neighborhood?
- How is the town/neighborhood different from other places you have lived?
- What qualities of the neighborhood/town do you most enjoy?
- Aesthetic?
- Social?
- Political?
- Cultural?
- What qualities of the neighborhood/town would you like to see change?
- How long do you imagine you'll live here?

- [If intends to stay:] Is there anything that would change your intention to stay?
- [If intends to leave:] Is there anything that would encourage you to stay?
- How likely do you think it is that your mind could be changed?

Neighborhood / Town Change

- How has the town/neighborhood changed in the time you have lived here?
- How do you feel about those changes?
- Tell me about how others talk about those changes.
- What is the best change you have witnessed?
- What is the worst change you have witnessed?
- What are your greatest concerns regarding the future of the neighborhood/town?
- If you could freeze the town/neighborhood in a particular time, which period would you choose?
- Why?
- Would you freeze the whole country/world in that moment or just the town/neighborhood?
- Why might other neighborhood/town residents agree or disagree with the time period you selected?

Political Values

[Introduce topics of local election or political issue (e.g., school referendum, new condominium development, election of alderman, streetscape). Ask for opinion on those issues. Ask how others they know fall on the issues.]

- Are you a registered voter?
- With which party do you tend to vote?
- Do most of your friends and family vote the same way that you do?
- Do you belong to, or donate to, any national, state, or city political groups such as the NRA, NOW, etc.?
- Which national or regional political issue do you feel strongly about?

Civic Engagement

- Do you currently belong, or have you ever belonged, to any local civic groups?
- Cultural groups (e.g., chorus, book club)?
- Local or extralocal?
- Religious or spiritual institutions or groups?
- Local or extralocal?
- Block clubs?
- Chamber of Commerce?
- Do you currently participate, or have you ever participated, in town/neighborhood politics?
- Elected or voluntary positions?
- Do you currently volunteer, or have you ever volunteered?
- If so, do you volunteer in the neighborhood or town?
- If not, have you ever?
- Which neighborhood/town residents participate in . . .
- Cultural groups?

- Religious groups?
- Block clubs?
- Politics?
- Voluntary work?
- Within those groups, who holds the leadership roles?
- Has that changed over time?
- Have you stopped participating, or recently started participating, in any of the above?
- If so, why?

Social Networks

- Do you socialize on a regular basis?
- With whom do you socialize?
- Where do you socialize?
- Where do the people you associate with live?
- Are you close to neighbors or others who reside in the neighborhood/town?
- How frequently do you interact with them?
- How did you meet most of your friends? (If prompting is needed, ask about school, work, church, civic groups, family, etc.).
- Are you content with your social circle/friendships?
- What would you change about your social circle/friendships?
- [If have children:] Are you content with your children's social circle? What would you change about it?
- [For newcomers:] Was it easy to make friends when you moved here?
- [For old-timers:] Is it easy to make friends with newcomers?
- Have any of your neighborhood friends or acquaintances moved away?
- If so, do you know why they did so?
- Do you know of any other residents who have moved away and why they did so?
- How would you describe the local community?
- What do you call your place of residence when referring to it in conversation with nonresidents?
- How do you describe it to friends and/or family members who do not live here?

Cultural Tastes (Questions for Newcomers)

- Tell me about the books you read [ask for both nonfiction and fiction].
- What books are you reading now?
- What newspapers or newsmagazines do you read?
- Which do you subscribe to or read online daily?
- Do you watch movies either in the theater or at home?
- Tell me about the movies you watch.
- What was the last movie you watched?
- What is your favorite movie?
- Tell me about the television you watch.
- What is your favorite television program?
- Do you vacation? If so, where do you do so? If not currently, have you ever?
- What is the place you would most like to visit?

Notes

Acknowledgments

1. All informants' names are pseudonyms.

Introduction

1. Mary told me that she found that she did not qualify as a member of a "vulnerable" population. Provincetown's Comprehensive Plan identifies "at-risk" populations as very low income, low income, families with children, single-parent heads of households, racial minorities, people with AIDS, the elderly, the homeless, the disabled, and others with special needs. It also offers discussion of the value of low-income artists to the local economy and "community character."

2. Mary generously allowed me to share her e-mail with readers.

3. On rural gentrification, see Connell (1978), Long (1980), Parsons (1980), Spain (1993), Bell (1994), Smith and Phillips (2001), Bridge (2003), Salamon (2003b), Macgregor (2005), and Phillips (2004).

4. On gentrifiers' culture, see, e.g., Zukin (1982/1989, 1987) and Long and DeAre (1980).

5. A few years after publishing an article identifying social preservation (Brown-Saracino 2004), I came across a 1973 master's thesis written by Yu Hui Yin, a Berkeley architecture student, entitled "Context in Architecture: Strengthening the Urban Fabric or Bridging the Gap between the Social Preservation and Urban Renewal." While Yin is not the only other individual to use the term *social preservation,* the meaning he assigns to it is closer to mine than most. Specifically, his thesis is something of a political treatise calling for methods for "bridging the gap that exists between social preservation and physical renewal" (2). Yin proposes a method for revitalizing a San Francisco rooming house "without displacing the existing tenants" (7). His commitment to social preservation is rooted in his dedication to civil rights, an appreciation for social heterogeneity, a desire to preserve community, and criticisms of gentrification. He writes of gentrification: "It is the poor area that is benefited, rather than the poor of the area" (5). He recommends community organizing to prevent displacement. In this sense, his thesis does not describe social preservation as a social phenomenon, as this book does, but is, rather, an early call to action by a social preservationist. This demonstrates that elements of social preservation, if not the full process, existed more than three decades ago.

6. This may have resulted from the fact that social preservationists construct their ideology and practices in opposition to that which they associate with the pioneer or speculator; in words and action, they stress the distinction between themselves and the pioneer. Therefore, because of the nature of the phenomenon I was studying, and because the literature led me to anticipate a pervasive frontier and salvation orientation, my early work was a study in contrasts.

7. For discussion of Rose, see Zukin (1987), Caulfield (1994), and Smith (1996/2000). For my initial argument about social preservation, see Brown-Saracino (2004).

8. On the notion of gentrifiers' single value structure, see London (1980) and Caulfield (1994). I depart from those like Butler and Robson (2003) who suggest that, as a group, gentrifiers have changed.

9. For similar arguments, see Allen (1984), Berry (1985), Caulfield (1994), Butler and Robson (2003), Bridge (2003), and Lees (2003).

10. For exceptions, see Berry (1985), Butler and Robson (2003), Caulfield (1994), Taylor (2002), and Pattillo (2007).

11. See http://idsachicago.org/fightclub/html/gentrification.html (accessed December 2008).

12. See http://prairiehome.publicradio.org (accessed February 2006).

13. Monique Taylor notes that Harlem's gentrifiers are also self-conscious about gentrification, likely because of popular and scholarly attention paid to the topic: "Speculation, displacement, and class warfare are loaded terms in community rhetoric and media attention" (2002, 96). On self-consciousness about gentrification, see also Smith (1996/2000, 30).

14. While I wholeheartedly agree with this notion, I depart from Butler and Robson (2003) in a few ways. First, while they identify a general shift among gentrifiers, my findings suggest that discourse about gentrification has not entirely shifted to a social preservation perspective. In each of my sites, some adhere to the frontier and salvation ideology (Spain 1993). Second, Butler and Robson suggest that some neighborhoods are much more likely than others to contain highly self-reflexive gentrifiers. In contrast, my findings suggest that preservationists and other gentrifiers intermingle in all my sites. Third, Butler and Robson do not link this growing self-consciousness to the practices of social preservation that I identify. Finally, some aspects of the perspective that they identify seem to align more closely with a commitment to multiculturalism or an appreciation for diversity writ large (as part of what it means to live on the urban frontier) expressed by social homesteaders, rather than with the identification of space with a particular group and the concern for the preservation of that group that social preservationists articulate.

15. For similar critiques, see Lees (1994), Podmore (1998), and Ley (2004).

16. For a more recent work that argues that culture—in the guise of appreciation for diversity—serves property owners' and developers' interests, see Berrey (2005).

17. There certainly is evidence that this is often the case. Indeed, much recent work in urban sociology and the sociology of culture attends to this. See, e.g., Florida (2002) and Clark (2003). For a discussion of this tradition, see Borer (2006).

18. On the symbolic consumption of diversity or appreciation for diversity writ large, see Allen (1980, 1984), Boyd (2000), Zukin (1995), Anderson (1990), Mele (2000), Butler and Robson (2003), Grazian (2003), Lloyd (2002, 2005), Berrey (2005), Davila (2004, 86, 87, 90), and Taylor (2002).

19. On awareness of the displacement of longtime residents, see Taylor (2002), Butler and Robson (2003), and Pattillo (2007). On appreciation for urban diversity, see Allen (1984), Boyd (2000), Caulfield (1994), Zukin (1995), Anderson (1990), Mele (2000), Butler and Robson

(2003), Grazian (2003), Lloyd (2002, 2005), Berrey (2005), Davila (2004, 86, 87, 90), and Taylor (2002).

20. Caulfield (1994, 204) and a few others (e.g., Mele 2000) identify newcomers' efforts to limit old-timers' displacement. However, they typically associate such efforts with first-wave gentrifiers (Caulfield 1994, 215), while other gentrifiers are nearly as likely as social preservationists to have participated in the first wave of gentrification.

Chapter One

1. These are the real names of the research sites. I chose not to rely on pseudonyms because Provincetown and the Chicago neighborhoods are easily identifiable and to mask their identities I would have had to disguise important place characteristics, hindering discussion of findings. Instead, I refer, as noted earlier, to all informants by pseudonym, and, in some cases, I have disguised defining traits to protect confidentiality, such as the particular college an individual attended.

2. Argyle and Andersonville are unofficial neighborhoods that extend into the official Chicago community areas of Uptown and Edgewater, while Provincetown and Dresden are small towns.

3. Discussion of Dresden's history draws from interviews as well as from Mary Bolté's *Portrait of a Woman Down East* (1983), Dresden's Web site (http://www.townofdresden.com/index .shtml), and *Maine: The Pine Tree State from Prehistory to the Present* (Judd et al. 1995).

4. The Comprehensive Plan is a state-mandated document produced each decade by a committee of residents who propose a plan for managing anticipated changes in population size, industry, and ecology. At a town meeting, citizens decide, by vote, whether to accept the plan.

5. Unless noted, population and economic data are from the 1980, 1990, or 2000 U.S. Census or Census Sample (available at www.census.gov).

6. For a careful discussion of how Provincetown became a gay resort, see Krahulik (2005). Importantly, Krahulik writes: "Although a similar phenomenon took place in Greenwich Village, gay resort areas do not spring automatically from art colonies" (16). On San Francisco's military history and the establishment of gay enclaves, see Castells (1983).

7. These figures reflect only the percentage of town residents reporting ancestry.

8. For work that does examine lesbians' role in gentrification, see Rothenberg (1995), Forsyth (1997b), and Smith and Holt (2005).

9. Lakewood Balmoral was nominated for placement on the National Register of Historic Places in 1999. See http://www.edgewaterhistory.org.

10. Property listings are from www.realtor.com (January 11, 2008).

11. As a further illustration of the neighborhood's nebulous boundaries, even though most residents readily distinguish between the two neighborhoods they have one 2000 census tract in common.

12. For such discussion, see Pattillo (2007, 20).

13. Recent work that shares a similar approach includes Pattillo (2007) and Taylor (2002).

14. My relationship to the sites studied aided the research. I either lived in or near or paid an extended visit to three of the research sites prior to or during the period of observation.

15. This is a count of the number of individuals I interviewed formally. It does not include informal interviews I conducted and occasionally draw from in my analysis.

16. The terminology that social preservationists use to refer to old-timers varies by site. For instance, in Provincetown, they refer to them as *townies,* and, in Argyle, they are simply *the*

Vietnamese. However, the meaning and esteem they award such population groups are constant across the sites, and, for ease, I refer to them as *old-timers.*

Chapter Two

1. On appreciation for the natural environment as a motivation for rural gentrifiers' relocation, see Bell (1994), Connell (1978, 105), and Salamon (2003b).

2. On gentrifiers' appreciation for community, both that which they establish through relationships with others like themselves and that which they establish with working-class old-timers, see Caulfield (1994).

3. In contrast, preservationists celebrate old-timers and reject the melting-pot model in favor of clusters of residents that they perceive to be homogeneous. Social preservationists are nostalgic for the ethnic enclave, adhering to the belief that "nationalized cultures loom as a menace to differentiation, seemingly obliterating all particularisms" (Zelizer 1999, 206). In fact, social preservation may have developed as part of a broader "romantic reaction to the success of assimilation (much like the adoration of nature that emerged from the centers of the urban-industrial order)" (Fischer 1999, 219) and the notion that "local cultures were weakening right along with ethnic cultures" (Higham 1999, 53). This latter criticism of assimilation is particularly central to social preservationists' ideology, for, despite their appreciation for the relationship between local communities and broader categories of identification (i.e., Andersonville old-timers' Swedish American identity), social preservationists concern themselves, not with broadscale cultural preservation, but with the preservation of particular local cultures and communities. After all, they celebrate Andersonville, not because it is a tourist destination for Swedish Americans, but, rather, because of the presence of Swedish residents with enduring and intimate relationships with one another.

Chapter Three

1. "Scholarship, journalism, and grassroots expressions celebrate white ethnics for their family loyalties and neighborhood ties. In fact, advertising in this period [i.e., the 1970s] began to exploit 'cute' white ethnic imagery—the pizza-baking grandmother, the extended family at the laden dinner table—in order to invest frozen and canned foods with the cachet of the gemeinschaft" (di Leonardo 1998, 94).

2. In his study of Chicago's Uptown neighborhood, Michael Maly reports a similar statement made several years earlier by a community organizer: "'People like being in a community where almost anybody walking down the street doesn't look like us'" (2005, 67).

3. For those with the privilege to select their place of residence, living in a particular locale is a mode of self-definition: "Community ideology provides a convincing rendering of varied social, moral, and other qualities of communities and their inhabitants, diverse qualities that can be appropriated for self-characterization" (Hummon 1990, 143; see also Hunter 1975, Ley 1996).

4. On gentrifiers who acknowledge that they are complicit in gentrification, see Caulfield (1994, 213).

Chapter Four

1. In his "Yankee City" series, W. Lloyd Warner focuses on Yankee City's tercentenary celebration. For Warner, the ritualization of the past is a collective statement of place identity. While he attends to tercentenary events, he also emphasizes the planning process through which

certain groups became the "custodians of tradition in Yankee City" (Warner 1959, 114–15, 147 [quote]). For contemporary festival analysis, see Horton (1995). On how "cultural work" can be "seen as a deterrent to the area's gentrification, specifically to the area's cultural gentrification," see Davila (2004, 90).

2. On resistance to heritage preservation, see Caulfield (1994).

3. On newcomers' practices aimed to prevent or minimize displacement, see Caulfield (1994, 204). Caulfield also notes, as I do in this chapter, that response to gentrification varies by neighborhood (1994, 211).

4. As Swidler writes: "[Weber] defined social action as action that is meaningfully oriented toward the action of others" (2001, 165). It is implausible that the actions of any group would not take others into account. Furthermore, given preservationists' intense interest in old-timers, it is not surprising that old-timers influence their practices.

5. In so doing, the discussion bridges four approaches to the study of gentrifiers' discourse and strategies. The first and most common is the identification of the frontier and salvation ideology, which suggests a monolithic model of gentrification (Spain 1993; Smith 1996/2000). The second suggests that gentrifiers' ideology and practices evolve over time with the stage of gentrification (Gale 1980; Berry 1985, 77–79; Ley 1996). The third explores deviation from the frontier model in a single site, often addressing the "surprising" finding that gentrifiers appreciate diversity (see Hunter 1975; Allen 1984; Berry 1985; and Lloyd 2005) or that, because of individual demographic characteristics or the wave of gentrification of which they are a part, some gentrifiers have a unique, complicated, or unusually supportive relationship with old-timers (see Rose 1984; Caulfield 1994; Ley 1996; Taylor 2002; and Pattillo 2007). Finally, the fourth finds that gentrifiers' perspectives on gentrification vary by place. For instance, Butler and Robson (2003) found that gentrifiers' ideology is closer in some places to the frontier model and in others to that of those I term *homesteaders*. According to Butler and Robson, gentrifiers' discourse and practices are intact on arrival, shaping residential choices. Like some of the findings detailed above, mine demonstrate variation from the frontier model (although, in each site, some articulate the frontier ideology) and also that stage of gentrification shapes dialogue. However, I argue that, while expressions of belief (practices, public discourse) vary by place, ideology does *not*.

6. Despite months of planning and much support from residents, PALISS's proposal to build studio and living space for artists on town land failed to pass at the annual town meeting.

7. See http://www.provincetowngov.org/cpa.htm (accessed March 10, 2005). For more on the CPA see http://commpres.env.state.ma.us/content/cpa.asp (accessed March 11, 2005).

8. See http://chrgroup.net/CHRHome.htm (accessed March 11, 2005).

9. Social preservationists are not the only residents interested in affordable housing. Many business owners support the creation of affordable housing because they hope it will encourage their workers (waiters, chambermaids, etc.) to stay in town, attract new laborers, and limit wage increases.

10. See http://www.iamprovincetown.com/PortugueseWomen/ (accessed March 10, 2005).

11. Housing prices were taken from www.realtor.com (accessed May 17, 2006).

12. All demographic data are taken from the U.S. Census (1990 or 2000) or the U.S. Census Sample unless otherwise noted.

13. The last count was in 2001. See http://www.provincetowngov.org/affordable/AffHsg Ch40Binfosheet.htm (accessed January 8, 2009).

14. For discussions of gays and gentrification, see Knopp (1990a, 1995), Lauria and Knopp (1985), Rothenberg (1995), Skeggs (1999), and Brown and Knopp (2003). For a discussion of gentrifiers' claims to neighborhood improvements, see Zukin (1987).

15. Of course, there are longtime residents who are not working- or middle-class, such as Portuguese gays and lesbians, a handful of wealthy Portuguese business owners, and a large number of working-class gays and lesbians who change bed linens and wait tables. However, most social preservationists exclude such individuals from the *old-timer* category.

16. For a discussion of the use of art by ACT-UP, see McDonnell (2002).

17. This particular social preservationist is Jewish. Likely as a result of his own identity, he sometimes included recent Jewish immigrants who reside in Argyle in the *old-timer* category. He was alone in this, and he typically emphasized Vietnamese old-timers' presence.

18. See http://www.onechicago.org/ (accessed July 7, 2005).

19. For a similar debate among residents of a gentrifying Chicago neighborhood about who "these people" are, see Pattillo (2003, 20). Importantly, in Pattillo's site, it was a public-housing resident who questioned another's generalizing statement about low-income neighborhood residents.

20. For discussion of name changes and gentrification, see Mele (2000). Argyle pioneers and homesteaders call the neighborhood "the bastard child of the alderwoman" because they believe that she concentrates revitalization resources in other parts of her ward.

21. I am not the first to use the term *paternalistic advocacy*. It has been used in many contexts by a variety of scholars. It is often used to discuss the relationship between caregivers or professional advocates and their clients. See, e.g., Reid (2000), Barnes (2007), Schwartz (2002), and Bayliss and Polk (2005).

22. The Chamber had hired the firm Civic Economics.

23. In 2007, one of the Swedish delis temporarily closed in preparation for relocating a few blocks south. At the time of writing, it had yet to reopen.

24. Property listings are from www.realtor.com (accessed January 2008).

25. In 2004, property tax increases led many middle- and upper-middle-class Andersonville homeowners to rally for reduced rates. As part of their effort, they placed "Endangered Homeowner" signs in their yards. Perhaps because the bulk of participants were middle-class and were *not* Swedish, their movement attracted little attention from social preservationists.

26. From the Web site of the Maine secretary of state: http://www.maine.gov/sos/cec/elec/enr/enr04.htm (accessed July 8, 2005).

27. In stark contrast to Dresden preservationists' abstinence, some homesteaders sought to build relationships with the snowmobile and ATV clubs (largely composed of old-timers) to encourage them to care for the environment and to recognize their common interest in Dresden's natural character. In this sense, they sought to encourage cultural change, albeit in a very cordial manner, while preservationists go to great lengths to avoid it.

28. Income data are taken from the 2000 Census Sample.

29. However, on risks for the elderly, see Atkinson (2002, 9) and Henig (1981).

30. Dresden's low population density may limit property values, at least in the near future, as developable land is abundant. The town's sixteen hundred residents live in an area that is "nine miles long, three to four miles wide, [and] contains some 25,000 acres" (www.townofdresden .com [accessed November 12, 2004]). Of course, today's young adults may not be able to carry on the tradition of land gifting with their children, for the parcels that their parents give them tend to be for a single home. In other words, this strategy may be a temporary fix should Dresden's gentrification continue.

31. The town did place temporary restrictions on trailer parks when one park, established in 1984, expanded from sixteen to thirty or more units. One official said: "We used two periods of 180 days to revamp the trailer park policy." The park eventually expanded to include forty-four

units. However, individual trailers, outside the town's two parks, are much more common. The rapid expansion of Dresden's trailer park was part of a national trend. Salamon writes: "Between 1980 and 1990 the nation experienced more than a 50 percent increase in mobile home numbers to 7.3 million units" (2003a, 925).

32. Salamon identifies this as a trend: "Rural areas, in particular, are characterized by individual mobile homes scattered alone or next to a 'stick' house, typically owned by a relative who owns the land" (2003a, 926).

33. See http://www.maine.gov/tools/whatsnew/index.php?topic=HouseDems+News&id= 1599&v=Article (accessed November 15, 2004).

34. On the codes, contexts, and institutions that shape the relationship between ideology and action, see Swidler (2001). She writes, e.g.: "The cultured capacities for action that individuals develop will be coherent not within an individual self but within a particular institutionally organized arena of action" (179). For a discussion of cultural coherence, see also this book's conclusion.

35. "Settled" and "unsettled" times (Swidler 1986) do not strictly determine which resources are available to preservationists. For, as we have seen, the measures that might come closest to determining whether a site is settled or unsettled—the stage of gentrification or the level of community conflict—do not alone explain the type of action that preservationists engage in or prevent or enable the establishment of new strategies.

Chapter Five

1. This chapter builds on Brown-Saracino (2007).

2. Notably, not all of Andersonville's Swedish establishments are owned by Swedes.

3. On the privilege to select one's own identity, see Gans (1979), Waters (1990), and Kibria (2000). On the power of community ideology, see Anderson (1991), Hunter (1975), and Ryle and Robinson (2006).

4. I follow a model that Howard Becker provides in his discussion of medical students' use of the term *crock* to refer to patients. He writes: "When members of one status category make invidious distinctions among the members of another status category . . . , the distinction will reflect the interests of the members of the first category in the relationship" (1993, 31). This chapter demonstrates that this is the case, not only for "invidious distinctions," but also for those that are complimentary.

5. Of course, the other actors from whom they receive input about "real" old-timers also construct authenticity.

6. This shares a relationship with Monique Taylor's argument that, for Harlem gentrifiers, "the Harlem that exists in their minds is both romantic and real" (2002, 79).

7. We can imagine social preservationists worrying about other processes: a factory closure, suburbanization, a hurricane, even downscaling. As with historic preservation, the impetus for social preservation may change.

8. The interview protocol included the question, "If you could freeze your neighborhood or town in a particular moment, which moment would you choose?" Nearly all gentrifiers and old-timers referred either to a historic period or to the future, while social preservationists spoke of the present state of community or the recent past. This is a key distinction between social preservationists and others who engage in the preservation of history, culture, or landscape. However, like other preservationists, social preservationists select objects for preservation; they do not work to preserve all artifacts. See Barthel (1989, 91, 92), Collins (1980, 86), Datel (1985,

128, 129), Francaviglia (2000, 68), Lowenthal (1999, 389), Milligan (2004, 2007), O'Laughlin and Munski (1979, 55), and Schuyler and O'Donnell (2000, 76).

9. In this way, they seek "representative characters" of strong community (Bellah et al. 1985, 39; see also MacIntyre 1984).

10. Contra those who argue that the myth of rugged individualism weakens community (Putnam 2000, 24), preservationists associate individuals' independence with community autonomy and, thus, with strong ties.

11. Note that they "saved" the community from "rowdy" Appalachians and African American "gang-bangers."

12. This admiration for child labor is not limited to this particular preservationist. Having offered to provide comments on this chapter, a young sociologist who fit the description of the preservationist quoted in the text asked me: "Did you quote me about the eight-year-old girl?" I told him that I had not and asked why he thought I had done so. He explained that he had made similar statements in the past and that, when traveling in Africa, had met another Chicago resident, who also spoke with great admiration for the restaurant and its young waitress.

13. On "community autonomy," see Warren (1970).

14. Appreciation for old-timers' independence is coupled with assessment of the tenuous nature of their livelihoods. Preservationists recognize that old-timers are dependent on broad economic factors. This is the basis for their concerns about gentrification: that old-timers' stores will fail in the face of chains or that rising property taxes will close farms. The continuation of such practices in the face of broad changes *enhances* old-timers' seeming self-determination.

15. Women's old-timer status is often derived through association with men's labor or through ethnic identity. For instance, the wife of a Portuguese fisherman or a Swedish deli owner is as much an old-timer as her husband.

16. This appreciation for artists may be a twenty-first-century adaptation of appreciation for *artisans* (Bellah et al. 1985, 35). It also relates to a historically enduring appreciation for those who work with their hands. See Stansell (1987, xii).

17. In the nineteenth century, some urban reformers romanticized qualities associated with certain impoverished individuals and groups, such as members of the "deserving poor," which typically included the rural poor and widows. See Riis (1971, 43, 77).

18. Most Chicago preservationists work outside their neighborhood, in other parts of the city or the surrounding suburbs. Some Dresden preservationists telecommute or commute one to three hours for work, in either Portland or Boston.

19. Similarly, Krahulik writes that Provincetown "artists perpetuated the notion that Portuguese fishermen were 'picturesque' and 'authentic' only when they appeared near the wharves and in proper fishing attire" (2005, 79).

20. Fine (2004, 226) notes that some folk artists sacrifice their authenticity by successfully marketing products.

21. See the related discussions of Horatio Alger and Benjamin Franklin in Weiss (1988) and Bellah et al. (1985), respectively.

22. Old-timer's length of residence varies by site. In Dresden, preservationists admire families associated with the town's settlement, while, in Argyle, they turn to Vietnamese residents who came to the United States in the 1970s.

23. However, membership in such a traditional family is not always enough to ensure old-timer status. For instance, social preservationists do not consider some gay Portuguese Provincetown residents old-timers, likely because they do not regard them as independent from newcomers' own networks or as traditional as their heterosexual counterparts.

24. While professionals "engage in complicated networks of intimate relationships, [they] are often not tied to a particular place" (Bellah et al. 1985, 186). This encourages preservationists' appreciation for old-timers' place-based community.

25. This demonstrates the occasional, albeit atypical, confluence of social and historic preservation. A member of the Provincetown historic preservation committee was a social preservationist, as was a Dresden Conservation Commission member. Otherwise, few informants were committed to both social and historic or landscape preservation.

26. On this notion that people and place lend each other meaning, see French (1995).

27. I observed most such preservationists at meetings and other events. They use political abstinence, not to avoid participation or to selectively participate (Janowitz 1967; Suttles 1972), but as a model for how to participate.

28. On the influence of "between-group contact" for local culture, see Fischer (1995, 545).

29. In 2000, assigning meaning to the term quite divergent from my own, the *LA Weekly* referred to Ralph Nader's "virtuous marginality" (see http://www.laweekly.com/2000-03-02/news/endorsements/3). On social distance, see the discussion of the patron's "formal and respectful distance" from "the folk" and of embarrassment about one's social privilege in Cantwell (1996, 365). See also Roy (2002).

30. On measures of community vitality, see Tönnies (1887), Durkheim (1897/1997), Zorbaugh (1929/1976), Wirth (1938), Redfield (1947), Hunter (1975), Oldenburg (1989), Schudson (1996), Paxton (1999), Rotolo (1999), Putnam (2000), Brint (2001), Hampton and Wellman (2003), and Monti et al. (2003).

31. Data were gathered on informants' personal and professional networks as well as on their "sense of community."

32. On elective communities, see Radway (1984), Fine (1979), Wuthnow (1996), and Jindra (1994). On identity subcultures, see Fischer (1995).

33. Connell also found that some newcomers to rural areas did not consider themselves real villagers: "Again most households (but certainly not all) valued the ethos of villager: 'At heart we're villagers but we haven't been here long enough to deserve that'" (1978, 151–52). Others worried that the space in which they lived could no longer be considered a village: "The most frequently cited reason ran along the lines 'You can't have a stockbroker suburb and call it a village,' or 'too pseudo-stockbrokerish to be a village,' to 'this is 'glorified suburbia' full of people playing at living in the country.' . . . Partly, therefore, it was not a village because the wrong sort of people were there and the right sort of people were not there." (152–53). (The reference here is to not enough craftsmen.) Some blamed themselves for destroying those characteristics of their place of residence that made it a village: "A company representative in Effingham Junction observed, somewhat sadly: 'We're suburbanites; we ruin what's left of the village atmosphere'" (153).

34. I depart from Ryle and Robinson (2006) by showing that it is not merely education level or social class that shapes assessment of community vitality as social preservationists share many demographic traits with other gentrifiers.

35. For similar searches for authentic community, see Bell (1997), Bendix (1997), Cantwell (1996), Davila (2001), di Leonardo (1998), Fine (2004), Grana (1971), Grazian (2003), Hobsbawm and Ranger (1983), Hummon (1990), Johnson (2003), Krahulik (2005), MacCannell (1976/1999), Orvell (1989), Peterson (1997), Roy (2002), Ryden (1993), and Trilling (1971). On the impermanent sampling of social authenticity, see Grazian (2003), Fine (2004), and Johnson (2003, 188). However, for a discussion of extended consumption of authenticity and gentrifiers as "resident tourists," see Allen (1984). On commercialized or marketed authenticity, see Beverland (2005), Graham (2001), Grayson and Martinec (2004), Grazian (2003, 17), Fine (2004, 57, 226), Judy

(2004), Lewis and Bridger (2000), and Peterson (1997). For an exception to work on the marketing of authenticity, see Johnson (2003). For work on taste for the "uncommercial," see Bendix (1997), Grazian (2003), Fine (2004), Lewis and Bridger (2000), Orvell (1989), and Peterson (1997). On a desire for personal authenticity, Grazian, e.g., writes: "The search for authenticity may not only stem from the fantasy of consuming that which is considered authentic, but from the desire to actually become authentic" (2003, 22). Finally, I do not wish to suggest that markets and products are not involved in preservationists' quest or that it is devoid of self-interest. Social preservationists' efforts are not without economic reward or consequence. Indeed, they may unintentionally benefit financially from their efforts (as inadvertent conduits for reinvestment). However, their immediate rewards are an affirmation of a moral identity, a sense of personal sincerity (Trilling 1971, 2), and the preservation of the authenticity they value.

36. On how the authentic necessitates the existence of the inauthentic, see Bendix (1997, 17). On how interest in the other is predicated on self-criticism—"a critique of one's own cultural habits"—rather than a straightforward quest for personal authenticity, see Bendix (1997, 34). See also Cantwell (1996) and Roy (2002).

37. However, it is notable that appreciation for independence, tradition, and relationship to place is constant across the urban and rural sites, perhaps because preservationists are themselves so transient.

38. However, Lamont finds that elites are not the only to hold such views; workers view "whites (as middle class people) [who] are less communal than blacks" and emphasize the "poorer quality of their interpersonal relationships" (2000, 124, 147).

Chapter Six

1. It is plausible that engagement relates to certain demographic and attitudinal characteristics—such as access to cultural, social, and economic resources that enable participation—that separate engaged from less active gentrifiers. Thus, the results speak most directly to the characteristics of preservationists, pioneers, and homesteaders in my sites engaged in local organizations and activities.

2. Because my sample contained only a handful of pioneers, I do not separate social homesteaders from other non–social preservationist gentrifiers. Interview data cited draw from the full sample of eighty gentrifiers, while the demographic analysis is limited to thirty-eight social preservationists and thirty-two pioneers and homesteaders, as some interviews were conducted early in the research process before the protocol included questions about demographic and cultural tastes. In addition, some interviews ended before I could ask for this information. Therefore, the number of informants from whom I was able to collect demographic data varies by measure, and I specify the sample size for each measure.

3. Issues of generalizability and validity may confront all who seek to document the relation between gentrifiers' demographic traits and their beliefs. On the one hand, survey research may best identify the social location of a random sample of gentrifiers, but it may not adequately uncover the ideological variation that interviews and observation make apparent. For instance, a closed survey question about attitudes toward displacement would likely generate more universal expression of concern than an open-ended interview that asks a gentrifier to reflect on local change. Furthermore, survey research does not allow a researcher to document gentrifiers' practices. A survey researcher might not categorize a social preservationist as such if he or she reports participating in a block club (a key agent of gentrification), but observation might uncover that the preservationist uses such meetings to combat renewal strategies and to advocate

for old-timers. However, while ethnographic and interview research offers a more nuanced picture of gentrifiers, it rarely relies on a random sample, raising questions about representativeness and generalizability.

4. This is not to suggest that all gentrifiers are white. Rather, it reflects the demographic characteristics of my sites.

5. Preservationists in my sample range in age from twenty-six to sixty-five, with an average age of forty-six, while other gentrifiers range from thirty to sixty-three, with an average age of forty-nine. That they have comparable lengths of residence suggests that appreciation for "a mixed ethnic neighborhood" (Berry 1985, 78) is not associated only with first-wave gentrifiers.

6. Data on social mobility were available for twenty-three social preservationists and twenty-one other gentriers. They were generated by comparing gentifiers' occupational status with that of their parents. Drawing on others' work on the topic, I measured occupational status according to educational requirements associated with a position as well as the estimated income bracket associated with the position. On occupational status, see Sorkin (1971), Duncan (1961), and Warren, Sheridan, and Hauser (2002).

7. I am indebted to Elizabeth Long for proposing this explanation.

8. For this measure, the sample consisted of twenty-five social preservationists and twenty-two other gentrifiers. Ninety-six percent of social preservationists attended college. A few artists in the sample did not complete their formal education, thus producing the small discrepancy between preservationists and other gentrifiers. Thirty-two percent of social preservationists and 36% of other gentrifiers have or are working toward advanced degrees.

9. This high number of English and education majors is not simply a measure for gender (i.e., it does not mean preservationists are disproportionately female). Fifty percent of social preservationists in my sample are female. However, women account for a little over half the social preservationists who majored in English and two-thirds of those who majored in education.

10. The concentration of English majors relates to Douglas Holt's finding that those in "the top quintile of cultural capital resources" "learn to emphasize and value metaphysical aspects of life," specifically "through informal and formal humanistic education" (1998, 6, 11).

11. This question was entered into the protocol after a third of the interviews were collected. For this reason, data on degree earned and type of school are not available for all informants.

12. School status is measured according to current (2005) *U.S. News and World Report* rankings. *U.S. News* ranks liberal arts colleges and national universities separately—listing the scores for all liberal arts colleges and only the top 120 universities (the rest are assigned either third- or fourth-tier status). When calculating the percentage of students who attended elite vs. nonelite schools, I combined the scores of top-ranked national universities and liberal arts colleges. For instance, a gentrifier who attended Harvard University would receive a score of 1, as would one who attended Williams College. The status of some schools may have changed since informants' graduation, and, for this reason, the measure is not precise.

13. A few scholars concur with this finding that course of study and institution shape ideology. For instance, Phelan, Link, Stueve, and Moore (1995, 137) note that those with graduate degrees have the most liberal attitudes toward homelessness. Knoke and Isaac (1976) found that those who attend elite colleges are more liberal than those who attend less selective schools.

14. I have included one preservationist who is a religious leader in the *public and voluntary sector* category.

15. Among gentrifiers, 31.54% of social preservationists and 41.1% of other types are members of creative professions (Florida 2002). However, this figure does not capture the number

of preservationists whose small businesses are arts related or who identify as artists but work in routine occupations (e.g., as waiters or clerks). Thus, substantial portions of preservationists as well as homesteaders and pioneers are likely members of the creative class.

16. In recent decades, many scholars have pointed to cultural differences between subgroups of the middle class, arguing that cultural tastes correspond with occupation (Bourdieu 1984, 14, 128; Savage et al. 1992, 109; Brint 1985; Brint 1984; Ehrenreich and Ehrenreich 1977; Gouldner 1979; Heath and Jowell 1991; Heath and Savage 1995, 281; Holt 1998; Johnstone et al. 1972–73; Kristol 1972a, 1972b; Ladd 1978a, 1978b; Ladd and Lipset 1975; Lipset 1979, 1983; Rothman 1979; Tunstall 1971). With few exceptions, such studies do not address causality. An exception is Bagguley's study of 1960s radicals, which concludes: "The relationship between class and social movement activism in the case of the middle classes is best seen as an emergent property of politicized job choice, enabled by educational qualifications, and structured by the wider labour market opportunities for educated labour" (1995, 308). Similarly, Fendrich (1993) found that African American civil rights activists sought careers that complemented their political beliefs.

17. Of course, social preservationists' relative economic privilege facilitates the sense of agency that many seem to possess regarding their courses of study, occupations, and residential choices. In other words, their (relative) privilege may allow their ideology to have greater influence or power over their life choices and, therefore, over some of their traits. Thus, even in this case, ideology cannot be completely divorced from actors' social locations.

18. Data on this measure were collected from a small portion of the sample: seventeen preservationists and thirteen other gentrifiers.

19. Social preservationists reported that they were reading an average of 1.94 books, while other gentrifiers currently were reading an average of 1.23 books.

20. On the relation between education, other markers of social position, and tastes, see Bourdieu (1984), Bryson (1996), DiMaggio (1987), Erickson (1996), Gans (1974), Peterson and Kern (1996), Savage et al. (1992), and Shively (1992).

21. Bryson (1996) suggests that the "tolerance" embedded in the consumption of a variety of cultural forms, products, or objects may now "separate high-status culture from other group cultures. This *tolerance line* recreates the pattern of high-status (cosmopolitan) culture in opposition to non-high-status (group-based) culture. Thus, it provides a new criterion of cultural exclusion" (897). Similarly, Holt (1998) found that some elites value products not of their own milieu: "They favor bluegrass and other much less popular traditional styles that are described as original, unique varieties of American music, rather than a music genre that speaks to their lives" (15). See also Johnston and Baumann (2007).

22. While social preservationists', homesteaders', and pioneers' tastes and practices vary significantly, there is not much variation among them in terms of education level. This complicates DiMaggio's proposition that, "the greater the degree of access to higher education, the more differentiated the [Artistic Classification System]" (1987, 447). My findings suggest that the *type* of educational institution or course of study is of great import.

23. This is further supported by Erickson's finding that weak ties (rather than intimate ties) encourage broad cultural consumption (1996, 237). Most relationships between social preservationists and old-timers could be considered weak, for the former consciously seek to maintain social distance from the latter.

24. As Elizabeth Long writes: "People construct themselves with materials culled from a vast cultural storehouse. These cultural materials come to us as commodities to be sure, but they are what we use (and are used by) as we imagine our futures and create our lives" (2003, xi). Connell specifically notes that cultural objects influence newcomers' relation to their place of residence:

"The new residents of central Surrey are literate people. Many have read their Jane Austen and can easily come to feel that they, too, are almost a part of the world she describes. They may even share some of her values: they too, may treat the agricultural workers with the contempt of total neglect" (1978, 213).

25. The preservationists in my sample are less likely than other gentrifiers to own a home. Specifically, 60% of preservationists own their current place of residence (and, in some cases, second homes elsewhere), compared to just over 90% of other gentrifiers. Notably, 92% of preservationists who rent live in either Provincetown or Argyle, the sites with the highest median home value and the lowest median rent, respectively. Some may hesitate to purchase property for fear it will contribute to gentrification, for my sample includes pioneers and homesteaders on limited incomes who nonetheless own homes as well as affluent preservationists who rent. In sum, given preservationists' relative status security, this difference is most likely a result of the fact that social preservationists do not emphasize homeownership as a mobility strategy.

26. On average, preservationists have lived in their place of residence for 12.85 years, while other gentrifiers have lived in theirs for 11.53 years. The mean number of places preservationists have lived is 3.53. For other gentrifiers it is 3.67.

27. Even when this is accounted for—and, frequently, foreign residence as a child was for a limited period of time—preservationists are twice as likely as other gentrifiers to have lived outside the United States as adults. (Residential data were available for thirty-eight social preservationists and twenty-seven other gentrifiers.)

28. On gentrifiers as "resident tourists," see Allen (1984).

29. On the imperialist fascination with the other, see Hammond and Jablow (1970, 160), Edwardes (1968, 10), Fulford (2004, 43), Guelke and Guelke (2004, 11), Murti (2001, 118), and Pratt (1992). Despite their fascination with the other and the fact that they rarely distinguish themselves from other gentrifiers, preservationists view themselves as anti-imperialists, as newcomers more self-conscious about the influence of their actions on old-timers than other gentrifiers are. In fact, social preservation borrows from popular and academic traditions of criticizing the practices of both imperialism and globalization (Said 1978/1994).

30. MacCannell identifies this as a broader orientation, suggesting that, for all modern tourists, "reality and authenticity are thought to be elsewhere; in other historical periods and cultures; in purer, simpler lifestyles" (1976/1999, 3).

31. On cosmopolitan's educational backgrounds, see Holt (1998). On how colleges and the media attract gentrifiers' attention to urban social problems, see Allen (1984, 30).

32. According to IIE (2005), between the 1998/1999 and 2003/2004 school years, U.S. college students studying the social sciences were more likely than those with other majors to study abroad.

33. The sample for this measure was composed of thirty-six social preservationists and thirty-one others. In all, 66.7% of nonheterosexual social preservationists and 70% of nonheterosexual other gentrifiers in my sample live in Provincetown or Andersonville (the two sites residents and the media most associate with gays and lesbians).

34. In all, 60% of nonheterosexual preservationists are gay men, 28% are lesbians, and 12% are bisexual, while 66% of nonheterosexual pioneers and homesteaders are gay men and the rest lesbians.

35. For discussion of gays and gentrification, see Bellafonte (2004), Castells (1983), Elari (2004), Forsyth (1997a, 1997b), Moss (1997), Sibalis (2004), and Smith (1987, 1996/2000). For work that questions the straightforward relation between gays and gentrification, see Knopp (1990a, 1990b). Gale suggests that gays' movement into neighborhoods populated by those un-

like themselves can be read as a method for avoiding "feelings of rejection likely in more conventional neighborhoods" (1980, 105). See also Pattison (1977).

36. On gay men, suburbia, and perspectives on gay ghettos, see Brekhus (2003). One of Brekhus's informants "determined he no longer wanted to live in a 'fortress' apart from the rest of the world. Indeed, the conformities of San Francisco's gay world eventually bothered him as much as the conformities of the straight world he had escaped. . . . Other gay commuters also referenced their ability to live in suburban space as a source of pride and . . . maturity" (68).

37. As a lesbian nonnative, and as a person who is neither Portuguese nor a struggling artist, the informant did not consider herself an old-timer—despite her extended residence in Provincetown—nor did I encounter any evidence that others considered her to be such.

38. In the same document, as well as in accompanying discussions, Q2L draws attention to the impact of gentrification on people living with AIDS.

39. Notably, the film was written by a gay couple who moved to a gentrifying Los Angeles neighborhood and is, therefore, akin to the use of art forms to criticize gentrification common among Provincetown preservationists.

40. Not only do my data suggest that heterosexuals often participate in first-wave gentrification; they also demonstrate that we have underestimated the role of lesbians as first-wave gentrifiers. This may be because lesbians have fewer resources for the construction of institutions that mark their presence or because they are less conspicuous than gay men or are often displaced before media and scholarly attention turns to the neighborhood's or town's gentrification.

41. However, it is also plausible that the ghetto's persistence facilitates a desire to live outside it. For instance, middle-class lesbians may move to more affordable neighborhoods whose cultural economies do not cater to lesbians with the assurance that they may visit Andersonville bars.

Chapter Seven

1. This chapter does not directly enter a debate about the costs and benefits of gentrification for longtime residents. On gentrification's costs and benefits, see Allen (1980), Atkinson (2000, 2002), Bailey and Robertson (1997), Berry (1985), Chernoff (1980), Freeman and Braconi (2004), Freeman (2006), Gale (1980), Henig (1981, 1984), Kain and Apgar (1985), Kasinitz (1983), Lee and Lodge (1984), Levy and Cybriwsky (1980), Logan and Molotch (1987), Lyons (1996), Marcuse (1986), Mele (2000), Quercia and Galster (1997), Robinson (1995), Schill and Nathan (1983), Smith (1996/2000), Spain (1993), Tobin and Anderson (1982), Vigdor (2002), and Zukin (1987).

2. I relied on social preservationists' definitions and categorizations to identify old-timer research subjects from among the population of longtime residents. Of course, as noted in chapter 5, it is not just preservationists who engage in the construction of old-timers. Their construction borrows from local logics and interactions, and, in each site, many residents who are *not* social preservationists concur with their definitions.

3. The precise number of longtime residents I interviewed whom preservationists do not consider to be old-timers is difficult to calculate with precision, for several reside on the borders of preservationists' definitions, such as gay and lesbian Portuguese residents of Provincetown and Chinese American residents of Argyle. I prioritized interviews with old-timers to ascertain their perspectives as the object of preservationists' and some homesteaders' admiration. While my sample includes several displaced over the course of fieldwork, it best captures the experiences of those who weathered gentrification's first stages (i.e., those who remained in my sites). Furthermore, most of those interviewed ostensibly benefit from preservationists' practices—at

least more so than those longtime residents overlooked—and, therefore, many of my informants may be less critical of gentrification than others.

4. On the implications of physical displacement for old-timers' community, see Henig (1984, 171), Logan and Molotch (1987), Zukin (1987, 133), Smith (1996/2000, 32–33), and Atkinson (2002, 10). On measures of community vitality, see Brint (2001), Durkheim (1897/1997), Hunter (1975), Putnam (2000), Oldenburg (1989), Redfield (1947), Tönnies (1887), Wirth (1938), and Zorbaugh (1929/1976).

5. On the consequences of loss of space for community identity, see Milligan (1998, 2003).

6. On the importance of a sense of turf for community identity and cohesion, see Suttles (1968).

7. Vidgor writes: "A possible negative consequence of [gentrification] is the loss of 'character,' as perceived by a neighborhood's initial residents. . . . Character might be determined by the demographic composition of the neighborhood or the amenities, commercial establishments, and local institutions present" (2002, 147).

8. Mele writes: "Minorities and the poor are also indirectly 'pushed out' [through gentrification] or made to feel like strangers in their own neighborhood by . . . new commercial spaces geared toward middle- and upper-class consumers. . . . Cultural dislocation . . . abounds as new shops, restaurants, and groceries open to serve more-affluent newcomers rather than the remaining underprivileged families" (2000, 304).

9. Sadly, a fire destroyed St. Peter's Catholic Church in 2005. However, the congregation continued to hold services as funds were raised for reconstruction, and construction of a new church has been completed.

10. On institutions as the building blocks of community, see Fowler (1991), Hawley (1971), Hunter (1975), Kornblum (1974), Oldenburg (1989), Putnam (2000), Warren (1963).

11. On the benefits of gentrification-induced commercial improvements, see Henig and Gale (1987) and Freeman and Braconi (2004). On how improvements better serve gentrifiers than they do old-timers, see Tobin and Anderson (1982).

12. My observations suggest that the gay and lesbian takeover of public space is less complete than some old-timers suggest. Specifically, gays' and lesbians' presence is most notable at particular times of the year (in the summer in Provincetown), particular times of the week (on the weekend in both Andersonville and Provincetown, when tourists and bar patrons are most likely to visit), and particular times of day (in the evening). While gays and lesbians were almost always visible in Andersonville and Provincetown, their population swelled at these times, and they were more likely to express affection (hold hands, kiss), display markers of gay identity (e.g., T-shirts or buttons), or engage in the displays of sexual attraction that displease some old-timers. Thus, it is not enough to suggest that open displays of gay identity and relationships occur in certain places (Castells 1983; Graham 1998; Forsyth 1997a, 1997b; Gieryn 2000); even within such places they occur at particular times and in particular spaces (see also Small 2004, 43).

13. Of course, the Dyke March occurs only once per year, so ordinarily there is little risk that her grandsons will encounter topless women. However, for her, the march symbolizes the changing character of the neighborhood's public space, which is of daily significance. On the Dyke March, see Brown-Saracino and Ghaziani (2009).

14. Some, but far from all, preservationists include Chinese business owners and their families in the *old-timer* category.

15. This contrasts with Oldenburg's (1989) notion that the "great good place" contains a diverse ensemble of community members as well as new urbanism's emphasis on the relation be-

tween diversity and community. On new urbanism and residential integration, see Bohl (2000, 762, 766) and Grant (2006, 188). While social preservationists seek to maintain the economic diversity of the places in which they live, it is not clear that they would advocate for new urbanists' economically integrated neighborhoods, for they fear the costs of integration for old-timers. Specifically, they share scholars' concern that integration "runs the risk of diluting the local community" (Bohl 2000, 780; see also Briggs 1997b). Furthermore, like some scholars, preservationists may regard integration as an attempt to encourage gentrification (Palen 1988; see also Bohl 2000, 779). Finally, preservationists are less interested in the intermingling of people that diversity or integration implies than in the concentration of what they regard as relatively homogeneous old-timers.

16. While they talk about sexual identity and other differences, old-timers likely also fear class differences. As Mary Pattillo writes: "Lay people express their own class standing and read others' class positions through signs of language, dress, demeanor, and other objects and behaviors that have social meaning" (2003, 2). In this sense, old-timers may also respond to physical representations of class difference in public spaces.

17. This may be, in part, because they are able to maintain connections with some close friends and family members whom gentrification displaced.

18. On institutions, see Fowler (1991), Hawley (1971), Hunter (1975), Kornblum (1974), Putnam (2000), and Warren (1963).

19. According to Hunter (1975, 170, 188), organizations provide a framework for social interaction and, therefore, increase primary bonds. They also serve as communication systems, increasing knowledge, particularly positive information, thus encouraging affirmative feelings about the local community. Block clubs and community-policing groups may increase local social control and social interaction.

20. Connell suggests that rural gentrification can strengthen civic participation (1978, 134).

21. This may be because, as business owners and, in some cases, as leaders of business associations, many Argyle old-timers may have mistakenly assumed that I was solely interested in their businesses and, therefore, in their business community, rather than in their more informal networks of friends, relatives, neighbors, etc.

22. On rising housing costs for the elderly, see Vigdor (2002, 147).

23. Atkinson also writes: "Few studies have alluded to increased property values as a benefit of gentrification though clearly it is for those who are homeowners" (2002, 14; see also Logan and Molotch 1987).

24. Mele notes a similar trend in New York: "Many of the poorest Lower East Siders equated social mobility with departure from the tenements" (2000, 27; see also Davila 2004).

25. Such words reflect the fact that "there are few year-round skilled job opportunities in town" (Gleason 1999, 106). They also reflect the belief that getting out is paramount to upward mobility (Davila 2004, 39).

26. On the economic impetus behind Provincetown old-timers' relocation, see Gleason (1999, 270). On some gentrifiers' recognition that old-timers may wish to sell for a profit, see Caulfield (1994, 102).

27. Arlene Davila (2004, 3, 13) also notes old-timers' support for gentrification, which is rooted in the hope that the neighborhood might improve without leading to their displacement.

28. See http://egov.cityofchicago.org/webportal/COCWebPortal/COC_EDITORIAL/04AR .pdf (March 28, 2006). The perception of high crime rates in Argyle is likely a product of several factors: racism, fear of the mentally ill population, and pockets of poverty along the neighborhood's Kenmore-Winthrop corridor.

29. Atkinson writes: "[The] most obvious upside to gentrification is the rehabilitation of the physical fabric of neighbourhoods" (2002, 14).

30. On self-preservation, see Davila (2004, 42), Mele (2000), Muñiz (1998), and Smith (1996/2000, 138).

31. On class and historic preservation, see Barthel (1996, 11–13) and Walton (2001, 257). On class and environmental preservation, see Bell (1998), Cotgrove (1982), and Kegan, Van Liere, and Dunlap (1980, 192). Davila (2004, 84) makes a more general point that middle-class old-timers are more likely to be involved in local politics.

Conclusion

1. On this ideology, see Spain (1993), Smith (1996/2000), and Zukin (1987).

2. On "ideological elements" that hide in plain sight, see di Leonardo (1998).

3. I am indebted to an anonymous *Theory and Society* reviewer for helping to clarify this point.

4. I intentionally use the male pronoun, for the popular image of the pioneer is of a man.

5. Others offer similar critiques of urban scholars' devotion to political economy. For instance, Michael Ian Borer writes that urban sociologists often regard culture "as a by-product of economic and politically interested decision and actions" (Borer 2006, 176; see also Podmore 1998; and Ley 2004).

6. Michael Ian Borer makes a similar point, arguing that, in urban scholarship, social actors are typically regarded as "somehow passive in relation to culture: they receive it, transmit it, express it, but do not create it" (2006, 180).

7. For a few exceptions—those who do attend to cultures that conflict with growth-machine politics—see Caulfield (1994) and Butler and Robson (2003).

8. While many will ultimately gain from gentrification when they sell their homes or use home equity as a resource for further investment, to the extent that preservationists' efforts are successful—that they block further gentrification—they limit or reduce their economic return. Furthermore, some reduce their rental income by renting affordably.

9. On the marketing of neighborhood difference and authenticity, see Grazian (2003) and di Leonardo (1998).

10. For instance, preservationists celebrate Argyle's Asian character and Andersonville's Swedish character. As such, they engage in efforts to mark the space as belonging to old-timers. This may attract tourists and even gentrifiers. However, it may be a less effective method of boosterism than others, such as framing Argyle as "Lincoln Park North" or Andersonville as an upscale gay neighborhood.

11. On conspicuous consumption and gentrification, see Beauregard (1986) and Ley (1985).

12. Of course, social preservationists may achieve status through inconspicuous consumption; indeed, their ability to abstain from conspicuous consumption may be a marker of their status (see Bourdieu 1984, 282). Nonetheless, I have found little evidence to suggest that preservationists predicate their practices on a desire to accumulate capital or status.

13. On culture as fractured or incoherent, see Swidler (1986, 2001), DiMaggio (1997), Martin (1992), and Tilly (1992). Michael Ian Borer suggests that this view of the social world as fragmented also shapes urban scholarship: "They all see the city as fractured and fragmented and, perhaps, inevitably so" (2006, 177).

14. For this argument, see Butler and Robson (2003). On revanchism, see Smith (1996/2000).

15. On culture as at once coherent and fragmented, see Ghaziani and Ventresca (2005).

16. On culture's coherence, see Geertz (1973), Bourdieu (1984), Alexander (1990), Alexander and Smith (1993), Biernacki (1995), and Mohr and Lee (2000).

17. I suggest that frames are not so much "situationally cued" (DiMaggio 1997, 265) as situationally articulated or translated into practice. This may have been difficult to discern without the combined methods of interviews and extended observation. On culture and context, see Swidler (2001, 184).

18. For instance, departures from the pioneer prototype are typically explained by turning to the demographic characteristics of these atypical gentrifiers or by suggesting that resistance to gentrification is but one orientation that all gentrifiers may sample from. See Berry (1985), Butler and Robson (2003), Caulfield (1994), Taylor (2002), and Pattillo (2007).

19. It should also encourage scholars to question whether the ethnic enclaves and other places they study are as homogeneous as residents and some scholars presuppose (e.g., does Little Italy's commercial district accurately reflect the demographic characteristics of its neighborhood?).

Appendix One

1. In 2001, I was unable to attend the Portuguese Festival and Blessing of the Fleet but sent two research assistants in my stead. I personally observed the festival in 2002.

References

Abu-Lughod, J. 1961. "Migrant Adjustment to City Life: The Egyptian Case." *American Journal of Sociology* 67, no. 1 (July): 22–32.

———. 1994. *From Urban Village to East Village: The Battle for New York's Lower East Side.* Cambridge: Blackwell.

Alanen, A., and R. Melnick. 2000. *Preserving Cultural Landscapes in America.* Baltimore: Johns Hopkins University Press.

Alexander, J. C. 1990. "Analytic Debates: Understanding the Relative Autonomy of Culture." In *Culture and Society: Contemporary Debates,* ed. J. C. Alexander and Steven Seidman, 1–30. Cambridge: Cambridge University Press.

Alexander, J. C., and N. J. Smelser. 1999. "Introduction: The Ideological Discourse of Cultural Discontent." In *Diversity and Its Discontents: Cultural Conflict and Common Ground in Contemporary American Society,* ed. N. J. Smelser and J. C. Alexander, 3–18. Princeton, NJ: Princeton University Press.

Alexander, J. C., and P. Smith. 1993. "The Discourse of American Civil Society: A New Proposal for Cultural Studies." *Theory and Society* 22:151–207.

Allen, I. 1980. "The Ideology of Dense Neighborhood Redevelopment: Cultural Diversity and Transcendent Community Experience." *Urban Affairs Quarterly* 15:409–28.

———. 1984. "The Ideology of Dense Neighborhood Redevelopment." In *Gentrification, Displacement, and Neighborhood Revitalization,* ed. J. J. Palen and B. London, 27–42. Albany: State University of New York Press.

Anderson, B. 1991. *Imagined Communities: Reflections on the Origins of Nationalism.* New York: Verso.

Anderson, E. 1990. *Streetwise: Race, Class, and Change in an Urban Community.* Chicago: University of Chicago Press.

———. 2004. "The Cosmopolitan Canopy." *Annals of the American Academy of Political and Social Science* 595, no. 1:14–31.

Anderson, J. 2003. "Kids Take Puppets on Parade to Make Point." *Chicago Tribune,* Metro section, September 9, 1.

Andersonville Streetscape Committee. 2002. Streetscape Memo. Chicago.

Andert, J. 2004a. "Washashores." *Provincetown Banner* 9, no. 40 (February 5): 24.

———. 2004b. "Washashores." *Provincetown Banner* 9, no. 41 (February 19): 25.

Antani, J. 2006. "Sweet Fifteen: Quinceanara Is Everything but Kitchen Sink Drama." *Los Angeles Alternative,* August 4.

Argyle Streetscape Task Force. 2001. *Argyle Streetscape Task Force Over-View and Survey.* Chicago.

Atkinson, R. 2000. "Measuring Gentrification and Displacement in Greater London." *Urban Studies* 37, no. 1:149–65.

———. 2002. "Does Gentrification Help or Harm Urban Neighborhoods? An Assessment of the Evidence-Base in the Context of the New Urban Agenda." CNR Paper 5. ESRC Centre for Neighborhood Research. Available at http://www.bris.ac.uk/sps/cnrpaperspdf/cnr5pap .pdf.

———. 2003a. "Domestication by Cappuccino or a Revenge on Urban Space? Control and Empowerment in the Management of Public Spaces." *Urban Studies* 40, no. 9:1829–43.

———. 2003b. "Introduction: Misunderstood Saviour or Vengeful Wrecker? The Many Meanings and Problems of Gentrification." *Urban Studies* 40, no. 12:2343–50.

Avila, O. 2005. "Argyle St. Burglar Bars Divide Old, New Cultures." *Chicago Tribune,* January 23, 1.

Bagguley, P. 1995. "Middle Class Radicalism Revisited." In *Social Change and the Middle Classes,* ed. T. Butler and M. Savage, 293–312. London: University College of London Press.

Bagguley, P., et al. 1990. *Restructuring: Place, Class and Gender.* London: Sage.

Bailey, N., and D. Robertson. 1997. "Housing Renewal, Urban Policy and Gentrification." *Urban Studies* 34, no. 4:561–78.

Barbaro, M. 2006. "In Wal-Mart's Home, Synagogue Signals Growth." *New York Times,* June 20, B1.

Barker, R. G., and H. F. Wright. 1971. *Midwest and Its Children: The Psychological Ecology of an American Town.* Hamden, CT: Archon.

Barnes, V. 2007. "Young People's Views on Children's Rights and Advocacy Services: A Case for 'Caring' Advocacy?" *Child Abuse Review* 16:140–52.

Barrell, J. 1991. *The Infection of Thomas De Quincy: A Psychopathology of Imperialism.* New Haven, CT: Yale University Press.

Barry, D. 2004. "Waiting, Camera in Hand, to Capture a Fish Market's Final Days." *New York Times,* April 3, A12.

———. 2005. "A Last Whiff of Fulton's Fish, Bringing a Tear." *New York Times,* July 10, 1, 21.

Barry, J., and J. Derevlany. 1987. *Yuppies Invade My House at Dinnertime: A Tale of Brunch, Bombs, and Gentrification in an American City.* Hoboken, NJ: Big River.

Barthel, D. 1989. "Historic Preservation: A Comparative Analysis." *Sociological Forum* 4, no. 1:87–105.

———. 1996. *Historic Preservation: Collective Memory and Historical Identity.* New Brunswick, NJ: Rutgers University Press.

Bayliss, C. M., and O. E. Polk. 2005. "Attorneys' Self-Reported Perspectives and Criteria for Requesting Competency Evaluations in Criminal Defense Cases." *Criminal Justice Review* 30:312–24.

Bearman, P. 2005. *Doormen.* Chicago: University of Chicago Press.

Beauregard, R. A. 1986. "Planning Practice." *Urban Geography* 7:172–78.

Bebow, J. 2004. "Residents, Builders Gird for Zoning War." *Chicago Tribune,* April 11, 1, 4.

Bechtel, R. 1977. *Enclosing Behavior.* Stroudsburg, PA: Dowden, Hutchinson & Ross.

Becker, H. 1993. "How I Learned What a Crock Was." *Journal of Contemporary Ethnography* 22:28–35.

Beggs, J. J., et al. 1996. "Revisiting the Rural-Urban Contrast: Personal Networks in Nonmetro-politan and Metropolitan Settings." *Rural Sociology* 61:306–25.

Bell, M. M. 1994. *Childerley: Nature and Morality in a Country Village.* Chicago: University of Chicago Press.

———. 1997. "The Ghosts of Place." *Theory and Society* 26, no. 6 (December): 813–36.

———. 1998. *An Invitation to Environmental Sociology.* Thousand Oaks, CA: Pine Forge.

Bell, W. 1961. "The City, the Suburb, and a Theory of Social Choice." In *The New Urbaniza-tion,* ed. W. Bell, S. Greer, D. C. McElrath, D. W. Minar, and P. Orleans, 132–68. New York: St. Martin's.

Bell, W., and M. Boat. 1957. "Urban Neighborhoods and Informal Social Relations." *American Journal of Sociology* 62, no. 4 (January): 391–98.

Bellafonte, G. 2004. "A Gay Boomtown Is More Mainstream and Less the Cliché." *New York Times,* May 15, A1.

Bellah, R. N., et al. 1985. *Habits of the Heart: Individualism and Commitment in American Life.* New York: Harper & Row.

Belluck, P. 2005. "To Preserve Its Way of Life, Island Seeks Independence." *New York Times,* August 25, A12.

Bendix, R. 1997. *In Search of Authenticity: The Formation of Folklore Studies.* Madison: University of Wisconsin Press.

Bennett, L. 1997. *Neighborhood Politics: Chicago and Sheffield.* London: Routledge.

Berger, J. 2002. "Sit in This Chair, Go Back in Time." *New York Times,* November 11, A15.

———. 2005. "Goodbye South Bronx Blight, Hello Trendy SoBro." *New York Times,* June 24, A1, A20.

Berrey, E. C. 2005. "Divided over Diversity: Political Discourse in a Chicago Neighborhood." *City and Community* 4, no. 2 (June): 143–70.

Berry, B. J. L. 1985. "Islands of Renewal in Seas of Decay." In *The New Urban Reality,* ed. P. E. Petersen, 69–96. Washington, DC: Brookings Institution.

Beverland, M. B. 2005. "Crafting Brand Authenticity: The Case of Luxury Wines." *Journal of Management Studies* 42, no. 5:1003–29.

Biernacki, R. 1995. *The Fabrication of Labor: Germany and Britain, 1640–1914.* Berkeley and Los Angeles: University of California Press.

Bobo, L., and F. C. Licari. 1989. "Education and Political Tolerance: Testing the Effects of Cogni-tive Sophistication and Target Group Affect." *Public Opinion Quarterly* 53:286–308.

Bohl, C. C. 2000. "New Urbanism and the City: Potential Applications and Implications for Distressed Inner-City Neighborhoods." *Housing Policy Debate* 11, no. 4:761–801.

Bolté, M. 1983. *Portrait of a Woman Down East: Selected Writings of Mary Bolté.* Edited by C. Bolté. Halifax: Nimbus.

Borer, M. I. 2006. "The Location of Culture: The Urban Culturalist Perspective." *City and Com-munity* 5, no. 2:173–97.

Bourdieu, P. 1984. *Distinction: A Social Critique of the Judgement of Taste.* Translated by R. Nice. Cambridge, MA: Harvard University Press.

Boyd, M. 2000. "Reconstructing Bronzeville: Racial Nostalgia and Neighborhood Redevelop-ment." *Journal of Urban Affairs* 22, no. 2 (April): 107–22.

Bragg, M. 2004. "Manor Board Recant on Motta Site." *Provincetown Banner,* March 25, 1.

Bragg, M., and K. R. Lum. 2005. "Nicolau Ekes Out Victory." *Provincetown Banner,* May 5, 1.

Bragg, R. 2002. "Driving the Blues Trail, in Search of a Lost Muse." *New York Times,* April 19, D1, D5.

Braham, J., and P. Peterson. 1998. *Starry Starry Night: Provincetown's Response to the AIDS Epidemic.* Cambridge, MA: Lumen.

Breen, T. H., and T. Kelly. 1996. *Imagining the Past: East Hampton Histories.* Reading, MA: Addison-Wesley.

Brekhus, W. 2003. *Peacocks, Chameleons, Centaurs: Gay Suburbia and the Grammar of Social Identity.* Chicago: University of Chicago Press.

Breton, R. 1964. "Institutional Completeness of Ethnic Communities and the Personal Relations of Immigrants." *American Journal of Sociology* 70:193–205.

Bridge, G. 2003. "Time-Space Trajectories in Provincial Gentrification." *Urban Studies* 40, no. 12:2545–56.

Briggs, J. E. 2005. "Historic Tree Studios Survives Redo, but Homes for Artists Don't." *Chicago Tribune*, sec. 2, August 24, 1, 10.

———. 2006. "Housing Boom's 2 Sides Detailed Study Defines Strain in Gentrifying Areas." *Chicago Tribune*, January 2, 1.

Briggs, X. de S. 1997a. "Comment on Sandra J. Newman and Ann B. Schnare's '". . . And a Suitable Living Environment': The Failure of Housing Programs to Deliver on Neighborhood Quality.'" *Housing Policy Debate* 8, no. 4:743–53.

———. 1997b. "Moving Up versus Moving Out: Neighborhood Effects in Housing Mobility Programs." *Housing Policy Debate* 8, no. 1:195–234.

Brint, S. G. 1984. "'New Class' and Cumulative Trend Explanations of the Liberal Political Attitudes of Professionals." *American Journal of Sociology* 90:30–71.

———. 1985. "The Political Attitudes of Professionals." *Annual Review of Sociology* 11:389–414.

———. 2001. "Gemeinschaft Revisited: A Critique and Reconstruction of the Community Concept." *Sociological Theory* 19, no. 1 (March): 1–23.

Brown, Michael, and Larry Knopp. 2003. "Queer Cultural Geographies—We're here! We're queer! We're over there, too!" In *Handbook of Cultural Geography*, ed. Kay Anderson, 313–24. London: Sage.

Brown, P. L. 2005. "A Man's Home May Be His Castle, but Only to a Point." *New York Times*, December 23, A13.

Brownell, B. 1950. *The Human Community: Its Philosophy and Practice for a Time of Crisis.* New York: Harper & Bros.

Brown-Saracino, J. 1999. "Newcomers v. Old-Timers: Symbolic Class Conflict in a Small Massachusetts Town." Honors thesis, Department of Sociology, Smith College.

———. 2004. "Social Preservationists and the Quest for Authentic Community." *City and Community* 3, no. 2 (June): 135–56.

———. 2006. "Social Preservation: The Quest for Authentic People, Place and Community." Ph.D. diss., Northwestern University.

———. 2007. "Virtuous Marginality: Social Preservationists and the Selection of the Old-Timer." *Theory and Society* 36, no. 5:437–68.

Brown-Saracino, J., and A. Ghaziani. 2009. "The Constraints of Culture: Evidence from the Chicago Dyke March." *Cultural Sociology* 3, no. 1:51–75.

Brown-Saracino, J., and C. Rumpf. 2008. "Nuanced Imageries of Gentrification: Evidence from Newspaper Coverage in Major U.S. Cities, 1986–2006." Paper presented at the meeting of the American Sociological Association, August 2008.

Bryson, B. 1996. "'Anything but Heavy Metal': Symbolic Exclusion and Musical Dislikes." *American Sociological Review* 61, no. 5:884–99.

Burgess, E. W. 1967. "The Growth of the City." In *The City*, ed. R. Park and E. W. Burgess, 47–62. Chicago: University of Chicago Press.

Butler, M. E. 1999. "Critchley Sees Chance to Cash in on Y2K." *Provincetown Banner*, Arts section, July 9, A1.

Butler, T., and G. Robson. 2003. *London Calling: The Middle Classes and the Re-Making of Inner London*. Oxford: Berg.

Buzzelli, M. 2001. "From Little Britain to Little Italy: An Urban Ethnic Landscape Study in Toronto." *Journal of Historical Geography* 27, no. 4:573–87.

Cantwell, R. 1996. *When We Were Good: The Folk Revival*. Cambridge, MA: Harvard University Press.

Castells, M. 1983. *The City and the Grassroots*. Berkeley and Los Angeles: University of California Press.

Caulfield, J. 1994. *City Form and Everyday Life: Toronto's Gentrification and Critical Social Practice*. Toronto: University of Toronto Press.

Chernoff, M. 1980. "Social Displacement in a Renovating Neighborhood's Commercial District: Atlanta." In *Back to the City: Issues in Neighborhood Revitalization*, ed. S. B. Laska and D. Spain, 204–19. New York: Pergamon.

Christenson, J. A. 1984. "Gemeinschaft and Gesellschaft: Testing the Spatial and Community Hypotheses." *Social Forces* 63, no. 1 (September): 160–68.

Clark, T. N. 2003. *The City as an Entertainment Machine*. New York: Elsevier.

Clarke, R. 2004. *Against All Enemies: Inside America's War on Terror*. New York: Free Press.

Collins, R. C. 1980. "Changing Views on Historical Conservation in Cities." *Annals of the American Academy of Political and Social Science* 451, no. 1 (September): 86–97.

Colman, D. 2005. "Rich Gay, Poor Gay." *New York Times*, sec. 9, September 4, 1.

Connell, J. 1978. *The End of Tradition: Country Life in Central Surrey*. London: Routledge & Kegan Paul.

Cooke, T. J., and M. Rapino. 2007. "The Migration of Partnered Gays and Lesbians between 1995 and 2000." *Professional Geographer* 59, no. 3:285–97.

Coontz, S. 1992. *The Way We Never Were: American Families and the Nostalgia Trap*. New York: Basic.

Cotgrove, S. F. 1982. *Catastrophe or Cornucopia: The Environment, Politics, and the Future*. Chichester: Wiley.

Cunningham, M. 2002. *Land's End: A Walk in Provincetown*. New York: Crown.

Datel, R. E. 1985. "Preservation and a Sense of Orientation for American Cities." *Geographical Review* 75, no. 2 (April): 125–41.

Davila, A. 2001. *Latinos, Inc.: The Marketing and Making of a People*. Berkeley and Los Angeles: University of California Press.

———. 2004. *Barrio Dreams: Puerto Ricans, Latinos, and the Neoliberal City*. Berkeley and Los Angeles: University of California Press.

Davis, F. 1979. *Yearning for Yesterday: A Sociology of Nostalgia*. New York: Free Press.

Davis, J. A. 1982. "Achievement Variables and Class Cultures: Family, Schooling, Job, and Forty-Nine Dependent Variables in the Cumulative GSS." *American Sociological Review* 47, no. 5 (October): 569–86.

Davis, M. 1992. *City of Quartz: Excavating the Future in Los Angeles*. New York: Vintage.

Desroches, S. 2003a. "Icons of P'town." *InNewsweekly* (Boston), September 3.

———. 2003b. "New Art Exhibit—Meadows Motel." *InNewsweekly* (Boston), October 29.

———. 2003c. "Townie Talk." *InNewsweekly* (Boston), August 6.

———. 2004. "P-town's Angry Clown: Comic and Playwright Ryan Landry Combines Nudity and Activism to Defend the Spirit of His Beloved Provincetown." *The Advocate,* June 24, 114.

Dewar, H. 2003. "Go with the Slow." *Utne Reader,* September/October, 34, 36.

di Leonardo, M. 1998. *Exotics at Home: Anthropologies, Others, American Modernity.* Chicago: University of Chicago Press.

DiMaggio, P. 1987. "Classification in Art." *American Sociological Review* 52, no. 4 (August): 440–55.

———. 1997. "Culture and Cognition." *Annual Review of Sociology* 23:263–87.

Doty, M. 1995. *My Alexandria.* London: Jonathan Cape.

Downey, D. J. 1999. "From Americanization to Multiculturalism: Political Symbols and Struggles for Cultural Diversity in Twentieth-Century American Race Relations." *Sociological Perspectives* 42, no. 2 (Summer): 249–78.

Duncan, J., and N. Duncan. 2003. *Landscapes of Privilege: The Politics of the Aesthetic in an American Suburb.* New York: Routledge.

Duncan, O. D. 1961. "Occupational Components of Educational Differences in Income." *Journal of the American Statistical Association* 56, no. 296:783–92.

Durkheim, E. 1897/1997. *Suicide.* Edited by J. A. Spaulding and G. Simpson. New York: Free Press.

Edgewater Development Corporation. *Edgewater Developments.* Chicago, Summer 2005.

Edwardes, M. 1968. *Glorious Sahibs: The Romantic as Empire-Builder, 1799–1838.* London: Eyre & Spottiswoode.

Ehrenreich, J., and B. Ehrenreich. 1977. "The Professional-Managerial Class." *Radical America* 11:7–31.

Eight Forty-Eight. 2002. *Interview with Maxwell Street Activist.* February 28. Copy in author's files, available on request.

Elari, J. 2004. "Upper West Side Story." *Gay and Lesbian Review Worldwide* 11, no. 6 (November/December): 30–33.

Eliasoph, N. 1998. *Avoiding Politics: How Americans Produce Apathy in Everyday Life.* Cambridge: Cambridge University Press.

———. 1997. "'Close to Home': The Work of Avoiding Politics." *Theory and Society* 26, no. 5:605–47.

Erickson, B. H. 1996. "Culture, Class, and Connections." *American Journal of Sociology* 102, no. 1 (July): 217–51.

Erikson, K. T. 1976. *Everything in Its Path: Destruction of Community in the Buffalo Creek Flood.* New York: Simon & Schuster.

Etzioni, A. 1993. *The Spirit of Community: The Reinvention of American Society.* New York: Simon & Schuster.

———. 1996. "The Responsive Community: A Communitarian Perspective." *American Sociological Review* 61, no. 1 (February): 1–11.

Evans, G. 2003. "Hard-Branding the Cultural City—from Prado to Prada." *International Journal of Urban and Regional Research* 27, no. 2 (June): 417–40.

Everson, J. G. 1970. *Tidewater Ice of the Kennebec River.* Freeport, ME: Bond Wheelwright.

Fabrikant, G. 2005. "Old Nantucket Warily Meets the New." *New York Times,* June 5, 1, 16–17.

Faiman-Silva, S. 2004. *The Courage to Connect: Sexuality, Citizenship, and Community in Provincetown.* Urbana: University of Illinois Press.

Fantasia, R. 1988. *Cultures of Solidarity: Consciousness, Action and Contemporary American Workers.* Berkeley and Los Angeles: University of California Press.

Fendrich, J. M. 1993. *Ideal Citizens: The Legacy of the Civil Rights Movement.* Albany: State University of New York Press.

Feuer, A. 2006. "Can't Get a Beer Anymore, but Soon You Can Get a Condo." *New York Times,* January 22, 23.

Fine, G. A. 1979. "Small Groups and Culture Creation: The Idioculture of Little League Baseball Teams." *American Sociological Review* 44:733–45.

———. 2004. *Everyday Genius: Self-Taught Art and the Culture of Authenticity.* Chicago: University of Chicago Press.

Finke, R., et al. 1996. "Mobilizing Local Religious Markets: Religious Pluralism in the Empire State, 1855–1865." *American Sociological Review* 61:203–18.

Fischer, C. S. 1975. "Toward a Subcultural Theory of Urbanism." *American Journal of Sociology* 80, no. 6 (May): 1319–41.

———. 1982. *To Dwell among Friends: Personal Networks in Town and City.* Chicago: University of Chicago Press.

———. 1995. "The Subcultural Theory of Urbanism: A Twentieth-Year Assessment." *American Journal of Sociology* 101, no. 3 (November): 543–77.

———. 1999. "Uncommon Values, Diversity, and Conflict in City Life." In *Diversity and Its Discontents: Cultural Conflict and Common Ground in Contemporary American Society,* ed. N. J. Smelser and J. C. Alexander, 213–28. Princeton, NJ: Princeton University Press.

Florida, R. 2002. *The Rise of the Creative Class: And How It's Transforming Work, Leisure, Community and Everyday Life.* New York: Basic.

Forsyth, A. 1997a. "NoHo: Upscaling Main Street on the Metropolitan Edge." *Urban Geography* 18, no. 7:622–52.

———. 1997b. "'Out' in the Valley." *International Journal of Urban Regional Research* 21:38–62.

Fowler, R. B. 1991. *The Dance with Community.* Lawrence: University Press of Kansas.

Francaviglia, R. 2000. "Selling Heritage Landscapes." In *Preserving Cultural Landscapes in America.,* ed. A. Alanen and R. Melnick, 44–69. Baltimore: John Hopkins University Press.

Freeman, L. 2006. *There Goes the 'Hood: Views of Gentrification from the Ground Up.* Philadelphia: Temple University Press.

Freeman, L., and F. Braconi. 2004. "Gentrification and Displacement: New York City in the 1990s." *Journal of the American Planning Association* 70, no. 1 (Winter): 39–52.

French, H. T., ed. 1995. *Maine, a Peopled Landscape: Salt Documentary Photography, 1978 to 1995.* Hanover: University Press of New England.

Friend, T. 2005. "The Parachute Artist: Have Tony Wheeler's Guidebooks Travelled Too Far?" *New Yorker,* April 18, 78–91.

Fulford, T. 2004. "The Taste of Paradise: The Fruits of Romanticism in the Empire." In *Cultures of Taste/Theories of Appetite: Eating Romanticism.,* ed. T. Morton, 41–58. New York: Palgrave.

Gale, D. E. 1980. *Neighborhood Revitalization and the Postindustrial City: A Multinational Perspective.* Lexington, MA: Lexington.

Gans, H. J. 1962. "Urbanism and Suburbanism as Ways of Life: A Re-Evaluation of Definitions." In *Human Behavior and Social Processes,* ed. A. M. Rose. Boston: Houghton Mifflin.

———. 1974. *Popular Culture and High Culture: An Analysis and Evaluation of Taste.* New York: Basic.

———. 1979. "Symbolic Ethnicity: The Future of Ethnic Groups and Cultures in America." *Ethnic and Racial Studies* 2:1–20.

———. 2005. "Gentrification's Victims." *New York Times,* Letter to the Editor, sec. 14, June 26, 9.

"Gay Urban Pioneers: Going against the Grain." 1995. *CQ Researcher* 5, no. 38:908.

Geertz, C. 1973. *The Interpretation of Cultures*. New York: Basic Books.

"Getting Real in Andersonville." 2006. *Chicago Free Press*, Freetime section, March 8, 3.

Ghaziani, A., and M. Ventresca. 2005. "Keywords and Cultural Change: Frame Analysis of Business Model Public Talk, 1975–2000." *Sociological Forum* 20, no. 4 (December): 523–59.

Gieryn, T. F. 2000. "A Space for Place in Sociology." *Annual Review of Sociology* 26:463–96.

Gilderbloom, J. I. 2004. "University Partnerships to Reclaim and Rebuild Communities." *Practicing Planner*, Winter 2004. http://myapa.planning.org/affordablereader/pracplanner/univpartnersvol2no4.htm.

Gleason, D. S. 1999. "Becoming Dominant: Shifting Control and the Creation of Culture in Provincetown, MA." Ph.D. diss., University of Pennsylvania.

Goldberg, L. 1991. *Here on This Hill: Conversations with Vermont Neighbors*. Marshfield, VT: Vermont Folklife Center.

Gotham, K. 2005. "Tourism Gentrification: The Case of New Orleans' Vieux Carre (French Quarter)." *Urban Studies* 42, no. 7:1099–1121.

Gouldner, A. W. 1979. *The Future of Intellectuals and the Rise of the New Class*. New York: Seabury.

Graham, C. 2001. "'Blame It on Maureen O'Hara': Ireland and the Trope of Authenticity." *Cultural Studies* 15:58–75.

Graham, S. 1998. "The End of Geography or the Explosion of Place? Conceptualizing Time and Information Technology." *Progress in Human Geography* 22:165–85.

Grana, C. 1971. *Fact and Symbol*. New York: Oxford University Press.

Grant, J. 2006. *Planning the Good Community: New Urbanism in Theory and Practice*. London: Routledge.

Grayson, K., and R. Martinec. 2004. "Consumer Perceptions of Iconicity and Indexicality and Their Influence on Assessments of Authentic Market Offerings." *Journal of Consumer Research* 31:296–312.

Grazian, D. 2003. *Blue Chicago: The Search for Authenticity in Urban Blues Clubs*. Chicago: University of Chicago Press.

Greenhouse, C. J., et al. 1994. *Law and Community in Three American Towns*. Ithaca, NY: Cornell University Press.

Griswold, W. 1986. *Renaissance Revivals: City Comedy and Revenge Tragedy in the London Theatre, 1576–1980*. Chicago: University of Chicago Press.

———. 1992. "The Writing on the Mud Wall: Nigerian Novels and the Imaginary Village." *American Sociological Review* 57, no. 6:709–24.

———. 1994. *Cultures and Societies in a Changing World*. Thousand Oaks, CA: Pine Forge.

———. 2000. *Bearing Witness: Readers, Writers, and the Novel in Nigeria*. Princeton, NJ: Princeton University Press.

———. 2001. "The Ideas of the Reading Class." *Contemporary Sociology* 30, no. 1:4–6.

Griswold, W., and N. Wright. 2004. "Cowbirds, Locals, and the Endurance of Regionalism." *American Journal of Sociology* 109, no. 6 (May): 1411–51.

Guelke, L., and J. K. Guelke. 2004. "Imperial Eyes on South Africa: Reassessing Travel Narratives." *Journal of Historical Geography* 30:11–31.

Gusfield, J. R. 1975. *Community: A Critical Response*. New York: Harper & Row.

Hackworth, J., and J. Rekers. 2005. "Ethnic Packaging and Gentrification: The Case of Four Neighborhoods in Toronto." *Urban Affairs Review* 41, no. 2 (November): 211–36.

Hall, A., and B. Wellman. 1983. *Social Structure, Social Networks, and Social Support*. Toronto: Centre for Urban and Community Studies.

Hall, P. 2006. "Quinceañera." *Edge* (Boston), August 4. http://www.edgeboston.com/index.php?ch=entertainment&sc=movies&sc2=reviews&sc3=features&id=45942.

Hammond, D., and A. Jablow. 1970. *The Africa That Never Was: Four Centuries of British Writing about Africa.* New York: Twayne.

Hampton, K., and B. Wellman. 2003. "Neighboring in Netville: How the Internet Supports Community and Social Capital in a Wired Suburb." *City and Community* 2, no. 4 (December): 277–312.

Hannerz, U. 1990. "Cosmopolitans and Locals in World Culture." In *Global Culture: Nationalism, Globalization, and Modernity,* ed. M. Featherstone, 237–52. London: Sage.

Hannigan, J. 2004. "Boom Towns and Cool Cities: The Perils and Prospects of Developing a Distinctive Urban Brand in a Global Economy." Paper presented at "Leverhulme International Symposium 2004: The Resurgent City," London School of Economics, April 19–21.

Harden, B. 2002. "More Greens? A Red Light on the Vineyard." *New York Times,* October 18, 1.

Hardesty, D. L. 2000. "Ethnographic Landscapes: Transforming Nature into Culture." In *Preserving Cultural Landscapes in America.,* ed. A. Alanen and R. Melnick, 169–85. Baltimore: John Hopkins University Press.

Harrison, S. 2000. "Charnick Creates 'Popeye' from Cans." *Provincetown Banner,* Arts section, September 7, 35.

———. 2002a. "Critchley Parlays Arrest into Art." *Provincetown Banner,* May 30, 40.

———. 2002b. "Year in Review: Building Up and Tearing Down." *Provincetown Banner,* December 26, 29–30.

Hartmann, D., and J. Gerteis. 2005. "Dealing with Diversity: Mapping Multiculturalism in Sociological Terms." *Sociological Theory* 23, no. 2 (June): 218–40.

Hawley, A. H. 1971. *Urban Society: An Ecological Approach.* New York: Ronald.

Heath, A. F., and R. Jowell. 1991. *Understanding Political Change: The British Voter, 1964–1987.* Oxford: Pergamon.

Heath, A. F., and S. Michael. 1995. "Political Alignments within the Middle Classes, 1972–1989." In *Social Change and the Middle Classes,* ed. T. Butler and M. Savage, 275–92. New York: Routledge.

Hebdige, D. 1979. *Subculture: The Meaning of Style.* London: Methuen.

Hebert, E. 1979. *The Dogs of March.* Hanover, NH: University Press of New England.

Henig, J. R. 1981. "Gentrification and Displacement of the Elderly: An Empirical Analysis." *Gerontologist* 21, no. 1:67–75.

———. 1984. "Gentrification and Displacement of the Elderly: An Empirical Analysis." In *Gentrification, Displacement, and Neighborhood Revitalization,* ed. J. J. Palen and B. London, 170–84. Albany: State University of New York Press.

Henig, J., and D. Gale. 1987. "The Political Incorporation of Newcomers to Racially Changing Neighborhoods." *Urban Affairs Quarterly* 22:399–419.

Higham, J. 1999. "Cultural Responses to Immigration." In *Diversity and Its Discontents: Cultural Conflict and Common Ground in Contemporary American Society,* ed. N. J. Smelser and J. C. Alexander, 39–62. Princeton, NJ: Princeton University Press.

Hirsch, A. R. 1998. *Making the Second Ghetto: Race and Housing in Chicago, 1940–1960.* Chicago: University of Chicago Press.

Hobsbawm, E., and T. Ranger. 1983. *The Invention of Tradition.* Cambridge: Cambridge University Press.

Hoffman, L. M. 2003. "The Marketing of Diversity in the Inner City: Tourism and Regulation in Harlem." *International Journal of Urban and Regional Research* 27, no. 2 (June): 286–99.

Holstein, J. A., and G. Miller. 1993. "Social Constructionism and Social Problems Work." In *Constructionist Controversies: Issues in Social Problems Theory,* ed. G. Miller and J. A. Holstein, 131–52. New Brunswick, NJ: Aldine/Transaction.

Holt, D. B. 1998. "Does Cultural Capital Structure American Consumption?" *Journal of Consumer Research* 25, no. 1 (June): 1–25.

Horton, J. 1995. *The Politics of Diversity: Immigration, Resistance, and Change in Monterey Park, California.* Philadelphia: Temple University Press.

Hummon, D. M. 1990. *Commonplaces: Community Ideology and Identity in American Culture.* Albany: State University of New York Press.

Hunter, A. 1972. "The Expanding Community of Limited Liability." In *The Social Construction of Communities,* by G. D. Suttles, 44–81. Chicago: University of Chicago Press.

———. 1974. *Symbolic Communities.* Chicago: University of Chicago Press.

———. 1975. "The Loss of Community: An Empirical Test through Replication." *American Sociological Review* 40, no. 5 (October): 537–52.

———. 1985. "Private, Parochial and Public Social Orders: The Problem of Crime and Incivility in Urban Communities." In *The Challenge of Social Control: Citizenship and Institution Building in Modern Society,* ed. G. D. Suttles and M. N. Zald, 230–42. Norwood, NJ: Ablex.

Institute of International Education (IIE). 2005. *Open Doors 2005: Report on International Educational Exchange.* New York: IIE.

Jacobs, A. 2006. "There Goes the Neighborhood: The Last of the Machinery District Is Packing Up." *New York Times,* February 7, A21.

Janowitz, M. 1967. *The Community Press in an Urban Setting.* Chicago: University of Chicago Press.

Jaster, R. S. 1999. *Russian Voices on the Kennebec: The Story of Maine's Unlikely Colony.* Orono: University of Maine Press.

"Jay Critchley on 'P-Town Inc.'" 2002. *Gay and Lesbian Review,* September/October, 40.

Jeffers, G., and R. Osterman. 2004. "Pilsen Uneasy with Development: Some Residents Fear Gentrification." *Chicago Tribune,* Metro section, January 31, 3.

Jindra, M. 1994. "Star Trek Fandom as a Religious Phenomenon." *Sociology of Religion* 55:27–51.

Johnson, E. P. 2003. *Appropriating Blackness: Performance and the Politics of Authenticity.* Durham, NC: Duke University Press.

Johnson, J. 2004. "Two Camps in Big Sur." *Los Angeles Times,* March 15, 1.

Johnston, J., and S. Baumann. 2007. "Democracy versus Distinction: A Study of Omnivorousness in Gourmet Food Writing." *American Journal of Sociology* 113, no. 1 (July): 165–204.

Johnstone, J. W. C., et al. 1972–73. "The Professional Values of American Newsmen." *Public Opinion Quarterly* 36, no. 4 (Winter): 522–40.

Jones, R. E., and R. E. Dunlap. 1992. "The Social Bases of Environmental Concern: Have They Changed over Time?" *Rural Sociology* 57, no. 1:28–47.

Judd, R. W., et al. 1995. *Maine: The Pine Tree State from Prehistory to the Present.* Orono: University of Maine Press.

Judy, R. A. T. 2004. "On the Question of Nigga Authenticity." In *That's the Joint! The Hip Hop Studies Reader,* ed. M. Forman and M. A. Neal, 105–18. New York: Routledge.

Kain, J. F., and W. C. Apgar Jr. 1985. *Housing and Neighborhood Dynamics: A Simulation Study.* Cambridge, MA: Harvard University Press.

Kasarda, J. D., and M. Janowitz. 1974. "Community Attachment in Mass Society." *American Sociological Review* 39, no. 3 (June): 329–39.

Kasinitz, P. 1983. "Gentrification and Homelessness: The Single Room Occupant and the Inner City Revival." *Urban and Social Change Review* 17, no. 1:9–14.

———. 1988. "The Gentrification of Boerum Hill: Neighborhood Change and Conflicts over Definitions." *Qualitative Sociology* 11, no. 3 (September): 163–82.

Kefalas, M. 2003. *Working-Class Heroes: Protecting Home, Community, and Nation in a Chicago Neighborhood*. Berkeley and Los Angeles: University of California Press.

Kegan, P., K. D. Van Liere, and R. E. Dunlap. 1980. "The Social Bases of Environmental Concern: A Review of Hypotheses, Explanations and Empirical Evidence." *Public Opinion Quarterly* 44, no. 2 (Summer): 181–97.

Keller, C. 2001. *Frommer's Chicago, 2001*. Hoboken, NJ: Frommer.

Kennedy, R. F. 2005. "An Ill Wind off Cape Cod." *New York Times*, December 16, A35.

Kibria, Nazli. 2000. "Race, Ethnic Options, and Ethnic Binds: Identity Negotiations of Second-Generation Chinese and Korean Americans." *Sociological Perspectives* 43, no. 1 (Spring): 77–95.

Kiecolt, K. J. 1988. "Recent Developments in Attitudes and Social Structure." *Annual Review of Sociology* 14:381–403.

Kilgannon, C. 2005. "Be It Ever So Humble. O.K., It's Shabby." *New York Times*, August 25, A20.

Kirby, D. 2003. "My Brooklyn: Racing Upscale at Full Throttle." *New York Times*, February 14, B37, B46.

Kitchin, R., and K. Lysaght. "Sexual Citizenship in Belfast, Northern Ireland." *Gender, Place and Culture* 11, no. 1 (March): 83–103.

Klawans, S. 2006. "The Magic Bus." *The Nation*, August 28, 42–44.

Kleinman, S. 1996. *Opposing Ambitions: Gender and Identity in an Alternative Organization*. Chicago: University of Chicago Press.

Knoke, D., and L. Isaac. 1976. "Quality of Higher Education and Sociopolitical Attitudes." *Social Forces* 54, no. 3 (March): 524–29.

Knopp, L. 1990a. "Exploiting the Rent Gap: The Theoretical Significance of Using Illegal Appraisal Schemes to Encourage Gentrification in New Orleans." *Urban Geography* 11:48–64.

———. 1990b. "Some Theoretical Implications of Gay Involvement in an Urban Land Market." *Political Geography Quarterly* 9:337–52.

———. 1995. "Sexuality and Urban Space: A Framework for Analysis." In *Mapping Desire: Geographies of Sexuality*, ed. D. Bell and G. Valentine, 149–62. London: Routledge.

Kornblum, W. 1974. *Blue Collar Community*. Chicago: University of Chicago Press.

Krahulik, K. C. 2005. *Provincetown: From Pilgrim Landing to Gay Resort*. New York: New York University Press.

Kramer, M. 1987. *Three Farms: Making Milk, Meat, and Money from the American Soil*. Cambridge, MA: Harvard University Press.

Kristol, I. 1972a. "About Equality." *Commentary* 54:41–47.

———. 1972b. *On the Democratic Idea in America*. New York: Harper & Row.

Kugel, S. 2006. "Preservation: Sure, It's a Good Thing, but . . ." *New York Times*, January 15, 3, 13.

Ladd, E. C., Jr. 1978a. "The New Lines Are Drawn: Class and Ideology in America, Part I." *Public Opinion* 3:48–53.

———. 1978b. *Where Have All the Voters Gone? The Fracturing of America's Political Parties*. New York: Norton.

Ladd, E. C., Jr., and S. M. Lipsett. 1975. *The Divided Academy: Professors and Politics*. New York: McGraw-Hill.

Lamont, M. 1992. *Money, Morals, and Manners*. Chicago: University of Chicago Press.

————. 2000. *The Dignity of Working Men: Morality and the Boundaries of Race, Class, and Immigration.* Cambridge, MA: Harvard University Press.

Lane, K. B. 2003. *Andersonville: A Swedish-American Landmark Neighborhood.* Chicago: Swedish American Museum.

Larson, C. 1998. "Bitter Blend: Will Starbucks Suck the Soul out of Andersonville." *The Reader* (Chicago), sec. 1, December 4, 6, 8, 10.

Laumann, E., et al. 2004. *The Sexual Organization of the City.* Chicago: University of Chicago Press.

Lauria, M., and L. Knopp. 1985. "Toward an Analysis of the Role of Gay Communities in the Urban Renaissance." *Urban Geography* 6, no. 2:152–69.

Lee, B. A., and D. C. Lodge. 1984. "Spatial Differentials in Residential Development." *Urban Studies* 21, no. 3:219–32.

Lees, L. 1994. "Rethinking Gentrification: Beyond the Positions of Economics or Culture." *Progress in Human Geography* 18:137–50.

————. 2003. "Super-Gentrification: The Case of Brooklyn Heights, New York City." *Urban Studies* 40, no. 12:2487–2509.

Leff, L. 2007. "After AIDS, Gay Neighborhoods in U.S. May Be Victims of Own Success." *The Advocate,* March 13. http://www.advocate.com/news_detail_ektid42852.asp.

Lehane, D. 2003. *Mystic River.* New York: HarperTorch.

Lethem, J. 2003. *Fortress of Solitude.* New York: Vintage.

Levy, P. A., and R. A. Cybriwsky. 1980. "The Hidden Dimensions of Culture and Class: Philadelphia." In *Back to the City: Issues in Neighborhood Revitalization,* ed. S. B. Laska and D. Spain, 138–55. New York: Pergamon.

Lewis, D., and D. Bridger. 2000. *The Soul of the New Consumer: Authenticity, What We Buy and Why in the New Economy.* London: Nicholas Brealey.

Ley, D. 1985. "Gentrification in Canadian Inner Cities: Patterns, Analysis, Impacts and Policy." Report prepared for the Canada Mortgage and Housing Corp. Vancouver: University of British Columbia, Department of Geography.

————. 2004. "Transnational Spaces and Everyday Lives." *Transactions of the Institute of British Geographers* 29, no. 2:151–64.

Lindstrom, B. 1997. "A Sense of Place: Housing Selection on Chicago's North Shore." *Sociological Quarterly* 39, no. 1:19–39.

Lipset, S. M. 1960. *Political Man.* Garden City, NJ: Anchor.

————. 1979. *The First New Nation: The United States in Historical and Comparative Perspective.* New York: Norton.

————. 1983. "Radicalism or Reformism: The Sources of Working-Class Politics." *American Political Science Review* 77, no. 1:1–18.

Lipton, S. G. 1977. "Evidence of Central City Revival." *Journal of the American Institute of Planners* 43:136–47.

Lloyd, R. 2002. "Neo-Bohemia: Art and Neighborhood Redevelopment in Chicago." *Journal of Urban Affairs* 24, no. 5:517–32.

————. 2005. *Neo-Bohemia: Art and Commerce in the Post-Industrial City.* New York: Routledge.

Lofland, L. H. 1998. *The Public Realm: Exploring the City's Quintessential Social Territory.* New York: Aldine de Gruyter.

Logan, J., and H. Molotch. 1987. *Urban Fortunes.* Berkeley and Los Angeles: University of California Press.

London, B. 1980. *Metropolis and Nation in Thailand: The Political Economy of Uneven Development.* Boulder, CO: Westview.

London, B., and J. J. Palen. 1984. Introduction to *Gentrification, Displacement, and Neighborhood Revitalization*, ed. J. J. Palen and B. London, 4–26. Albany: State University of New York Press.

Long, E. 2003. *Book Clubs: Women and the Uses of Reading in Everyday Life.* Chicago: University of Chicago Press.

Long, L. H. 1980. "Back to the Countryside and Back to the City in the Same Decade." In *Back to the City: Issues in Neighborhood Revitalization*, ed. S. B. Laska and D. Spain, 61–76. New York: Pergamon.

Long, L., and D. DeAre. 1980. *Migration to Metropolitan Areas: Appraising the Trends and Reasons for Moving.* Washington, DC: U.S. Department of Commerce.

Lowenthal, D. 1999. *The Past Is a Foreign Country.* Cambridge: Cambridge University Press.

Lynd, R., and H. Lynd. 1959. *Middletown: A Study in Modern American Culture.* San Diego: Harvest/HBJ.

Lyons, M. 1996. "Gentrification, Socioeconomic Change, and the Geography of Displacement." *Journal of Urban Affairs* 18, no. 1:39–62.

MacCannell, D. 1976/1999. *The Tourist: A New Theory of the Leisure Class.* Berkeley and Los Angeles: University of California Press.

Macgregor, L. 2005. "Habits of the Heartland: Producing Community in a Small Midwestern Town." Ph.D. diss., University of Wisconsin, Madison.

MacIntyre, A. 1984. *After Virtue: A Study in Moral Theory.* South Bend, IN: University of Notre Dame Press.

Makdisi, S. 1998. *Romantic Imperialism: Universal Empire and the Culture of Modernity.* Cambridge: Cambridge University Press.

Maly, M. T. 2005. *Beyond Segregation: Multiracial and Multiethnic Neighborhoods in the United States.* Philadelphia: Temple University Press.

Mansbridge, J. 1983. *Beyond Adversary Democracy.* Chicago: University of Chicago Press.

Manso, P. 2002. *Ptown: Art, Sex, and Money on the Outer Cape.* New York: Scribner's.

Marcuse, P. 1986. "Abandonment, Gentrification, and Displacement: The Linkages in New York City." In *Gentrification of the City*, ed. N. Smith and P. Williams, 153–77. Boston: Allen & Unwin.

Marketing Concepts, Inc. 1978a. *The Study of American Opinion, 1978 Report, Vol. 1.* New York: US News and World Report.

Martin, J. 1992. *Cultures in Organizations: Three Perspectives.* New York: Oxford University Press.

McDonnell, T. 2002. "The Trouble with Representation: ACT UP, Protest Art, and Collective Identity." Working paper, Northwestern University.

McLaughlin, A. T. 1995. "Ross Perot's Latest Conquest Maine." *Christian Science Monitor*, December 1, 2.

McNulty, E. 2000. *Chicago Then and Now.* Berkeley, CA: Thunder Bay.

Mele, C. 2000. *Selling the Lower East Side: Culture, Real Estate, and Resistance in New York City.* Minneapolis: University of Minnesota Press.

Merton, R. K. 1968. *Social Theory and Social Structure.* New York: Free Press.

Miller, G., and J. A. Holstein, eds. 1993. *Constructionist Controversies: Issues in Social Problems Theory.* New Brunswick, NJ: Aldine/Transaction.

Miller, S. L. 2002. "Ukrainian Village Gets Push as City Landmark." *Chicago Tribune*, sec. 2, November 20, 3.

Milligan, M. J. 1998. "Interactional Past and Potential." *Symbolic Interaction* 21, no. 1:1–33.

———. 2003. "Displacement and Identity Discontinuity: The Role of Nostalgia in Establishing New Identity Categories." *Symbolic Interaction* 26, no. 3:381–403.

———. 2004. "The House Made Me Do It: Viewing Dwelling as Social Actor." Working paper, Sonoma State University.

———. 2007. "Buildings as History: The Place of Collective Memory in the Study of Historic Preservation." *Symbolic Interaction* 3, no. 1:105–23.

Mitchell, D. 1997. "The Annihilation of Space by Law: The Roots and Implications of Anti-Homeless Laws in the United States." *Antipode* 29:303–35.

Mohr, J., and H. Lee. 2000. "From Affirmative Action to Outreach: Discourse Shifts at the University of California." *Poetics* 8:47–71.

Monti, D., et al. 2003. "Private Lives and Public Worlds: Changes in Americans' Social Ties and Civic Attachments in Late-20th Century America." *City and Community* 2, no. 2 (June): 143–63.

Moore, M. 2001. *Stupid White Men . . . and Other Sorry Excuses for the State of the Nation!* New York: HarperCollins.

Morgan, A. E. 1942. *The Small Community: Foundation of Democratic Life.* New York: Harper & Bros.

Moss, M. L. 1997. "Reinventing the Central City as a Place to Live and Work." *Housing Policy Debate* 8, no. 2:471–90.

Muñiz, V. 1998. *Resisting Gentrification and Displacement: Voices of Puerto Rican Women of the Barrio.* New York: Garland.

Murti, K. P. 2001. *India: The Seductive and Seduced "Other" of German Orientalism.* Westport, CT: Greenwood.

Nelson, K. P. 1988. *Gentrification and Distressed Cities: An Assessment of Trends in Intrametropolitan Migration.* Madison: University of Wisconsin Press.

Newman, A. 2006. "Hold the Mustard, Maybe Forever." *New York Times,* January 6, A19.

Newman, M. 2003. "A Town That's Quaint, Diverse, Convenient—and Costly." *New York Times,* March 22, A13.

Nisbet, R. 1976. *The Quest for Community.* Oxford: Oxford University Press.

Nyden, P., E. Edlynn, and J. Davis. 2006. *The Differential Impact of Gentrification on Communities in Chicago.* Chicago: Loyola University of Chicago Center for Urban Research and Learning.

Ogden, P. 2000. "Households, Reurbanisation and the Rise of Living Alone in the Principal French Cities, 1975–90." *Urban Studies* 37, no. 2 (February): 367–90.

O'Laughlin, J., and D. Munski. 1979. "Housing Rehabilitation in the Inner City: A Comparison of Two Neighborhoods in New Orleans." *Economic Geography* 55, no. 1 (January): 52–70.

Oldenburg, R. 1989. *The Great Good Place: Cafes, Coffee Shops, Bookstores, Bars, Hair Salons, and Other Hangouts at the Heart of a Community.* New York: Marlowe.

Oliver, M. 2002. *What Do We Know: Poems and Prose Poems.* Cambridge, MA: Da Capo.

Orvell, M. 1989. *The Real Thing: Imitation and Authenticity in American Culture, 1880–1940.* Chapel Hill: University of North Carolina Press.

Ouroussoff, N. 2006. "Outgrowing Jane Jacobs." *New York Times,* sec. 4, April 30, 1, 4.

Palen, J. 1988. "Gentrification, Revitalization, and Displacement." In *Handbook of Housing and the Built Environment in the United States,* ed. E. Huttman and W. van Vliet, 417–32. New York: Greenwood.

Paris, B. 1999. *Generation Queer: A Gay Man's Quest for Hope, Love, and Justice*. New York: Warner.

Parsons, D. 1980. "Rural Gentrification: The Influence of Rural Settlement Planning Policies." Research Paper no. 3. University of Sussex, Brighton, Department of Geography.

Parsons, T. 1951. *The Social System*. New York: Free Press.

Pascarella, E. T., and P. T. Terinzini. 1991. *How College Affects Students*. San Francisco: Jossey-Bass.

Pattillo, M. 2003. "Negotiating Blackness, for Richer or for Poorer." *Ethnography* 4, no. 1:1–33.

———. 2007. *Black on the Block: The Politics of Race and Class in the City*. Chicago: University of Chicago Press.

Pattison, T. J. 1977. "The Process of Neighborhood Upgrading and Gentrification: Aan Examination of Two Neighborhoods in Boston Metropolitan Area." M.A. thesis, Massachusetts Institute of Technology, Department of Urban Studies and Planning.

Paulsen, K. 2004. "Making Character Concrete: Empirical Strategies for Studying Place Distinction." *City and Community* 3, no. 3 (September): 243–62.

Paxton, P. 1999. "Is Social Capital Declining in the United States? A Multiple Indicator Assessment." *American Journal of Sociology* 105:88–127.

Payne, David. 2000. *Gravesend Light*. New York: Plume.

Perez, G. M. 2004. *The Near Northwest Side Story: Migration, Displacement, and Puerto Rican Families*. Berkeley and Los Angeles: University of California Press.

Peterson, R. A. 1997. *Creating Country Music: Fabricating Authenticity*. Chicago: University of Chicago Press.

———. 2005. "In Search of Authenticity." *Journal of Management Studies* 42, no. 5 (July): 1083–98.

Peterson, R. A., and R. M. Kern. 1996. "Changing Highbrow Taste: From Snob to Omnivore." *American Sociological Review* 61, no. 5:900–907.

Phelan, J., B. G. Link, A. Stueve, and R. E. Moore. 1995. "Education, Social Liberalism, and Economic Conservatism: Attitudes towards Homeless People." *American Sociological Review* 60, no. 1 (February): 126–40.

Phillips, M. 2004. "Other Geographies of Gentrification." *Progress in Human Geography* 28, no. 5:5–30.

Podmore, J. A. 1998. "(Re)Reading the 'Loft Living' Habitus in Montreal's Inner-City." *International Journal of Urban and Regional Research* 22, no. 2:283–302.

———. 2006. "Gone 'Underground'? Lesbian Visibility and the Consolidation of Queer Space in Montréal." *Social and Cultural Geography* 7, no. 4 (August): 595–625.

Pogrebin, R. 2006. "Culture Raises Its Head and Heart." *New York Times*, January 16, B1.

Porges, S. 2003. "Ukrainian Village: The Effects of Gentrification." *Citylink* (Chicago), April 25–May 1, 1.

Pratt, M. L. 1992. *Imperial Eyes: Travel Writing and Transculturation*. London: Routledge.

Putnam, R. 2000. *Bowling Alone: The Collapse and Revival of American Community*. New York: Simon & Schuster.

Queer to the Left. 2004. *Gentrification: Key Words*. Chicago.

Quercia, R. G., and G. C. Galster. 1997. "Threshold Effects and the Expected Benefits of Attracting Middle-Income Households to the Central City." *Housing Policy Debate* 8, no. 2: 409–35.

Quinn, B. 2005. *How Walmart Is Destroying America and the World: And What You Can Do about It*. Berkeley, CA: Ten Speed.

Rado, D. 2002. "Daley Touts Upgrades, Presses Flesh on Clark St." *Chicago Tribune,* Metro section, December 22, 3.

Radway, J. 1984. *Reading the Romance: Women, Patriarchy, and Popular Literature.* Chapel Hill: University of North Carolina Press.

Ray, A. 2006. "Dirt and Dead Ends." On Indigo Girls, *Despite Our Differences.* Hollywood Records, B000GG4SOC.

Redfield, R. 1947. "The Folk Society." *American Journal of Sociology* 52, no. 4 (January): 293–308.

Reed, C. 2003. "We're from Oz: Marking Ethnic and Sexual Identity in Chicago." *Environment and Planning D: Society and Space* 21, no. 4:425–40.

Reid, E. 2000. ""Understanding the Word 'Advocacy': Context and Use." In *Nonprofit Advocacy and the Policy Process* (vol. 1), ed. E. Reid, 1–8. Washington, DC: Urban Institute.

"Resident of Three Years Decries Neighborhood's Recent Gentrification." 2001. *The Onion,* June 20.

Richards, C. 2002. "Who'll Defend Real Pioneers? Nothing Short of City's Intervention Will Protect Longtime, Low-Income Residents of Gentrifying Neighborhoods." *Chicago Sun-Times,* May 15, 2002, 43.

Rieder, J. 1985. *Canarsie: The Jews and Italians of Brooklyn against liberalism.* Cambridge, MA: Harvard University Press.

Riis, J. 1971. *How the Other Half Lives: Studies among the Tenements of New York.* Harmondsworth: Penguin.

Robinson, T. 1995. "Gentrification and Grassroots Resistance in San Francisco's Tenderloin." *Urban Affairs Review* 30, no. 4:483–513.

Roistacher, E. A., and J. S. Young. 1980. "Working Women and City Structure: Implications of the Subtle Revolution." In "Women and the American City," suppl., *Signs* 5, no. 3 (Spring): 220–25.

Rose, D. 1984. "Rethinking Gentrification: Beyond the Uneven Development of Marxist Urban Theory." *Environment and Planning D: Society and Space* 2, no. 1:47–74.

————. 2004. "Discourses and Experiences of Social Mix in Gentrifying Neighbourhoods: A Montreal Case Study." *Canadian Journal of Urban Research* 13:278–316.

Rothenberg, T. 1995. "'And She Told Two Friends . . .': Lesbians Creating Urban Social Space." In *Mapping Desire: Geographies of Sexuality,* ed. D. Bell and G. Valentine, 165–81. London: Routledge.

Rothman, S. 1979. "The Mass Media in Post-Industrial America." In *The Third Century: America as Post-Industrial Society,* ed. S. M. Lipset, 245–88. Stanford, CA: Hoover Institution.

Rotolo, T. 1999. "Trends in Voluntary Association Participation." *Nonprofit and Voluntary Sector Quarterly* 28:199–212.

Roy, W. G. 2002. "Aesthetic Identity, Race, and American Folk Music." *Qualitative Sociology* 25, no. 3:459–69.

Runciman, W. G., ed. 1978. *Weber: Selections in Translation.* Cambridge: Cambridge University Press.

Ryden, K. 1993. *Mapping the Invisible Landscape: Folklore, Writing, and the Sense of Place.* Iowa City: University of Iowa Press.

Ryle, R. R., and R. V. Robinson. 2006. "Ideology, Moral Cosmology, and Community in the United States." *City and Community* 5, no. 1 (March): 53–69.

Said, E. W. 1978/1994. *Orientalism.* New York: Vintage.

Salamon, S. 2003a. "Mobile Home Communities." In *Encyclopedia of Community: From the*

Village to the Virtual World, ed. K. Christensen and D. Levinson, 3:925–29. Thousand Oaks, CA: Sage.

———. 2003b. *Newcomers to Old Towns: Suburbanization of the Heartland.* Chicago: University of Chicago Press.

Sanderson, E. D., and R. A. Polson. 1939. *Rural Community Organization.* New York: Wiley.

Saulny, S. 2005. "The Displaced: Cast from Their Ancestral Home, Creoles Worry about Culture's Future." *New York Times,* October 11, A15.

Savage, M., et al. 1992. *Property, Bureaucracy and Culture: Middle-Class Formation in Contemporary Britain.* London: Routledge.

Schill, M. H., and R. P. Nathan. 1983. *Revitalizing America's Cities: Neighborhood, Reinvestment and Displacement.* Albany: State University of New York Press.

Schmalenbach, H. 1922/1961. "The Sociological Category of Communion." In *Theories of Society: Foundations of Modern Sociological Theory,* ed. T. Parsons, 1:331–47. New York: Free Press.

Schmich, M. 2005. "Neighborhood Ghosts Roam a Changing City." *Chicago Tribune,* sec. 2, April 22, 1.

Schmitt, P. 1990. *Back to Nature: The Arcadian Myth in Urban America.* Baltimore: Johns Hopkins University Press.

Schudson, M. 1996. *The Good Citizen: A History of American Civic Life.* New York: Free Press.

Schuyler, D., and P. O'Donnell. 2000. "The History and Preservation of Urban Parks and Cemeteries." In *Preserving Cultural Landscapes in America.,* ed. A. Alanen and R. Melnick, 70–93. Baltimore: John Hopkins University Press.

Schwartz, L. 2002. "Is There an Advocate in the House? The Role of Health Care Professionals in Patient Advocacy." *Journal of Medial Ethics* 28:37–40.

Sciolino, E. 2004. "Jewish District Rallies to Save Its Soul from Renovation." *New York Times,* April 5, 4.

Scott, J. 2005. "In Fast-Growing New York, More Neighborhoods Are Seeking to Slow Things Down." *New York Times,* October 10, A18.

Seely, C. 2003. "Is Midtown Losing its Gay Appeal?" *Southern Voice* (Atlanta), April 15, 3, 13.

Severson, K. 2006. "Can New Orleans Save the Soul of Its Food?" *New York Times,* January 11, D1, D5.

Shively, J. 1992. "Cowboys and Indians: Perceptions of Western Films among American Indians and Anglos." *American Sociological Review* 57 (December): 725–34.

Shorr, K. 2002. *Provincetown: Stories from Land's End.* Beverly, MA: Commonwealth.

Sibalis, M. 2004. "Urban Space and Homosexuality: The Example of the Marais, Paris' 'Gay Ghetto.'" *Urban Studies* 41, no. 9 (August): 1739–58.

Simmel, G. 1971. *On Individuality and Social Forms.* Edited by D. Levine. Chicago: University of Chicago Press.

Skeggs, Beverly. 1999. "Matter Out of Place: Visibility and Sexualities in Leisure Spaces." *Leisure Studies* 18:213–32.

Slater, T. 2006. "The Eviction of Critical Perspectives from Gentrification Research." *International Journal of Urban and Regional Research* 30, no. 4:737–57.

Small, M. L. 2004. *Villa Victoria: The Transformation of Social Capital in a Boston Barrio.* Chicago: University of Chicago Press.

Smelser, N. J., and J. C. Alexander, eds. 1999. *Diversity and Its Discontents: Cultural Conflict and Common Ground in Contemporary American Society.* Princeton, NJ: Princeton University Press.

Smith, D. P., and L. Holt. 2005. "'Lesbian Migrants in the Gentrified Valley' and 'Other' Geographies of Rural Gentrification." *Journal of Rural Studies* 21:313–22.

Smith, D. P., and D. A. Phillips. 2001. "Socio-Cultural Representations of Greentrified Pennine Rurality." *Journal of Rural Studies* 17, no. 4:457–69.

Smith, N. 1987. "Gentrification and the Rent-Gap." *Annals of the Association of American Geographers* 77, no. 3: 462–65.

———. 1996/2000. *The New Urban Frontier: Gentrification and the Revanchist City*. London: Routledge.

———. 2002. "New Globalism, New Urbanism: Gentrification as Global Urban Strategy." *Antipode* 34, no. 3:427–50.

Solnit, R., and S. Schwartzenberg. 2000. *Hollow City*. New York: Verso.

Sontag, D. "Forced from New Orleans, but Neighbors Still." *New York Times*, October 12, 1, A14.

Sontag, S. 2001. *In America*. New York: Picador.

Sorkin, A. L. 1971. "Occupational Status and Unemployment of Nonwhite Women." *Social Forces* 49, no. 3:393–98.

Sowers, P. 2005. "School Hearing Draws Vocal Crowd." *Provincetown Banner*, March 9, 1, 24.

Spain, D. 1980. "Indicators of Urban Revitalization: Racial and Socioeconomic Changes in Central-City Housing." In *Back to the City: Issues in Neighborhood Revitalization*, ed. S. B. Laska and D. Spain, 27–41. New York: Pergamon.

Spain, D. 1993. "Been-Heres versus Come-Heres: Negotiating Conflicting Community Identities." *Journal of the American Planning Association* 59, no. 2 (Season): 156–71.

Stansell, C. 1987. *City of Women: Sex and Class in New York, 1780–1860*. Urbana: University of Illinois Press.

Steinhauer, J. 2005. "New York, Once a Lure, Is Slowly Losing the Creative Set." *New York Times*, New York/Region, December 18, 35.

Stephan, G. E., and D. R. McMullin. 1982. "Tolerance of Sexual Nonconformity: City Size as a Situational and Early Learning Determinant." *American Sociological Review* 47:411–15.

Stoneall, L. 1983. *Country Life, City Life: Five Theories of Community*. New York: Praeger.

Stouffer, S. A. 1955. *Communism, Conformity, and Civil Liberties*. New York: Wiley.

Stratton, Jim. 1977. *Pioneering in the Urban Wilderness*. New York: Urizen.

Strauss, A., and B. Glaser. 1967. *The Discovery of Grounded Theory: Strategies for Qualitative Research*. Chicago: Aldine.

Suttles, G. D. 1968. *The Social Order of the Slum*. Chicago: University of Chicago Press.

———. 1972. *The Social Construction of Communities*. Chicago: University of Chicago Press.

Swidler, A. 1986. "Culture in Action: Symbols and Strategies." *American Sociological Review* 51 (April): 273–86.

———. 2001. *Talk of Love: How Culture Matters*. Chicago: University of Chicago Press.

Talen, E. 1999. "Sense of Community and Neighbourhood Form: An Assessment of the Social Doctrine of New Urbanism." *Urban Studies* 36, no. 8:1361–79.

Tavernise, S. 2003. "To Young, a Russian Enclave Is Too Much the Old Country." *New York Times*, Travel section, sec. A, October 8, 1.

Taylor, E. 2006. "There Goes the Neighborhood: L.A. Gentrifiers Pay Tribute to a Community under Siege." *Village Voice*, August 1.

Taylor, M. M. 2002. *Harlem: Between Heaven and Hell*. Minneapolis: University of Minnesota Press.

Thompson, C. J., and S. K. Tambyah. 1999. "Trying to Be Cosmopolitan." *Journal of Consumer Research* 26, no. 3 (December): 214–41.

Tilly, C. 1992. *Coercion, Capital, and European States, AD 990–1992.* Cambridge: Blackwell.

Tobin, G. A., and B. D. Anderson. 1982. "Will Public Schools Benefit from Urban Redevelopment?" *Urban Education* 17, no. 1:73–96.

Tönnies, F. 1887. *Community and Organization.* London: Routledge & Kegan Paul.

Town of Dresden. 1991a. *Dresden Comprehensive Plan, Volume I.* Dresden, ME.

———. 1991b. *Dresden Comprehensive Plan, Volume II.* Dresden, ME.

Trebay, G. 2004. "The Last Lord of Gardiner's Island." *New York Times,* sec. 9, August 29, 1.

Trilling, L. 1971. *Sincerity and Authenticity.* New York: Harcourt.

Tunstall, J. 1971. *Journalists at Work: Specialist Correspondents: Their News Organizations, News Sources, and Competitor-Colleagues.* London: Constable.

Urquia, N. 2004. "'Doin' It Right': Contested Authenticity in London's Salsa Scene." In *Music Scenes: Local, Transnational, and Virtual,* ed. A. Bennett and R. A. Peterson, 96–114. Nashville: Vanderbilt University Press.

Urry, J. 2002. *The Tourist Gaze.* Thousand Oaks, CA: Sage.

U.S. Census Bureau. 2000. DP-2. Profile of Selected Social Characteristics: 2000. Washington, D.C.

Van Gelder, L., and P. R. Brandt. 1997. *The Girls Next Door: Into the Heart of Lesbian America.* New York: Simon & Schuster.

Van Liere, K. D., and R. E. Dunlap. 1980. "The Social Bases of Environmental Concern: A Review of Hypotheses, Explanations and Empirical Evidence." *Public Opinion Quarterly* 44, no. 2 (Summer): 181–97.

Venkatesh, S. 2006. *Off the Books: The Underground Economy of the Urban Poor.* Cambridge, MA: Harvard University Press.

Vidich, A., and J. Bensman. 1958. *Small Town in Mass Society: Class, Power, and Religion in a Rural Community.* Urbana: University of Illinois Press.

Vigdor, J. L. 2002. "Does Gentrification Harm the Poor?" *Brookings-Wharton Papers on Urban Affairs* 34 (June): 133–82.

Vorse, M. H. 1942. *Time and the Town: A Provincetown Chronicle.* Provincetown, MA: Cape Cod Pilgrim Memorial Association.

Walker, A. 1973. "Everyday Use." In *In Love and Trouble: Stories of Black Women,* 47–59. San Diego: Harcourt, Brace.

Walton, J. 2001. *Storied Land: Community and Memory in Monterey.* Berkeley and Los Angeles: University of California Press.

Warner, W. L. 1959. *The Living and the Dead: A Study of the Symbolic Life of Americans.* New Haven, CT: Yale University Press.

Warren, J. R., J. T. Sheridan, and R. M. Hauser. 2002. "Occupational Stratification Across the Life Course: Evidence from the Wisconsin Longitudinal Study." *American Sociological Review* 67, no. 3:432–55.

Warren, R. L. 1963. *The Community in America.* Chicago: Rand McNally.

———. 1970. "Toward a Non-Utopian Normative Model of Community." *American Sociological Review* 35, no. 2 (April): 219–28.

Waters, M. 1990. *Ethnic Options: Choosing Ethnic Identities in America.* Berkeley and Los Angeles: University of California Press.

Weber, M. 1968/1978. *Economy and Society: An Outline of Interpretive Sociolocy.* Edited by

G. Roth and C. Wittich. Translated by E. Fischoff et al. 2 vols. Berkeley and Los Angeles: University of California Press.

Weiner, T. S., and B. K. Eckland. 1979. "Education and Political Party: The Effects of College or Social Class?" *American Journal of Sociology* 84, no. 4 (January): 911–28.

Weiss, R. 1988. *The American Myth of Success: From Horatio Alger to Norman Vincent Peale.* Urbana: University of Illinois Press.

Wellman, B. 1977. "The Community Question: Intimate Ties in East New York." Research Paper no. 90. Toronto: University of Toronto, Centre for Urban and Community Studies.

White, K. I. C., and A. M. Guest. 2003. "Community Lost or Transformed? Urbanization and Social Ties." *City and Community* 2, no. 3 (September): 239–59.

Wilgoren, J. 2005. "At Center of a Clash, Rowdy Children in Coffee Shops." *New York Times,* November 9, A12.

Williams, B. 1988. *Upscaling Downtown: Stalled Gentrification in Washington DC.* Ithaca, NY: Cornell University Press.

Williams, M. 2000. "Harlem Journal; Gay White Pioneers, on New Ground." *New York Times,* sec. 1, November 19, 49.

Williams, R. 1973. *The Country and the City.* New York: Oxford University Press.

Williams, T. 2005. "Now Booming, Not Burning, the Bronx Fears a Downside." *New York Times,* March 19, 1, 27.

Wirth, L. 1938. "Urbanism as a Way of Life." *American Journal of Sociology* 44:3–24.

Wordsworth, W. 2000. *William Wordsworth: The Major Works: Including "The Prelude."* Edited by S. Gill. Oxford: Oxford University Press.

Wuthnow, R. 1989. *Communities of Discourse: Ideology and Social Structure in the Reformation, the Enlightenment, and European Socialism.* Cambridge, MA: Harvard University Press.

———. 1996. *Sharing the Journey: Support Groups and the Quest for a New Community.* New York: Free Press.

———. 1999. "The Culture of Discontent." In *Diversity and Its Discontents: Cultural Conflict and Common Ground in Contemporary American Society,* ed. N. J. Smelser and J. C. Alexander, 19–36. Princeton, NJ: Princeton University Press.

Yardley, J. 1998. "Park Slope, Reshaped by Money; As Rents and Prices Rise, Some Fear for Neighborhood's Soul." *New York Times,* March 14, 1998, B1.

Yednak, C., and C. Flynn. 2005. "Affordable Housing Law Deadline Looms." *Chicago Tribune,* Metro section, April 1, 1, 6.

Yin, Y. H. 1973. "Context in Architecture: Strengthening the Urban Fabric or Bridging the Gap between the Social Preservation and Urban Renewal." Master's thesis, University of California, Berkeley.

Zelinsky, W. 1991. "Seeing beyond the Dominant Culture." *Places* 7:32–35.

Zelizer, V. 1999. "Multiple Markets: Multiple Cultures." In *Diversity and Its Discontents: Cultural Conflict and Common Ground in Contemporary American Society,* ed. N. J. Smelser and J. C. Alexander, 193–212. Princeton, NJ: Princeton University Press.

Zernike, K. 2003. "The New Couples Next Door, Gay and Straight." *New York Times,* August 24, 12.

Zimmerman, C. C. 1938. *The Changing Community.* New York: Harper & Bros.

Zipp, J. F. 1986. "Social Class and Social Liberalism." *Sociological Forum* 1, no. 2 (Spring): 301–29.

Zorbaugh, H. W. 1929/1976. *The Gold Coast and the Slum: A Sociological Study of Chicago's Near North Side.* Chicago: University of Chicago Press.

Zukin, S. 1982/1989. *Loft Living: Culture and Capital in Urban Change.* New Brunswick, NJ: Rutgers University Press.

———. 1987. "Gentrification: Culture and Capital in the Urban Core." *Annual Review of Sociology* 13:129–47.

———. 1993. *Landscapes of Power: From Detroit to Disney World.* Berkeley and Los Angeles: University of California Press.

———. 1995. *The Cultures of Cities.* Cambridge: Blackwell.

Index

old-timers (*continued*)
farmers, 84, 136, 141, 147–48, *154*, 163–65,
198, 288n14; and economic struggle, 156–60;
ethnicity of, 146, 150; exclusion from category,
150–52, 288n23; food, 121; and gender, 288n15;
heterogeneity of, 223–24; independence and
definition of, 154–60; influence of, 133; institutions, 99, 120, 169–70, 217, 244–45; land-use
traditions, 62, 99; language barriers in Argyle,
124–25; length of residence, 288n22; local
networks, 157–58; and local organizations,
200; longevity of residence, 149–50, 166–67;
merchants, 154; middle class, 133, 226; mobility
of, 226; natives, 150; perceived isolation, 227; as
place-based category, 177; Portuguese, 150, 151,
156–57; preservation of, 19, 84; as Provincetown fishermen, 23, 105, 112, 147–48, *154*, 166;
and race, 150, 177, 261; recognition of social
preservation, 240–41; as reflective agents, 226;
resentment of political intrusions, 84, 138–39;
role of local culture in defining, 152; romanticization of, 75, 158; sampling of, 294n3; selection
of, 145–79; self-preservation, 241–47; and
sexual orientation, 150; struggling artists, 154,
156–57, 158–59; symbols of, 120; terminology
used to identify, 283n16; tradition and definition of, 160–67; townies, 149–50; variability
of attitudes, 213–49; views of public space and
community decline, 215–19; working-class, 156
Oliver, Mary: "On Losing a House," 34
Onion, 18
"On Losing a House" (Oliver), 34
Orvell, Miles, 289n35

Palen, J. John, 296n15
Parsons, D. J., 281n3
Parsons, Talcott, 160
Pascarella, Ernest T., 192
paternalistic advocacy, 117, 286n21; as context
specific, 124–25; defined, 117, 124; explained,
124–26
Pattillo, Mary, 17, 177, 211, 282n10, 282n19,
283nn12–13, 285n5, 286n19, 296n16, 298n18
Pattison, Timothy James, 294n35
Paulsen, Krista, 142
Paxton, Pamela, 289n30
Payne, David: *Gravesend Light*, 191
Perez, Gina, 4
perspectival variation, 103, 250
Peterson, Pamela, 32, 33
Peterson, Richard A., 148, 192, 211, 289n35, 292n20
Phelan, Jo, 291n13
Phillips, D. A., 281n3
Phillips, Martin, 281n3
pioneer(s), 5–6; and affordable housing, 52,
122, 124; attitudes toward displacement, 52,

53–54; defined, 12, 48, 51; economic success,
52; gentrification and common good, 53, 54;
images of, 8; investment, 53; literature on, 8;
priorities, 53; progentrification sentiments, 53,
114, 117–18, 122–24; reaction to social preservation discourse, 124; rhetoric, 124; as ruthless, 4,
250; safety and beautification concerns, 53, 55,
122, 124; and transformation, 53–55, 115; view of
gentrification as natural, 53, 122; views of social
preservationists, 123
place character, 142, 167, 255; associated with old-
timers, 92, 105, 167; and authenticity, 56, 95,
132; claims about, 255; and class, 70; influence
of on old-timers' views of gentrification, 239;
institutionalization of, 243; marketing of, 255;
in Provincetown, 35, 65; as mediating expression of beliefs, 143; and residents, 146; and
social homesteaders, 56; social preservationists'
concern for, 127
place-names, 122, 130, 170, 286n20; as gentrification strategy, 122, 286n20; as signifiers of
gentrification, 97–98; as social preservation
strategy, 130–31
Podmore, Julie A., 282n15, 297n5
political abstinence, 134, 174; consequences of, 138;
defined, 134; and Dresden, 134–42; influence
of town policies on, 141; and political avoidance, 106, 110, 134; and Provincetown, 108–17;
relationship to other ideologies, 135; source
of, 138
political avoidance, 106, 110, 134
political economy, 251; and culture, 253–54, 255,
297nn5–7; in Dresden, 141; scholars' attention
to, 251, 253–54, 297n5
Polk, O. Elmer, 286n21
Portrait of a Woman Down East (BoltÉ), 26, 283n3
Pratt, Mary Louise, 293n29
preservation: and class, 246–47; community, 245;
cultural, 86, 147, 284n3; historic, 56–61; landscape, 55–56, 61–65, 105, 172; of old-timers,
146; selection of residents for, 145–79; self,
144, 160, 241, 243, 246–47; symbolic, 11, 13, 94,
105, 132, 133, 260. *See also* conservation; social
preservation
property taxes, 166; increases, 60, 107, 118, 224–26,
286n25; resistance to increases, 107
property values, 89, 228; rising, 129, 224–25, 228,
296n23
protest, 118–19, 127–28; decreased, 129
Provincetown Banner: and anti-gentrification
artwork, 112; gentrification coverage, 113, 116,
232–33; property values reported in, 33–34
Provincetown, Massachusetts: affordable housing, 34, 52, 83, 108–10, 113, 114, 167; Affordable
Housing task force, 271; AIDS, 33, 69, 117;
artists, 32, 33; as arts colony, 32, 100; Blessing